Water Science and Application 4

Geomorphic Processes and Riverine Habitat

Water Science and Application 4

Geomorphic Processes and Riverine Habitat

Joseph M. Dorava
David R. Montgomery
Betty B. Palcsak
Faith A. Fitzpatrick
Editors

American Geophysical Union
Washington, DC

Geomorphic Processes and Riverine Habitat
Water Science and Application 4

Library of Congress Cataloging-in-Publication Data
Geomorphic processes and riverine habitat / Joseph M. Dorava ...[et al.].
p.cm.--(Water science and application ; 4)
Includes bibliographical references (p.).
ISBN 0-87590-353-3
1. Stream ecology--Congresses. 2. Geomorphology--Congresses. I. Dorava, Joseph M. II. American Geophysical Union. Meeting (1998 : San Francisco, Calif. III. Series.

ISBN 0-87590-353-3
ISSN 1526-758X

2000 Florida Avenue, N.W.
Washington, DC 20009

Front Cover: Mosaic of land cover in the Encampment River watershed, approximately 15 kilometers northeast of Two Harbors, Minnesota along Lake Superior's North Shore. Infrared air photographs reproduced with permission, © State of Minnesota, Department of Natural Resources. **Inset:** Photograph of Peshekee River, Fall 1994, near Champion, Upper Peninsula of Michigan by Kevin Richards, U.S. Geological Survey, Middleton, Wisconsin.

Back Cover: A small channel with an alluvial morphology forced by logs and log jams on Chichagof Island in Southeast Alaska (Buckhorn Creek). Photograph by David R. Montgomery.

Printed in the United States of America.

CONTENTS

CONTENTS

PREFACE

The study of the physical and biological aspects of our surroundings has interested scientists for centuries. Many physical processes, some as subtle as weathering of rock into soil, changes in sea level, or global warming and others as violent as volcanic eruptions, floods, or hurricanes, are influenced by and actively change the surrounding landscape. Geomorphic processes and the landscape and biological changes that result from their interactions are of interest to society as well as scientists because of the effects these processes potentially have on the economic, aesthetic, and health-related aspects of our lives. When the processes are life threatening, as in the case of volcanic eruptions or floods, scientists and society are highly motivated to improve their understanding of the magnitude, frequency, and predictability of future occurrences. When the consequences of the processes are not as severe, as is the case of weathering, scientific study has been less urgent. Quite recently, and driven in part by the decline in salmon populations throughout the Pacific Northwest and a near continuous loss of valuable riverine habitat resulting from human demands on the environment, there has been increasing interest in improving our understanding of how geomorphic processes affect riverine habitat and the biotic communities (including humans) that inhabit them.

This monograph describes results of recent research in fluvial geomorphology related to the assessment and characterization of riverine and riparian habitat, and the response of biota to changes in their environment. The contributions to the monograph are grouped into five sections. The first section, "Geomorphic and Habitat Assessments," presents methods and means to undertake geomorphic habitat investigations, ranging from routine characterizations or classifications through detailed process models. The next two sections, "Pristine Undisturbed Systems" and "Anthropogenically Altered Systems," compare and contrast active geomorphic processes and human disturbances. The fourth section, "Restoration Strategies," features discussion on various restoration methods and mitigation strategies by way of distinctive case studies. In the monograph's closing paper, "Geomorphology, River Ecology, and Ecosystem Management," issues involved in integrating an understanding of geomorphology and riverine habitat into landscape-scale environmental management are discussed.

A diverse group of researchers presented studies of geomorphic processes and riverine habitat at the 1998 fall meeting of the American Geophysical Union in San Francisco, California. The enthusiastic participation of researchers in that meeting and the reception of the research results motivated preparation of a monograph containing recent studies on the subject. This monograph thus brings together researchers who participated in the 1998 meeting with other contributors whose collective background and research breadth captures the diverse nature of developments on the interface of geomorphic processes and riverine ecology. In assembling this monograph, interdisciplinary points of view are brought together through the common focus of understanding the processes that form riverine habitat and the interactions among these processes and humans as both occupants and shapers of the environment.

We acknowledge, with sincere thanks, all the contributors, the numerous reviewers, and those that provided encouragement and support for this project, including AGU staff, our colleagues, and especially our families.

Joseph M. Dorava
Vierbicher Associates, Inc.

INTRODUCTION

Understanding Geomorphic Processes and Riverine Habitat

Joseph M. Dorava[1], David R. Montgomery[2], Betty B. Palcsak[3], and Faith A. Fitzpatrick[4]

INTRODUCTION

It is increasingly recognized that many interesting research opportunities lie on the interface between traditional scientific disciplines. In particular, the influence of geomorphic processes on riverine habitat is central to contemporary programs of river restoration. The role of geomorphologists in ecological studies ranges from understanding habitat-forming mechanisms to defining disturbance regimes and restoration or rehabilitation objectives. It has proven difficult to integrate the efforts of multiple scientific disciplines when evaluating the effects of humans on the environment and understanding the complex interactions of natural processes constantly changing the landscape. This monograph brings together the work of numerous researchers primarily focused on the study of fluvial geomorphic processes that create or alter riverine habitats in an effort to improve our understanding and assessment of geomorphic processes, to enhance communication among scientists, and to integrate the results of ongoing and future studies.

FROM ASSESSMENT TO RESTORATION

Landscape management for ecosystem restoration or rehabilitation purposes should follow a relatively simple logic for how to rebuild a complex system without an instruction manual: assess the ecosystem's condition; compare its condition to systems in an undisturbed state; determine the desired condition; hypothesize actions necessary to rehabilitate, restore, or modify key aspects of the system to attain the desired condition(s); take those actions; monitor and assess the results; and modify the approach if the results prove less than desired. This general framework for guiding landscape management from assessment to restoration includes studies of the interactions between geomorphology and ecology that address geomorphic and habitat assessments, undisturbed and anthropogenically altered systems, and restoration strategies.

Geomorphic and Habitat Assessments

A wide array of geomorphic and habitat assessments are presently in use for various applications. These assessments range from interpretive check lists of features, to score cards on which system traits are tallied to assess overall system condition, to classifications of objects (such as channel reaches) in manners considered to have interpretive power, each of which has advantages and disadvantages for practical applications. As a result, in this monograph, *Fitzpatrick* describes various assessment techniques and their use in various studies. *Potter and Gaffield* show how synoptic baseflow surveys can be used to identify channels within a watershed that have very different habitat characteristics and hydrology to reveal a pattern that would not be identified by typical channel assessment protocols. *Gore* describes several examples of the application of habitat simulation models to stream restoration. *Beechie et al.* demonstrate how post-glacial landscape evolution and recent human activity have changed valley-bottom landforms and salmonid habitats. These studies illustrate the wide range of scales over which geomorphic and habitat assessments can provide structure to the understanding of interactions between geomorphology and river ecology.

[1]Water Resources Department, Vierbicher Associates Inc., Reedsburg, Wisconsin

[2]Department of Earth & Space Sciences, University of Washington, Seattle, Washington

[3]Water Resources Division, U.S. Geological Survey, Lakewood, Colorado

[4]Water Resources Division, U.S. Geological Survey, Middleton, Wisconsin

Geomorphic Processes and Riverine Habitat
Water Science and Application Volume 4, pages 3-4

Undisturbed and Anthropogenically Altered Systems

Once an initial assessment of the system of interest is complete, restoration or rehabilitation programs require evaluation of current conditions against some benchmark expectations, objectives, or reference conditions, which are typically based on comparison with a similar undisturbed system. Evaluation of undisturbed landscapes is essential for defining the degree to which anthropogenically altered landscapes have been disturbed ecologically. Such an assessment is central to ecosystem management and in particular for developing landscape rehabilitation and restoration plans. Hence, the second and third sections of this volume address studies of undisturbed and anthropogenically altered systems.

A common theme throughout this monograph concerns the need to understand the particular landscape or watershed in question within the context of its geological, geomorphic, and cultural history. Compare, for example, the dominant processes controlling habitat diversity in the warm-water, gravel-bed stream studied by *McKenney* with those influencing channel island formation in the Snake River, as studied by *Osterkamp et al.* A clear example of the need to understand the dominant processes acting in a watershed and how wood debris that forms stable log jams mediates sediment export from small, debris-flow prone headwater streams in the Oregon Coast Range is provided by *Lancaster et al.* Their simulations show the importance of understanding both anthropogenically altered (wood-poor) and undisturbed (wood-rich) channel conditions in trying to connect geomorphological processes to ecological systems. In addition, *Larsen and Román* show how the legacy of past disturbances can dominate sediment storage in a channel system. *Frothingham et al.* report how changes in channel reach morphology, and in particular, the associated habitat complexity, can influence fish community structure. *Nakamura and Shin* explore the striking effects of flow regulation by dams on the disturbance ecology, and therefore stand composition and age structure, of riparian forests in Japan. *McDowell* provides an example of how local variability in valley width structures channel morphology and valley-bottom ecosystems, a finding that contrasts with the patterns anticipated by the popular River Continuum Concept. *Pasternack* describes how the actions of animals can counteract anthropogenic alteration of valley-bottom hydrologic conditions and landforms. In total, these papers show the rich variety of geomorphological processes that influence ecological systems, and thereby underscore the importance of incorporating a high level of understanding into landscape and watershed management.

Restoration Strategies

The primary goal of ecosystem management is often restoration of a disturbed system. The final section of this monograph describes various aspects of developing and implementing restoration strategies. Just as for geomorphic and habitat assessments, restoration strategies can be cast over a wide range of scales. Perhaps one of the most important, and overlooked, elements of any restoration strategy is clearly articulated goals. *Pitlick and Wilcock* address how to set what in many western U.S. river systems is a particularly contentious issue—how much flow does a river need? In addressing this basic attribute of aquatic habitat, *Pitlick and Wilcock* use the flow conditions associated with the onset of different styles of sediment transport to define minimum flow regimes necessary to maintain channel morphology and characteristics. *Jacobson et al.* explore passive approaches to channel restoration—the attractive idea of letting the river do the work. Their documentation of progressive changes in an evolving side-channel along the Missouri River illustrates the inherent (although not inexhaustible) resilience of river systems. *Dorava* shows how simple changes in human actions or behavior can dramatically reduce erosion of riverbanks and disturbance of aquatic ecosystems, if people are willing to accept changes in how they conduct their recreational activities. *Collins and Montgomery* show how process studies and archival methods are complementary approaches for generating an understanding of a place adequate for evaluating restoration plans and options.

As discussed in the final synthesis chapter by *Montgomery*, there are many exciting research opportunities on the frontier between geomorphology and ecology. These opportunities and new approaches can be integrated into the development of landscape management, recovery, and restoration plans with an improved understanding of the interactive processes. We hope that this monograph will help catalyze further interest in the study of the influences of geomorphic processes on riverine habitat, as well as other ecological systems.

Joseph M. Dorava, Water Resources Department, Vierbicher Associates Inc., 400 Viking Drive, P.O. Box 379, Reedsburg, WI 53959.

Faith A. Fitzpatrick, Water Resources Division, U.S. Geological Survey, 8505 Research Way, Middleton, WI 53562.

David R. Montgomery, Department of Earth & Space Sciences, University of Washington, Seattle, WA 98195-1310.

Betty B. Palcsak, Water Resources Division, U.S. Geological Survey, Box 25046, MS 415, Denver Federal Center, Lakewood, CO 80225.

Section 1

Geomorphic and Habitat Assessments

A Comparison of Multi-Disciplinary Methods for Measuring Physical Conditions of Streams

Faith A. Fitzpatrick

U.S. Geological Survey, Water Resources Division, Middleton, Wisconsin

A wide range of scientists, including fluvial geomorphologists, ecologists, fisheries managers, and river engineers, study human impacts on the physical characteristics of streams. Data gathered for stream physical conditions are useful for water- and fish-resource management, water-quality monitoring, and stream restoration. The physical condition of streams reflects a variety of natural and anthropogenic effects and multiple spatial and temporal scales. Thus, systematic approaches among scientific disciplines differ in terms of spatial and temporal scales of measurement and are heavily dependent on the academic background and field experience of the investigators. Data-collection efforts range from rapid, qualitative descriptions and categorizations to more time-consuming quantitative measurements. Most geomorphic assessments, stream classifications, and habitat characterizations measure similar key hydrologic and geomorphic features, even though individual methodologies may differ substantially. These key features include channel morphology and planform, valley and channel gradient or slope, sediment characteristics and movement, and bank characteristics. This paper examines several commonly used habitat characterizations, stream classifications, and geomorphic assessments in terms of how they effectively measure the physical condition of streams that are most likely to change in response to land-cover perturbations. Advantages and disadvantages of each type of assessment are discussed in the context of potential study goals and outcomes, and time, labor, and expertise constraints. The major goal of this paper is to promote communication among ecologists, geomorphologists, fisheries managers, and engineers, thereby reducing duplication of effort and increasing the utility of methods used to describe the physical characteristics of streams. By improving the understanding of the geomorphic processes within a watershed and their effect on physical features in a stream channel, various scientific disciplines will make better interpretations of the development of riverine habitat. Adjustments in sampling routines for each discipline may then create a complementary and more uniform data set from which decision makers can develop management strategies.

INTRODUCTION

Streams are dynamic features that adjust their physical characteristics to prevailing climate, geology, topography, vegetation, and base level (the level, or elevation, at which a stream enters a major water body). A change in any of these independent factors may alter streamflow and sediment movement. Human-induced changes in vegetation for agriculture, silviculture, or urban development have occurred in almost every watershed in North America. Conversion of land from forest or agriculture to urban is a common occurrence across much of the United States. Increases in impervious surfaces from parking lots, roads, and buildings result in a decrease in the time to peak and

Geomorphic Processes and Riverine Habitat
Water Science and Application Volume 4, pages 7-18

an increase in the volume and magnitude of floods [for example *Morisawa and LaFlure,* 1979; *Schueler,* 1994; *Sovern and Washington,* 1997]. In addition, urbanization can increase or decrease low flows by increasing effluent discharge or decreasing groundwater contributions [for example *Snodgrass et al.*, 1997; *Sovern and Washington,* 1997]. Sediment supply may increase during the construction phase but eventually decrease after grass, pavement, and buildings cover most of the urban landscape [*Wolman,* 1967].

In the process of responding to changes in flood characteristics and sediment supply and transport, stream channels alter their physical characteristics of channel geometry, planform, gradient, or base level [*Leopold and Maddock,* 1953; *Wolman,* 1967; *Wolman and Schick,* 1967; *Hammer,* 1972; *Gregory and Madew,* 1982; *Booth and Jackson,* 1997]. Channel adjustments in response to land-cover perturbations are dependent on the position of the stream within the stream network and can occur over time periods as brief as individual storms to as long as thousands of years [*Schumm,* 1977; *Fitzpatrick et al.*, 1999; *Fitzpatrick and Knox,* 2000]. Channels in an erosional zone of a stream network may become steeper and have eroding stream banks and channel beds, whereas channels in the depositional zone of a stream network will become filled with sediment as the additional material from upstream is deposited. Over time, continued erosion eventually reduces the stream gradient until deposition becomes the dominant sediment process (*Hey,* 1979). For management purposes, channels experiencing form adjustments are viewed as being unstable. These hydrogeomorphic adjustments may have substantial effects on aquatic life (for example *Wang et al.,* 1997; *Finkenbine et al.*, 2000).

Interest in measuring the effects of changes in land cover on the physical characteristics of streams draws scientists together from many different backgrounds, including fluvial geomorphologists, ecologists, fisheries managers, and river engineers. Each discipline has developed field techniques suitable for its area of expertise. Geomorphologists tend to specialize at multiple scale studies of channel and flood-plain characteristics with an important historical or time component [*Knox,* 1977; *Patrick et al.*, 1982; *Fitzpatrick et al.*, 1999]. Ecologists typically focus on physical features that may be limiting factors for aquatic communities, typically fish [*Simonson et al.*, 1994a]. River engineers more commonly concentrate on measurements of current hydraulic conditions for purposes of flood control, navigation, irrigation, remediation, and channel stabilization [*Chang,* 1992].

This paper attempts to bring together the disciplines of fluvial geomorphology, ecology, and river engineering by providing some background information on several widely used field techniques. These techniques are examined in terms of how they effectively measure the physical condition of streams and the geomorphic processes responsible for their formation. Five major characteristics are discussed here: channel geometry or morphology, channel planform, channel gradient or slope, bank conditions, and sediment characteristics such as size and movement. By closing the gap between specific disciplines and improving their understanding of geomorphic processes each will have a better interpretation of the development of aquatic habitat. Minor adjustments in data collection may also lead to a more complementary and uniform geomorphic data set that will assist natural resources managers.

CHARACTERIZATION OF A STREAM'S PHYSICAL CONDITION

Geomorphic principles are useful for determining a stream's response to changes in water discharge or sediment transport caused by watershed land-cover change. *Schumm's* [1969] qualitative model of channel metamorphosis describes the geomorphic response of a channel to changes in discharge and sediment load. For increasing discharge and sediment loads, the following proportional relations are possible:

$$Q^+Q_s^+ \sim W^+D^\pm F^+\lambda^+S^\pm P^-$$

where Q is discharge, Q_s is sediment load, W is channel width, D is channel depth, F is width/depth ratio, λ is meander wavelength, S is slope, and P is sinuosity. If discharge and sediment load increase, as typically the case during the early stages of urbanization, the expected geomorphic responses will be increases in channel width, width/depth ratio, and meander wavelength; a decrease in sinuosity; and increases or decreases in depth and slope.

Lane's [1955] description of equilibrium also gives some indication of possible outcomes for sediment size and slope:

$$Q_s d \propto QS.$$

A stream will adjust its slope (S) and sediment size (d) in response to changes in discharge and sediment load. These relations are important for quantifying data and determining direction of change and possible causes for geomorphic response [*Patrick, et al.*, 1982]. *Lane's* [1955] proportionality has a theoretical base—for a fixed width and straight channel, equations can be derived from solution of continuity, resistance, and bed material transport equations. Therefore, physical measurements of channel morphology, planform, gradient, and sediment load are important for characterizing the physical changes that occur in streams following changes in land cover (Table 1).

Typical aquatic habitat assessments may not be adequate to reveal the inherent relations among land use, subsequent

Table 1. Physical variables that adjust to changes in streamflow and sediment supply and transport

Channel morphology	Channel planform	Gradient	Sediment characteristics and movement	Bank measurements
Bankfull width	Sinuosity	Channel gradient	Suspended load	Height
Bankfull depth	Meander radius	Channel bed elevation	Bed load	Angle
	Meander belt width	Valley gradient	Suspended sediment size	Erosion
	Meander wavelength		Bed sediment size	Vegetation
			Fine sediment depth	Shape
			Embeddedness	
			Scour	
			Particle size and volume of overbank deposits	

changes in streamflow and sediment movement, and resulting physical characteristics of streams. Several studies have shown that urbanization or conversion of land from forest to agriculture lowers biological integrity or index scores; however, relations between habitat quality indexes and land cover appear to be more complex [*Dreher*, 1996; *Richards et al.*, 1996; *Roth et al.*, 1996; *Wang et al.*, 1997]. For example, *Wang et al.* [1997] found correlations between the index of biotic integrity (IBI) and the amount of agricultural or urban land in a watershed in Wisconsin. However, the habitat quality index for the same sites related to the amount of agricultural land in a watershed but not to urban land. For eastern Wisconsin streams, *Fitzpatrick et al.* [2000] found no statistically significant correlation between the Wisconsin and Michigan habitat quality indexes and the percentage of watershed agricultural land. The Wisconsin habitat quality index is based on semi-quantitative field methods and was created to rate the quality of warm-water fish habitat [*Simonson et al.*, 1994a]. However, *Wang et al.* [1997] adapted the habitat quality index as an indication of the health of warm-water stream ecosystems as affected by watershed land use. The Michigan habitat quality index was originally based on qualitative field methods and was specifically created for evaluation of nonpoint-source impacts [*Michigan Department of Natural Resources*, 1991]. It is not known whether the lack of correlation is due to the qualitative methods used to summarize the data into indexes or due to the methods used to collect the habitat data. Either the habitat indexes or the habitat measurements are not sensitive enough to distinguish changes in channel morphology, sediment movement, and bank conditions that are occurring, especially from urbanization.

MEASUREMENT OF A STREAM'S PHYSICAL CONDITION

A comparison of the types of measurements collected for selected habitat assessments, stream classifications, and geomorphic assessments to define the physical condition of a stream is shown in Table 2. The table does not include every measurement taken for each assessment but rather focuses on physical characteristics most likely to change in response to changes in streamflow and sediment movement. The measurements are grouped into five categories: channel morphology, channel planform, gradient or slope, sediment characteristics and movement, and bank measurements. A brief summary of common sampling approaches also is included (Table 2). In this paper, the term "transect" refers to a line established perpendicular to the channel along which measurements are made at a number of points along the line. The term "cross section" refers to a line established perpendicular to the channel along which distances and elevations are surveyed with an automatic level or total station.

Habitat Assessments

Qualitative habitat assessments are commonly used for rapid determination of habitat characteristics, such as during stream reconnaissance at a large number of sites. The U.S. Environmental Protection Agency's (USEPA) rapid bioassesment protocol [*Plafkin et al.*, 1989] contains a qualitative habitat assessment that is based on a Wisconsin qualitative model [*Ball*, 1982]. The Ohio EPA also developed a similar approach [*Rankin*, 1989]. The main goal of the USEPA qualitative habitat assessment is to determine the quality of available habitat for biota, primarily fish. The assessment is based on categorical determination of nine parameters. Scores are assigned for each category and a final score is determined for a stream reach. The rapid qualitative assessments generally take an hour or less to complete at each site but have subjective results because scores are based on visual estimation rather than quantitative measurements.

Several semi-quantitative habitat assessments that combine visual observations with measurements have been developed by federal and state agencies. Many are based on the transect line intercept method [*Platts et al.*, 1983; *Kaufmann and Robison*, 1994; *Simonson et al.*, 1994a;

Table 2. Methods comparison for habitat assessments, stream classifications, and geomorphic assessments
[--, Not measured; USEPA, U.S. Environmental Protection Agency; USFS, U.S. Forest Service; USGS, U.S. Geological Survey; NAWQA, National Water-Quality Assessment; USFWS, U.S. Fish and Wildlife Service]

Method name	Common sampling approach	Channel morphology measurements	Channel planform measurements	Gradient or slope measurements	Sediment characteristics and movement	Bank measurements
			HABITAT ASSESSMENT—QUALITATIVE			
USEPA [*Plafkin et al.*, 1989]	Reach—visual summaries	Pool/riffle or run/bend ratio Degree of channel alteration and bar formation	--	--	Gravel, percent Embeddedness, percent Scouring and deposition, percent	Amount of erosion, failures, or steep slopes Percent of vegetated surface
			HABITAT ASSESSMENTS—SEMI-QUANTITATIVE			
(1) USFS [*Platts et al.*, 1983]	Reach—transect method, sampling verticals at 1-ft intervals along transect	Pool/riffle/run/glide, length Channel cross section surveys	Sinuosity	Reach gradient	Dominant substrate categories Embeddedness categories Sediment cores Particle size analysis, laboratory or field	Angle Length of undercut Soil alteration rating Vegetative stability rating
(2) USGS NAWQA [*Fitzpatrick et al.*, 1998]	Reach—transect method, sampling verticals at 3 points along transect Segment—topographic maps Watershed—GIS	Pool/riffle/run, length Channel cross section surveys Bankfull width and depth	Sinuosity	Basin gradient Segment gradient Reach gradient	Dominant substrate categories Embeddedness, percent Particle size analysis, laboratory Silt presence/absence Pebble counts	Angle Height Substrate categories Presence/absence of erosion Percent vegetative cover Bank stability index
			HABITAT ASSESSMENT—QUANTITATIVE			
USFWS Instream Flow Incremental Methodology and Physical Habitat Simulation System [*Bovee*, 1982]	Reach—transect method Segment—topographic maps Watershed	Channel cross section surveys	Pattern categories	Segment gradient	Substrate categories	

Table 2 (Continued)

Method name	Common sampling approach	Channel morphology measurements	Channel planform measurements	Gradient or slope measurements	Sediment characteristics and movement	Observations and measurements of bank conditions
			STREAM CLASSIFICATION			
Rosgen channel classification, Levels I through IV [*Rosgen*, 1996]	Reach— visual summaries; aerial photographs; cross section surveys, cross section surveys over time Segment— topographic maps, aerial photography	Channel shape categories Bankfull width and depth categories Entrenchment ratio categories Debris/blockage categories Modified Pfankuch channel stability rating [*Pfankuch*, 1975] Width/depth ratio categories Lateral and vertical stability checks	Pattern categories Sinuosity categories Meander lengths Radius of curvature Meander belt width Sinuosity	Reach gradient categories Valley slope/sinuosity Water surface gradient Valley gradient	Substrate categories Pebble counts Depositional pattern categories Sediment supply categories Stream bed stability categories Bed form features Suspended and bed load measurements Substrate particle size distribution Embeddedness Pebble count Scour chains	Erosion potential categories Near bank stress categories Bank pins
			GEOMORPHIC ASSESSMENT—SEMI-QUANTITATIVE			
U.K. geomorphic assessment [*Thorn*, 1998]	Reach—visual summaries and measurements Sub reach—visual summaries and measurements Watershed and valley maps	Bankfull width Channel depth Instability status	Bend radius Meander belt width Wavelength Sinuosity Lateral activity Present status Instability status Problem severity Problem extent Width controls Width control types	Reach gradient Bed controls Bed control types	Presence, thickness, and type of overbank deposits Dominant substrate Presence of bed armor Loose sediment depth Surface size Substrate Bed form types Presence of islands/bars Island /bar substrate size Island/bar surface size Manning's n value	Substrate stratigraphy Height Angle Shape Degree of cracking Vegetation type; location; health; age; distribution Erosion location; status; rate; severity; extent; processes Geotechnical failure location; status; severity, extent, failure mode; distribution Toe sediment accumulation and de-

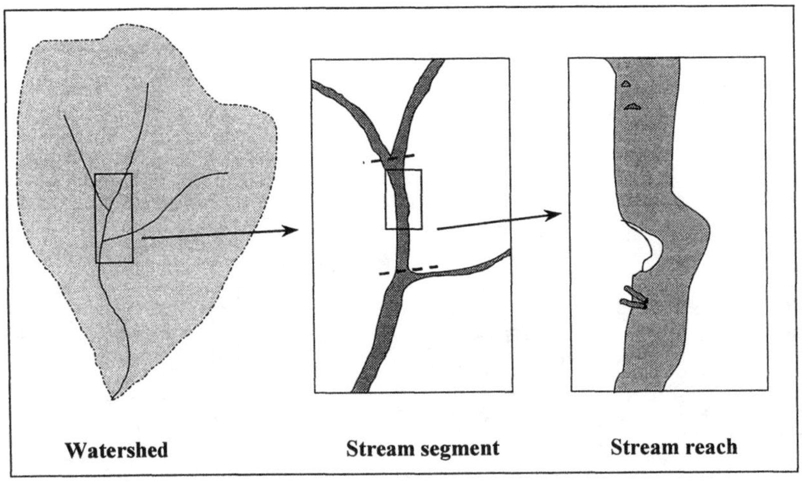

Figure 1. Spatial hierarchy of watershed, stream segment, and stream reach (modified from Frissell et al., 1986).

Fitzpatrick et al., 1998]. Examples of two semi-quantitative habitat assessments, the U.S. Forest Service (USFS) [*Platts et al.*, 1983] and the U.S. Geological Survey (USGS) National Water-Quality Assessment Program (NAWQA) [*Fitzpatrick et al.*, 1998], are described in Table 2. Transects, typically 11 to 20 or more depending on which assessment is selected, are usually spaced equidistantly along the sampled reach. Channel measurements of depth, velocity, substrate, and embeddedness are collected at several points, or verticals, along the transect. *Simonson et al.* [1994b] recommend an optimum of 20 transects to adequately describe the variability within a reach. The length of the sampling reach is dependent on the size of the stream and increases with stream width. Semi-quantitative assessments generally take 4 hours or more to complete at each site and are useful for studies comparing habitat at several sites. Proper training is essential for providing consistency among observers [*Roper and Scarnecchia*, 1995].

Several of the semi-quantitative assessments collect additional measurements of channel morphology and substrate. For example, the USFS assessment [*Platts et al.*, 1983] and the USGS NAWQA assessment [*Fitzpatrick et al.*, 1998] recommend surveying channel cross sections to further define channel shape, estimate bankfull discharge, and document changes in erosion and deposition over time. Both protocols suggest collecting channel substrate samples for laboratory particle-size analysis (Table 2). The NAWQA protocol also recommends pebble counts for characterizing coarse substrate as described by *Wolman* [1954]. Roughly speaking, cross section surveys and streambed-sediment sampling by use of NAWQA methods usually takes 2 to 3 people about 4 to 8 hours to complete for a stream with a drainage area of about 50 km^2 [*Fitzpatrick et al.*, 1998].

Frissell et al. [1986] first proposed the spatial hierarchical approach for habitat assessments and a modified hierarchical approach was included in the NAWQA habitat protocol (Table 2) [*Fitzpatrick et al.*, 1998]. Geomorphic features at the watershed, segment, and reach scale are collected as part of the NAWQA protocol (Fig. 1). Watershed measurements include basin length, basin relief, drainage shape, stream length, cumulative stream length, drainage density, stream order, and basin gradient. A segment is a section of stream bounded by confluences or physical or chemical discontinuities, such as major waterfalls, landform features, significant changes in gradient, or point-source discharges. Segment geomorphic measurements include gradient and sinuosity.

The U.S. Fish and Wildlife Service's (USFWS) Instream Flow Incremental Methodology (IFIM) and Physical Habitat Simulation System (PHABSIM) are used in combination for determining the instream flow and resulting habitat required for a given aquatic species, mainly fish (Table 2) [*Bovee*, 1982]. One of the many uses of this quantitative method is to explore the effects of watershed-scale changes in land cover on the instream flow and resulting habitat of a stream segment. *Gore* [this volume] provides a detailed overview of IFIM and PHABSIM. Habitat, as determined by IFIM or PHABSIM, may be computed by incorporating factors of longitudinal changes in channel characteristics, streamflow, water quality, and temperature [*Bovee*, 1982]. A series of hydraulic simulation models are used to evaluate the availability of depths, velocities, and substrates at a variety of flows [*Orth and Maughan*, 1982]. The final output is a time series of effective useable habitat for a given species for different disturbance scenarios.

Extensive data collection at a variety of spatial scales is required for IFIM and PHABSIM, thus limiting them to studies of one or two streams. At the watershed scale, an understanding of the water, sediment, and chemical yields is needed to characterize the stream. Streamflow time-series data are needed; either historical data from a streamflow gaging station or estimated data from rainfall/runoff models. Existing channel morphology is quantified by cross section surveys at numerous transects. Transects are established in the center of each geomorphic unit (riffle, run, pool, etc.) and at the transition between units [*Bovee*, 1982]. Depth, velocity, substrate type, and cover are measured at several points, or verticals along each transect. Multiple verticals are needed to quantify the combination of these features across the stream width. The number of verticals and transects are determined by the accuracy needed to adequately describe the habitat requirements for the species under study. A limitation of these methods is the lack of quantitative information on the habitat suitability criteria of target species and the regional applicability of existing criteria [*Orth and Maughan*, 1982; *Simonson et al.*, 1994a]. Other assumptions of the method are described in *Orth and Maughan* [1982].

Another group of habitat assessments not listed in Table 2 focus on descriptions and aerial extent of habitat units (also called geomorphic units) such as riffles, runs, and pools [*Bisson et al.*, 1982; *Hankin and Reeves*, 1988; *Hawkins et al.*, 1993]. The main goal of habitat-unit classifications is to quantify available fish habitat; however, a secondary goal is to evaluate the effects of environmental change [*Hawkins et al.*, 1993]. *Poole et al.* [1997] found that habitat-unit classifications are not useful for quantifying or monitoring aquatic habitat or geomorphic conditions. The classifications are limited by observer bias and a lack of statistical strength. In addition, anthropogenic disturbances in the watershed may not affect the frequency or aerial extent of habitat units [*Poole et al.*, 1997].

Stream Classifications

Hierarchical stream classifications, such as those by *Rosgen* [1994, 1996] and *Montgomery and Buffington* [1997, 1998], provide a consistent semi-quantitative means for describing and comparing the geomorphic characteristics of channels. Channel measurements are focused on geomorphic features that reflect hydrologic conditions and sediment supply and transport capacity (Table 2).

The Rosgen classification system is based on channel form and contains four levels of classification, ranging from a broad classification of channel pattern, slope, and shape (Level I) to a detailed description of hydrologic, sediment, and channel characteristics that are used to help identify geomorphic process relations (Level IV) [*Rosgen*, 1996]. Levels I and II data are used to classify stream types. Level III data are used to describe the geomorphic condition of each stream type. The Rosgen classification system is used by several federal agencies, including the USFS, USFWS, and the USEPA. The channel stability evaluation included in Rosgen's Level III assessment has been used by the USFS and the Bureau of Land Management in the central and western U.S. to describe the potential for erosion and changes in sediment supply due to changes in streamflow or watershed condition [*Pfankuch*, 1975; *Rosgen*, 1996]. Level IV data are used to develop empirical relations for a given stream type under a given condition and possibly extrapolate empirical relations of hydrologic and sediment characteristics to other stream types under similar conditions. Level IV data are collected from time series of streamflow, bedload and suspended sediment monitoring, multiple surveys at permanent cross sections, and multiple pebble counts and embeddedness measurements at permanent transects.

Criticisms of the Rosgen classification system have arisen when classified stream types are used as predictors of geomorphic processes [*Miller and Ritter,* 1996; *Goodwin*, 1999; *Harmel et al.*, 1999; *Montgomery*, 1999; *Doyle and Harbor,* 2000]. In addition, empirical relations may be too site specific to be useful regionally, and temporal and size scales are lacking. For example, an objective of the Rosgen classification is to provide a context for linking driving forces and response variables at multiple scales [*Rosgen*, 1996]. However, similar characteristics of channel form can result from different geomorphic processes [*Schumm*, 1984]. In addition, Rosgen's bank erosion potential ratings and near-bank stress estimates, when integrated into indices based on the stream classification system, were poor predictors of bank erosion for Oklahoma streams [*Harmel et al.*, 1999].

Geomorphic Studies And Assessments

The preferred method for identifying channel response to watershed disturbance is to monitor hydrologic conditions, sediment transport, and changes in channel morphology and pattern over long periods of time. However, very few studies have the luxury of collecting data for more than one or two years. Long-term hydrologic data collected by the USGS are probably the most readily available data collected on a routine, long-term basis. However, the number of streamflow gaging stations operated by the USGS has been greatly reduced over the last 15 to 20 years because of budget cuts related to basic data collection. Sediment transport data are less common. Long-term data on channel morphology are sparse.

Geomorphic assessments are considered reconnaissance surveys because they only provide data related to current geomorphic conditions, whereas more detailed studies enable current geomorphic conditions to be placed in a historical perspective. Many geomorphic studies are undertaken to provide a better understanding of current and historical changes in watershed land cover on aquatic habitat when long-term hydrologic, sediment, and channel morphology data are not available [*Knox*, 1977; *Patrick et al.*, 1982; *Knox*, 1987; *Jacobson and Primm*, 1997; *Fitzpatrick et al.*, 1999]. In many instances, longer time scales are examined by studying historical documents, flood-plain deposits, and characteristics of abandoned channels. Geomorphic studies may draw from various data sources such as historical aerial photography, descriptions of flood-plain and channel stratigraphy, age dating through sediment coring and geophysical techniques, dendrochronology, soil geomorphology, rainfall/runoff modeling, indirect measurements of discharge, quantification of historical land-use change, and quantification of historical climatic and tectonic changes. Results from geomorphic studies that include these elements can be used to predict likely channel responses to future watershed disturbances, and to help distinguish between climatic- and anthropogenic responses [*Knox et al.*, 1975; *Fitzpatrick et al.*, 1999; *Knox*, 1999; *Fitzpatrick and Knox*, 2000]. Detailed geomorphic studies require trained geomorphologists with ample field experience. The large amount of field data needed for these types of studies usually limits them to one or two streams and requires one to two field seasons or more of data collection.

Geomorphic assessments based on field reconnaissance have been developed to quantify channel conditions and relate them to geomorphic processes and to enable the impacts of both past and present watershed disturbance to be assessed [*Brookes*, 1995; *Thorne*, 1998]. Geomorphic assessments require a range of geomorphic training and experience but generally take less time and data collection than a geomorphic study and can be conducted on several streams in a field season.

Geomorphic assessment methods, such as that described by *Thorne* [1998], have been used in studies of United Kingdom and U.S. streams and are designed to work hand-in-hand with other engineering, ecological, and managerial studies (Table 2). *Thorne's* [1998] assessment, designed to characterize channel conditions, as well as indicate processes related to watershed disturbance, consists of region and valley description, channel description, left bank survey, and right bank survey. The assessment consists of a combination of visual qualitative observations and several quantitative measurements, which are summarized into categories or ranks. At the most detailed level, observations and measurements are recorded at the sub-reach level or for each geomorphic unit, usually 5 to 10 times the channel width. One of the most powerful aspects of the *Thorne* [1998] assessment is that it examines flood-plain and valley characteristics, as well as the channel itself, to evaluate past changes in channel planform and stability.

Comparison of Thorne's methodology with the stream classification and habitat assessment techniques summarized in Table 2 show a greater emphasis of this geomorphic assessment on extensively describing bank conditions and channel planform and morphology compared to the other methods. Thorne suggests that channel measurements need not be detailed or precise, but must be representative of the reach. Thus, the investigator draws on knowledge and experience to determine how many measurements need to be taken to determine typical characteristics. Bank and channel characterization are highly interpretive and require a background understanding of bank erosion processes and channel forming processes. The interpretive aspects of this assessment appear to require more background knowledge in geomorphology than other habitat assessments and stream classifications. However, several U.S. Army Corps of Engineers District Offices and the U.K. Environment Agency have used this method successfully [*Thorne*, 1998]. The widespread use of this method most likely results because Thorne's assessment focuses more on channel and bank features specifically related to channel instability and geomorphic processes. Thorne's emphasis on measurements and observations directly related to geomorphic processes might provide users with a better understanding of the effects of changes in land cover on aquatic habitat than the physical measurements included in habitat assessments and stream classifications.

DISCUSSION

Many aquatic habitat assessments do not measure channel dimensions in a manner useful for determination of geomorphic processes [*Ball*, 1982; *Platts et al.*, 1983; *Plafkin et al.*, 1989; *Rankin*, 1989; *Michigan Department of Natural Resources*, 1991; *Simonson et al.*, 1994a]. The geomorphic relations among channel morphology, discharge, and sediment load are based on dimensions of the channel at bankfull stage, not during low-flow or average conditions. Water depth and mean wetted channel width during low flow are required parameters for measuring aquatic habitat, but do not relate to geomorphic condition of the channel.

Measurements related to determining channel morphology at bankfull stage seem to be the most problematic for non-geomorphologists. Some of the problems relate to multiple definitions for bankfull or lack of field experience in identifying bankfull indicators. For example, four geomorphic measurements—bank-to-bank width, high-water stream width, bank rock content, and channel stability—were dropped from the USFS habitat assessment because of high observer error, problems with subjectivity, and poor accuracy [*Platts et al.,* 1983]. Training in fluvial geomorphology, combined with a clear definition of bankfull indicators, is needed to measure channel geometry [*Fitzpatrick et al.*, 1998]. Two of the more recent semi-quantitative habitat assessments include measurements of bankfull channel dimensions: USEPA's Environmental Monitoring and Assessment Program (EMAP) and USGS's NAWQA habitat assessments [*Kaufmann and Robison*, 1994; *Fitzpatrick et al.*, 1998]. The NAWQA assessment and IFIM methods also include cross section surveys to quantify channel dimensions.

Habitat-unit assessments may describe only specific flow conditions and do not usually quantify geomorphic conditions unless the assessment is designed to measure a specific watershed process. In contrast, quantitative measurements of geomorphic conditions generally can be used successfully to assess habitat availability and distribution [*Poole et al.*, 1997].

In many multi-disciplinary studies, the task of geomorphic data collection falls under the responsibility of an ecologist. Thus, the time available for geomorphic characterization of streams is shared with the time needed for biological characterizations and aquatic habitat measurements. In other multi-disciplinary studies, ecologists and engineers may be collecting similar geomorphic measurements as part of separate data-collection efforts. Geomorphic data collection efforts are somewhat specialized; however, all methods require some level of training, and relatively small changes in training and field methods for ecologists and engineers can potentially yield substantial improvements in both habitat characterizations and channel designs.

Empirical relations that define the dimensions and planform of stable channels can be used to compare observed morphology with predicted stable dimensions [*Hey and Thorne*, 1986]. Departures from stable dimensions could be used to identify potential sites of erosion or deposition. For example, *Hey and Thorne* [1986] developed stable channel characteristics for alluvial gravel channels based on empirical relations with bankfull discharge, sediment-bedload discharge, and median and 85-percent finer bed material size for streams with four types of riparian vegetation. Although empirical equations can be developed for different channel types, these equations cannot be used to predict unstable conditions because the new equilibrium Q and Q_s cannot be adequately predicted because of potential geomorphic feedback mechanisms. A clear understanding of the complex interaction of historical land-cover change and low-frequency climatic events and their long-term effects on geomorphic conditions of channels is still needed. Simple empirical relations do not provide enough insight into the temporal and spatial manifestations of long-term geomorphic processes.

Geomorphic studies tend to be highly interdisciplinary and to include time scales of centuries and millennia. Fluvial geomorphologists tend to have skills in a variety of disciplines, including hydrology, ecology, statistics, engineering, soils, meteorology, remote sensing, and geology. Many geomorphologists doing research are hesitant to get involved with the applied side of fluvial geomorphology. Because of the interdisciplinary nature of geomorphology, there also is a lack of standard methodologies [*Brookes*, 1995]. However, the multi-disciplined geomorphologist is in the best position to bridge the communication gap between ecologists and engineers [*Brookes*, 1995]. A geomorphologist is like a detective, drawing together evidence from any and all possible sources.

CONCLUSIONS

Some geomorphic training is necessary to adequately measure the physical aspects of streams for interpreting changes in channel form and aquatic habitat in response to changes in land cover. Quantitative measurements of aquatic habitat will be more useful if geomorphic measurements, such as bankfull width and depth, are measured correctly and incorporated into the methodology. Stream classifications are fast and easy to use and serve as a good starting point for describing physical conditions. However, some stream classification systems tend to oversimplify complex temporal and spatial geomorphic processes and may result in incorrect conclusions if applied outside the proper context. Interpretive summaries of geomorphic conditions at multiple scales provide information on the processes responsible for altering habitat conditions. Physical measurements need to focus on geomorphic variables that are most influenced by changes in runoff and sediment supply and transport capacity and those that reflect channel stability: channel morphology, channel planform, gradient, sediment characteristics, and bank conditions. Geomorphic assessments are useful for reconnaissance of several watersheds but do not replace detailed geomorphic studies. Creating a common understanding of geomorphic processes and their long-term influence on channel form will improve the interpretation of aquatic habitat by multi-disciplinary scientists. A more uniform and complementary geomorphic data set collected by ecologists, geomorphologists, and engineers would assist decision makers in developing appropriate management strategies.

Acknowledgements. The author wishes to thank those people that have given her a better understanding of the disciplines of geomorphology, civil engineering, and ecology and how they might better communicate among each other: James Knox of the University of Wisconsin-Madison; Richard Hey of University of East Anglia, England; Joseph Dorava of Vierbicher Associates, Inc.; Michael Meador, Waite Osterkamp, Cliff Hupp, Barbara Scudder, Ian Waite, Marty Gurtz, and Robb Jacobson of the U.S. Geological Survey; Ken Potter and John Hoopes of the University of Wisconsin-Madison; and Dennis Pratt of the Wisconsin Department of Natural Resources. Technical reviews by David Montgomery, Betty Palcsak, Joseph Dorava, Richard Hey, and Janet Heiny were appreciated and greatly improved the content of this paper. Errors or omissions in describing and summarizing the different methods presented here are the sole responsibility of the author.

REFERENCES

Ball, J., Stream classification guidelines for Wisconsin, *Wisconsin Dept. of Nat. Res. Technical Bulletin*, 1982.

Bisson, P.A., Nielsen, J.L., Palmason, R.A., and L.E. Grove, A system of naming habitat types in small streams, with examples of habitat utilization by salmonids during low streamflow, *Proc. Symp. Acquisition and Utilization of Aquatic Habitat Inventory Information*, pp. 62-73, 1982.

Booth, D.B., and C.R. Jackson, Urbanization of aquatic systems: degradation thresholds, stormwater detection, and the limits of mitigation. *J. of the Am. Water Res. Assoc.* 33(5):1077-1090, 1997.

Bovee, K.D., A guide to stream habitat analysis using the instream flow incremental methodology, *US. Fish and Wildlife Service, Instream flow Information Paper 12*, FWS/OBS-82/26, Washington, D.C., 248 pp., 1982.

Brookes, A., Challenges and objectives for geomorphology in U.K. river management, *Earth Surface Processes and Landforms* 20, pp. 593-610, 1995.

Chang, H.H., *Fluvial Processes in River Engineering*, Malabar, Fla., Krieger Publishing Co., 432 pp., 1992.

Doyle, M.W., and J.M. Harbor, Discussion of "Evaluation of Rosgen's streambank erosion potential assessment in northeast Oklahoma" by R.D. Harmel, C.T. Haan, and R.C. Dutnell. *J. of the Am. Water Res. Assoc.* 36(5):1191-1192, 2000.

Dreher, D.W., Watershed urbanization impacts on stream quality indicators in northeastern Illinois, *Proceedings of a National Symposium, Assessing the Cumulative Impacts of Watershed Development on Aquatic Ecosystems and Water Quality*, Chicago, Ill., pp. 129-135, 1996.

Finkenbine, J.K., J.W. Atwater, and D.S. Mavinic, Stream health after urbanization. *J. of the Am. Water Res. Assoc.* 36(5):1149-1160, 2000.

Fitzpatrick, F.A., E.M.P. Giddings, B.C. Scudder, D.J. Sullivan, and B.N. Lenz, Effects of riparian corridors and hydrology on agricultural stream biota, *Proceedings, International Conference on Riparian Ecology and Management in Multi-land Use Watersheds,* American Water Resources Association, pp. 215-219, 2000.

Fitzpatrick, F.A., and J.C. Knox, Spatial and temporal sensitivity of hydrogeomorphic response and recovery to deforestation, agriculture, and floods, *Phys. Geog.* 21(2), pp. 89-108, 2000.

Fitzpatrick, F.A., J.C. Knox, and H.E. Whitman, Effects of historical land-cover changes on flooding and sedimentation, North Fish Creek, Wisconsin, *U.S. Geol. Surv. Water-Resour. Invest. Rep.* 99-4083, 12 pp., 1999.

Fitzpatrick, F.A., I.R. Waite, P.J. D'Arconte, M.R. Meador, M.A. Maupin, and M.E. Gurtz, Revised methods for characterizing stream habitat in the National Water-Quality Assessment Program, *U.S. Geol. Surv. Water-Resour. Invest. Rep.* 98-4052, 67 pp., Raleigh, N.C., 1998.

Frissell, C.A., W.J. Liss, C.E. Warren, and M.D. Hurley, A hierarchical framework for stream habitat classification: viewing streams in a watershed context, *Environmental Management* 10(2), pp. 199-214, 1986.

Goodwin, C.N., Fluvial classification: Neanderthal necessity or needless normalcy, in *Wildland Hydrology*, edited by D.S. Olsen and J.P. Potyondy, American Water Resources Association, Herndon, Virginia, pp. 229-236, 1999.

Gore, J.A., Models of habitat use and availability to evaluate anthropogenic changes in channel geometry, in *Geomorphic Processes and Riverine Habitat*, edited by J.B. Dorava, D.R. Montgomery, B. Palcsak, and F.A. Fitzpatrick, American Geophysical Union Monograph xx, Washington, D.C., this volume.

Gregory, K.J., and J.R. Madew, Land use change, flood frequency and channel adjustments. in *Gravel-Bed Rivers*, edited by R.D. Hey, J.C. Bathurst, and C.R. Thorne, John Wiley and Sons, pp. 757-781, 1982.

Hammer, T.R., Stream channel enlargement due to urbanization. *Water Resources Research* 8(6):1530-1540, 1972.

Hankin, D.G. and G.H. Reeves, Estimating total fish abundance and total habitat area in small streams based on visual estimation methods, *Can. J. Fish. Aquat. Sci.* 45, pp. 834-844, 1988.

Harmel, R.D., C.T. Haan, and R.C. Dutnell, Evaluation of Rosgen's streambank erosion potential assessment in northeast Oklahoma. *J. of the Am. Water Resour. Assoc.* 35(1):113-121, 1999.

Hawkins, C.P., J.L. Kershner, P.A. Bisson, M.D. Bryant, L.M. Decker, S.V. Gregory, D.A. McCullough, C.K. Overton, G.H. Reeves, R.J. Steedman, and M.K. Young, A hierarchical approach to classifying stream habitat features, *Fisheries* 18(6), pp. 3-11, 1993.

Hey, R.D. Dynamic process-response model of river channel development. *Earth Surf. Processes* 4, pp. 59-72, 1979.

Hey, R., and C.R. Thorne, Stable channels with mobile gravel beds. *J. of Hydraul. Div.*, American Society of Civil Engineers 112(8), 671-689, 1986.

Jacobson, R.B., and A.T. Primm, Historical land-use changes and potential effects on stream disturbance in the Ozark Plateaus, *U.S. Geol. Surv. Water-Sup. Pap.* 2484, 85 pp., 1997.

Kaufmann, P.R., and E.G. Robison, Physical habitat assessment, in *Environmental Monitoring and assessment program 1994 pilot field operations manual for streams*, edited by Klemm, D.J., and Lazorchak, J.M., U.S. Environmental Protection Agency, Environ. Monit. Syst. Lab., Office of Research and Development, EPA/620/R-94/004, Cincinnati, Ohio, p. 6-1 to 6-38, 1994.

Knox, J.C., Human impacts on Wisconsin stream channels, *Annals of the Assoc. of Am. Geog.* 67(3), pp. 323-342, 1977.

Knox, J.C., Historical valley floor sedimentation in the upper Mississippi Valley, *Annals of the Assoc. of Am. Geog.* 77(2), pp. 224-244, 1987.

Knox, J.C., Long-term episodic changes in magnitudes and frequencies of floods in the Upper Mississippi River Valley, in *Fluvial Processes and Environmental Change*, edited by A.G. Brown and T.A. Quine, John Wiley and Sons, N.Y., pp. 255-282, 1999.

Knox, J.C., P.J. Bartlein, K.K. Hirschboeck, and R.J. Muckenhirn, The response of floods and sediment yields to climate variation and land use in the upper Mississippi valley, Center for Geographic Analysis, *Institute of Environmental Studies Rep.* 52, Univ. of Wis., Madison, Wis., 76 pp., 1975.

Lane, E.W., The importance of fluvial geomorphology in hydraulic engineering, *Proc. ASCE 81*, paper 745, pp. 1-17, 1955.

Leopold, L.B., and T. Maddock, The hydraulic geometry of stream channels and some physiographic implications, *U.S. Geol.Surv. Prof. Pap.* 252, 57 pp., 1953.

Michigan Department of Natural Resources, *Qualitative biological and habitat survey protocols for wadable streams and rivers*, Michigan Dept. of Natural Resources, Surface Water Quality Division, Great Lakes Environmental Protection Assessment Section Procedure 51, 40 pp., 1991.

Miller, J.R., and J.B. Ritter, An examination of the Rosgen classification of natural rivers, *Catena* 27, pp. 295-299, 1996.

Montgomery, D.R., Process domains and the river continuum, *J. of the Am. Water Res. Assoc.* 35(2), pp. 397-410, 1999.

Montgomery, D.R., and J.M. Buffington, Channel processes, classification, and response potential, in River Ecology and Management, edited by R.J. Naiman and R.E. Bilby, Springer-Verlag, Inc., New York, pp. 13-42, 1998.

Montgomery, D.R., and J.M. Buffington, Channel reach morphology in mountain drainage basins, Geological Society of America Bulletin, v. 109, pp. 596-611, 1997.

Morisawa, M., and E. Laflure, Hydraulic geometry, stream equilibrium, and urbanization, in *Adjustments of the Fluvial System*, edited by D.D. Rhodes and G.P. Williams, Kendal Hunt, Iowa, pp. 333-350, 1979.

Orth, D.J., and Maughan, O.E., Evaluation of the incremental methodology for recommending instream flows for fishes, *Transactions of the American Fisheries Society* 111, pp. 413-445, 1982.

Pfankuch, D.J., Stream reach inventory and channel stability evaluation, USDA Forest Service, R1-75-002, Government Printing Office #696-260/200, Washington D.C., 26 pp., 1975.

Patrick, D.M., L.M. Smith, and C.B. Whitten, Methods for studying accelerated fluvial change, in *Gravel-bed Rivers*, edited by Hey, R.D., J.C. Bathhurst, and C.R. Thorne, John Wiley and Sons, N.Y., pp. 783-815, 1982.

Plafkin, J.L., M.T. Barbour, K.D. Porter, S.K. Gross, and R.M. Hughes, Rapid bioassesment protocols for use in streams and rivers: benthic macroinvertebrates and fish, U.S. Environmental Protection Agency, EPA/444/4-89-001, 1989.

Platts, W.S., W.F. Megahan, and G.W. Minshall, Methods for evaluating stream, riparian, and biotic conditions, *U.S. Dept. Agriculture Forest Service General Technical Report* INT-138, 1983.

Poole, G.C., C.A. Frissell, and S.C. Ralph, In-stream habitat unit classification: inadequacies for monitoring and some consequences for management, *J. of the Am. Water Res. Assoc.*, 33 (4), pp. 879-896, 1997.

Rankin, E.T., The qualitative habitat evaluation index (QHEI): Rationale, methods, and application, Ohio Environmental Protection Agency, *Ecol. Assess. Sect. Rep.*, 1989.

Richards, C., L.B. Johnson, and G.E. Host, Landscape-scale influences on stream habitats and biota, *Can. J. Fish. Aquat. Sci.* (Suppl. 1), pp. 295-311, 1996.

Roper, B.B., and D.L. Scarnecchia, Observer variability in classifying habitat types in stream surveys, *North Amer. J. of Fish. Manage.* 15(1), pp. 49-53, 1995.

Rosgen, D.L., A classification of natural rivers, *Catena* 22, pp. 69-199, 1994.

Rosgen, D.L., *Applied River Morphology*, Wildland Hydrology, Pagosa Springs, Colo., 1996.

Roth, N.E., J.D. Allan, and D.L. Erickson, Landscape influences on stream biotic integrity assessed at multiple scales, *Landscape Ecology* 11(3), pp. 141-156, 1996.

Schueler, T., The importance of imperviousness, *Watershed Protection Techniques* 1(3): 100-111, 1994.

Schumm, S.A., River metamorphosis, *J. Hydraul. Div. ASCE, 95(Hyl), Paper* 6352, pp. 255-273, 1969.

Schumm, S.A., *The fluvial system*, Wiley and Sons, N.Y., 338 pp., 1977.

Schumm, S.A., River morphology and behavior: problems with extrapolation, in *River Meandering*, edited by C.M. Elliot, Amer. Society Civil Eng., New York, NY, pp. 16-29, 1984.

Simonson, T.D., J. Lyons, and P.D. Kanehl, Guidelines for evaluating fish habitat in Wisconsin streams, *U.S. Dept. Agric Forest Service General Technical Rep.* NC-164, 1994a.

Simonson, T.D., J. Lyons, P.D. Kanehl, Quantifying fish habitat in streams: transect spacing, sample size, and a proposed framework, *North Am. J. of Fish. Manage.* 14, pp. 607-615, 1994b.

Snodgrass, W.J., B.W. Kilgour, M. Jones, J. Parish, and K. Reid, Can environmental impacts of watershed development be measured, in Effects of Watershed Development and Management on Aquatic Ecosystems, edited by Roesner, L.A., *Proceedings of an Engineering Foundation Conference*, N.Y., American Society of Civil Engineers, pp. 351-385, 1997.

Sovern, D.T., and P.M. Washington, Effects of urban growth on stream habitat, in Effects of Watershed Development and Management on Aquatic Ecosystems, *Proceedings of an Engineering Foundation Conference*, edited by Roesner, L.A.,

N.Y., Am. Soc. of Civil Eng., pp. 163-177, 1997.

Thorne, C.R., *Stream reconnaissance handbook*, John Wiley and Sons, Chichester, 1998.

Wang, L., J. Lyons, P. Kanehl, and R. Gatti, Influence of watershed land use on habitat quality and biotic integrity in Wisconsin streams, *Fisheries* 22(6), pp. 6-12, 1997.

Wolman, M.G., A method for sampling coarse river-bed material, *Trans. of the Am. Geophys. Union* 35, pp. 951-956, 1954.

Wolman, M.G., A cycle of sedimentation and erosion in urban river channels, *Geografiska Annaler* 49, pp. 385-395, 1967.

Wolman, M.G., and A.P. Schick, Effects of construction on fluvial sediment, urban and suburban areas of Maryland, *Water Resour. Res.* 3(2), pp. 451-464, 1967.

Faith A. Fitzpatrick, U.S. Geological Survey, 8505 Research Way, Middleton, WI 53562

Watershed Assessment With Synoptic Base-Flow Surveys

Kenneth W. Potter

Department of Civil and Environmental Engineering, University of Wisconsin – Madison, Madison, Wisconsin

Stephen J. Gaffield

Wisconsin Geological and Natural History Survey, Madison, Wisconsin

A synoptic survey provides a nearly instantaneous observation of how base-flow discharge varies throughout a watershed. Discharge is measured at a number of watershed locations during a period when groundwater discharge is the only contribution to streamflow. Surveys can be repeated to estimate temporal variability, and an index method can be used to estimate parameters, such as mean annual base-flow discharge. The index method is based on dividing each synoptic base-flow discharge measurement by the concurrent discharge observed at a nearby gaging station. Five synoptic base-flow surveys were done in the Hillsboro Lake watershed in southwestern Wisconsin from 1995 to 1999. At many of the sites, the temporal variability of base-flow discharge is relatively low (coefficient of variation less than 0.2). For sites with higher variability, the measured discharges were low (about 0.05 cubic meters per second or less). Measurement error, which is proportionally larger for low-discharge sites than high-discharge sites, is probably responsible for part of this variability. Variability in discharge is probably also related to physical characteristics of the watersheds, such as soil type. For two streams, the pattern of unit-area discharge was as we expected: low in the headwaters and increasing downstream. However, in two other streams, unit-area discharge was about 5 times higher in the headwaters than farther downstream. This indicates that groundwater recharge rates also decrease downstream. This significant spatial variability in base-flow discharge has important habitat implications. In southwestern Wisconsin, upland sites with high unit-area base-flow discharges are likely to be important for coldwater habitat, because they are likely to have steeper gradients, greater flow depths, and more stable water temperatures than sites with low unit-area base-flow discharges.

INTRODUCTION

Flow in a stream can result from episodic events, such as rainfall or snowmelt, or from the drainage of groundwater, lakes, wetlands, or other sources that yield water at nearly constant rates. Streamflow derived from the latter sources is commonly referred to as base flow. When base flow is mainly derived from groundwater, as is commonly the case, it provides a steady source of generally high quality water at a relatively constant temperature. Base flow is essential to most stream ecosystems; without it a stream would be dry most of the time. Hence, habitat assessment in a watershed should include the estimation of the spatial distribu-

Geomorphic Processes and Riverine Habitat
Water Science and Application Volume 4, pages 19-25

tion of base-flow discharge. This paper explores the use of synoptic surveys as a means for estimating how base-flow discharge varies throughout a watershed.

In a synoptic base-flow survey, discharge is measured at a number of watershed locations during a period when groundwater discharge is the only contribution to streamflow. Sites are selected on the basis of the location of tributary confluences, known springs, and reaches where base-flow information is needed.

A synoptic survey thus provides instantaneous observations of the spatial variation of base-flow discharge. Surveys can be repeated to estimate temporal variability and base-flow parameters, such as mean annual base-flow discharge. However, because base-flow discharges are highly correlated in time, surveys should not be repeated until after the occurrence of a significant groundwater recharge event.

Even with multiple synoptic surveys, estimates of base-flow parameters may be subject to considerable uncertainty due to temporal variability. *Potter* [2001] presents a simple index method for reducing this uncertainty by using data from a nearby gaging station.

The index method is based on dividing each synoptic base-flow discharge measurement by the concurrent discharge observed at the gaging station. If base-flow discharges are approximately lognormally distributed, then the geometric mean of this ratio is equal to the ratio of the median discharges. If the variance of the log-transformed base-flow discharges at a given synoptic site is about equal to that of the base-flow discharges at the gaged site, then the geometric mean of the ratios is also equal to the ratios of the mean discharges. This relation suggests the following estimator of the median base flow at a site of interest:

$$\hat{y}_{0.5} = \exp\left[\left(\frac{1}{n}\right)\sum_{i=1}^{n}\left(\frac{\ln(y_i)}{\ln(x_i)}\right)\right] x_{0.5} \qquad (1)$$

where:

$y_1, y_2, \ldots, y_n$ is a set of base-flow measurements at the site of interest;

$x_1, x_2, \ldots, x_n$ are concurrent base-flow measurements at the gaging station;

$x_{0.5}$ is the median base-flow discharge at the gaging station for any period of interest within the period of record; and

$$\exp\left[\left(\frac{1}{n}\right)\sum_{i=1}^{n}\left(\frac{\ln(y_i)}{\ln(x_i)}\right)\right] \qquad (2)$$

is the geometric mean of the discharge ratios.

This equation can be used to estimate long-term or short-term parameters. For example, if the interest is in knowing the "population" median, then the period of record would be used to compute the median at the gaging station. On the other hand, if the interest is in knowing the median flow for a particular period of time (such as a particular year), then only the gaging station data for that period would be used.

The uncertainty in these estimators of median and mean base-flow discharge depend on the following factors:

- Correctness of the distributional assumptions;
- Degree of linear correlation between the log-transformed base-flow discharges at the synoptic site and the gaged site;
- Variance of the estimator of the median or mean at the gaging station;
- Uncertainty in the discharge measurements;
- Serial correlation in the base-flow ratios.

The last factor makes it difficult to develop good estimates of the standard errors of the estimators, as the serial correlation structure in base-flow discharges can be complex.

Potter [2001] tested this index method on base-flow discharge data from four streamflow-gaging stations in southwestern Wisconsin. For one pair of stations, the assumptions of lognormality, equal log-variance, and high log-correlation held reasonably well. As a result, the method worked very well. For the other pair of stations, the assumptions did not hold, and the application of the method to the estimation of long-term parameters suffered. However, for both station pairs, the method worked very well for estimating annual parameters.

The drainage areas of the watersheds used by *Potter* [2001] ranged from 368 to 707 km^2, larger than drainage areas relevant to many likely applications. This paper reports on the application of this method to sites in the Hillsboro Lake watershed, a 91 km^2 watershed also in southwestern Wisconsin.

STUDY SITE

The Hillsboro Lake watershed is located in the unglaciated part of southwestern Wisconsin, known as the Driftless Area (Figure 1). This region is characterized by well developed stream networks incised more than one hundred meters into nearly flat-lying sedimentary bedrock. Resistant Ordovician carbonate rocks form broad plateaus that are dissected by steep-sided valleys cut into underlying clastic rocks of the Cambrian period. Soils are developed in loess (deposits of windblown silt) and mass-movement deposits on hillslopes consisting of a mixture of loess and weathered bedrock. This study focuses on the watershed draining into Hillsboro Lake, a constructed impoundment fed by the

Figure 1. Location of the Hillsboro Lake watershed and Kickapoo River gaging station.

West and South Branches of the Baraboo River and Beaver Creek (Figure 2). This area forms the headwaters of the Baraboo River, a major tributary of the Wisconsin River.

Much of the groundwater discharge in this region comes from shallow, unconfined flow systems and occurs within a few kilometers of recharge areas [*Gaffield et al.*, 1998; *Young*, 1992; *Zaporozec and Cotter*, 1985]. The Cambrian rocks include regionally important sandstone aquifers [*Young*, 1992] as well as confining units that are commonly observed to produce springs. In places, the Ordovician carbonates are highly permeable due to fracture networks and solution features [*Young*, 1992].

The Driftless Area has a high concentration of coldwater streams that support both brook trout (*Salvelinus fontinalis*) and brown trout (*Salmo trutta*) and have long been valued for recreational fishing [*Becker*, 1983]. Agricultural development, that began in about 1850, has severely impacted area watersheds by increasing soil erosion, runoff and peak flows [*Johnson*, 1976], while decreasing groundwater recharge and base flows [*Potter*, 1991]. Consequently, the trout fishery was dramatically reduced by the middle of the 1900's [*Scott and Hoveland*, 1951]. Although the watershed is still dominated by agriculture, these trends have reversed over the past few decades [*Kent*, 1999; *Trimble and Lund*, 1982], probably due to the increase in organic matter as a result of improved land management practices [*Sampson*, 1981]. Today the watershed provides marginal trout habitat, and maintaining and improving the quality of the fishery is an important management issue.

METHODS

This study continues the work of *Desai* [1996], who measured base flow at 17 locations in the Hillsboro Lake watershed on three dates in 1995 and 1996. We conducted two additional base-flow surveys at most of these sites in 1998 and 1999. Sites are located on each of the main tributaries to Hillsboro Lake, with incremental contributing areas that are approximately the same size (Figure 2).

In choosing synoptic survey dates, we monitored the discharge of the Kickapoo River at the LaFarge gaging station (Figure 1) to ensure that base-flow conditions existed. We waited several days after any major precipitation event until the hydrograph formed a smooth recession curve. Each survey was completed in one day to minimize variability caused by base-flow recession between measurements. We measured discharge with Price "mini" current meters by wading and using the standard technique of the U. S. Geological Survey [*Buchanan and Somers*, 1969].

Synoptic base-flow measurements were indexed using the concurrent discharge of the Kickapoo River at the LaFarge gaging station (Figure 1). The indexed base flow equals the discharge at the synoptic measurement site divided by the discharge at the gaging station. In interpreting the results, however, it is important to keep in mind the possibility of groundwater underflow from one surface basin to another.

RESULTS

Measured discharges at each study site and concurrent discharges at the Kickapoo River gaging station are given in Table 1. Also given are summary statistics for the measured and indexed discharges. Figure 3 shows the discharge measurements at each site.

At many of the sites, the temporal variability of base-flow discharge is relatively low. For example, at seven of the sites, the coefficient of variation (CV) of the measured discharges is less than 0.2 (Figure 4). For the sites where the CV is high (greater than 0.2), the measured discharge is low (about 0.05 cubic meters per second or less). In these cases, the high values of CV are likely to be in part due to the errors associated with measuring very low discharges, although it is also likely that discharge variability increases as discharge decreases.

The indexed discharges are a bit more variable than the measured discharges (Table 1). This is contrary to the findings of *Potter* [2001]. Apparently, for these small drainage areas, indexing on a concurrent regional discharge does not reduce temporal variability.

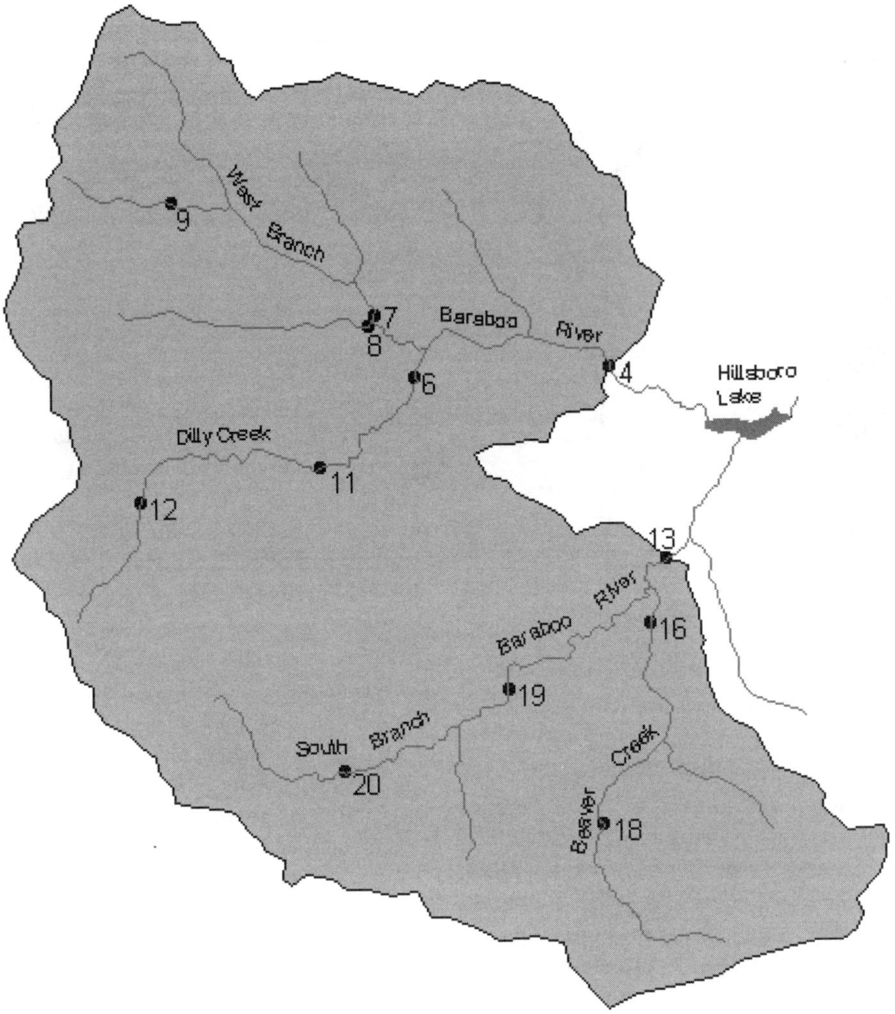

Figure 2. Synoptic survey site locations in the Hillsboro Lake watershed.

Median base-flow discharges for the Kickapoo River gaging station for the periods 1938-99 (long-term) and 1995-99 (short-term), computed using the modified Lyne-Hollick recursive digital filter [*Chapman*, 1991; *Nathan and McMahon*, 1990; *Lyne and Hollick*, 1979], are given in Table 2. Table 3 gives estimates of median base-flow discharge for each site for the same periods. These estimates are computed from equation 1, using the data in Table 2.

Base-flow discharge varies as a function of drainage area (Figure 5). The discharge values shown are the geometric mean of the indexed base-flow ratios, scaled by drainage areas. (Hence, a value of 1.0 means that the median unit-area base flow at the site is the same as that at the gaging station.) Note that stations on the same flow path are connected by line segments. The pattern shown for connected sites 12-11-6-4 and 18-16-13 is what one would normally expect. In the upper reaches of the watershed (e.g., site 12), unit-area base-flow discharge is low because groundwater tends to move downward and under the stream. Farther downstream, groundwater tends to flow upward toward the stream, and unit-area base-flow discharge increases.

The pattern shown by sites 9-7-4 and 8-4 was not expected. Unit-area base-flow discharges decrease significantly along these flow paths. Also, they are about 5 times higher at sites 7, 8, and 9 than at sites 12 and 18. The former observation implies that virtually all of the groundwater recharge above sites 7, 8, and 9 finds its way into the stream, perhaps as a result of an impermeable layer. The latter observation indicates that groundwater recharge rates decrease downstream in these watersheds.

Limited water temperature and fish data for sites 8, 11 and 16 (*L. Wang*, Wisc. Dept. Nat. Res., written communi-

Table 1. Summary of synoptic base-flow surveys and application of index method. Measured discharge is in cubic meters per second. Indexed discharge and coefficient of variation (CV) are dimensionless.

Site (Fig. 2)	Drainage area (km^2)	May 1995	Nov. 1995	July 1996	June 1998	July 1999	Arithmetic mean	CV	Indexed discharge		
									Arithmetic mean	CV	Geometric mean
4	42.5	0.27	0.25	0.24	0.23	0.22	0.24	0.08	0.80	0.13	0.80
6	12.4	0.084	0.074	0.054	0.071	0.068	0.070	0.15	0.78	0.15	0.77
7	10.9	0.086	0.085	0.085	0.082	0.065	0.081	0.11	1.04	0.16	1.03
8	7.0	0.060	0.052	0.065	0.062	0.048	0.057	0.13	1.15	0.14	1.14
9	1.8	0.013	0.026	0.017	0.008	0.008	0.014	0.52	1.19	0.61	1.03
11	9.1	0.057	0.054	0.025	0.042	0.045	0.045	0.28	0.69	0.30	0.66
12	2.6	0.0037	0.0037	0.0020		0.0031	0.0031	0.26	0.17	0.25	0.17
13	37.0	0.20	0.16	0.18	0.22	0.16	0.19	0.13	0.70	0.07	0.69
16	12.7	0.064	0.059	0.044	0.083	0.040	0.058	0.30	0.63	0.23	0.62
18	4.1	0.005	0.010	0.010	0.005		0.007	0.36	0.25	0.46	0.23
19	16.6	0.076	0.078	0.082			0.078	0.04	0.69	0.11	0.69
20	8.3	0.042	0.037	0.031			0.037	0.15	0.65	0.11	0.65
Kickapoo River gage	689	5.2	4.4	4.6	5.7	5.0	5.0	0.10			

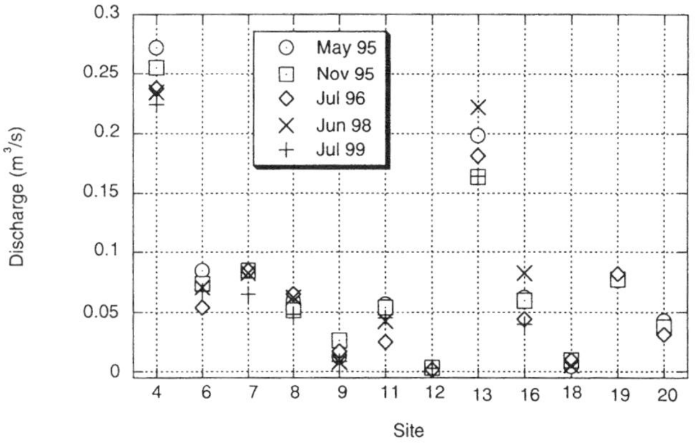

Figure 3. Measured base-flow discharges.

cation 2000) provide an indication that these patterns in base flow are significant for aquatic habitat. Higher unit-area base flow generally corresponds with cooler water temperatures and a greater percentage of coolwater and coldwater fish species (Table 4). One exception to this pattern is that site 11 is slightly cooler than site 8, even though site 8 has higher unit-area base flow. This difference in temperature may be due to other factors, such as shading by vegetation or channel shape, or the distribution of groundwater discharge upstream of the measurement sites.

DISCUSSION

The fact that indexing on a concurrent regional discharge did not reduce variability is disappointing. Variability due

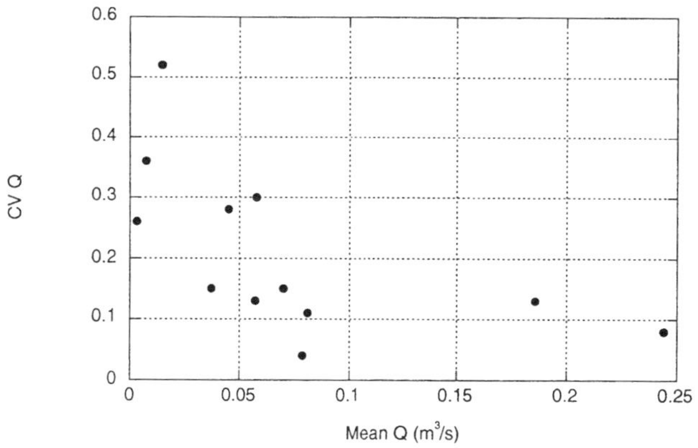

Figure 4. Coefficient of variation of measured base-flow discharges vs. the mean discharge.

Table 2. Base-flow discharge of the Kickapoo River at the LaFarge gaging station.

	Mean (m^3/s)	CV	Median (m^3/s)
1938-99	3.68	0.4	3.34
1995-99	4.05	0.2	3.77

Table 3. Estimated short- and long-term base-flow discharge for synoptic survey sites.

	Estimated median (m^3/s)	
Site	1938 - 99	1995 - 99
4	0.16	0.19
6	0.047	0.053
7	0.054	0.061
8	0.039	0.043
9	0.009	0.010
11	0.029	0.033
12	0.002	0.002
13	0.12	0.14
16	0.038	0.042
18	0.005	0.005
19	0.056	0.063
20	0.026	0.029

to measurement errors, which cannot be reduced by indexing, may partially explain this failure. For about one-half of the sites, the variability of the measured discharges was already very low, and may have been largely due to measurement error. At the sites most suitable for gaging, our discharge measurements are probably accurate within 5 to 10 percent. For the sites with high variability, the measured discharges were low. In these cases, measurement errors are likely to be significantly higher.

A second factor is the fact that the discharge measurements were made during a relatively wet period with no significant episodes of drought. This fact is reflected in the base flow at the Kickapoo gaging station. The mean base flow for 1995-99 was 10% higher than the period-of-record mean. Also, the coefficient of variation for 1995-99 was only 0.2, one-half of the period-of-record CV. This may mean that the temporal variability observed in base-flow discharges was due more to local rather than regional variations in groundwater recharge.

On the bright side, application of the method in the Hillsboro Lake watershed uncovered significant spatial variability in base-flow discharges, particularly at the upland sites. This finding has important habitat implications. Sites with relatively high unit-area base-flow discharges have greater average and minimum flow depths and more stable water temperatures, properties that are critical to the well-being of desired species such as brown trout and the native brook trout. Upland sites with high unit-area base-flow discharges are especially important. These sites have

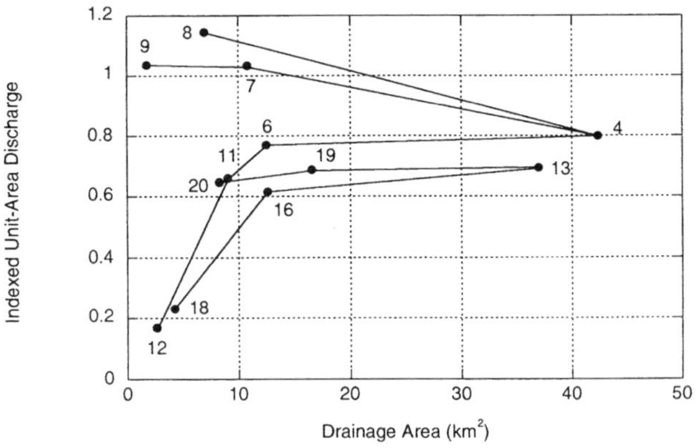

Figure 5. Indexed unit-area base-flow discharge. (Numbers on figure refer to measurement site.)

Table 4. Fish and temperature data for Hillsboro Lake watershed sites in summer 1998 (*L. Wang*, Wisc. Dept. Nat. Res., written communication 2000).

Stream	Site	Indexed arithmetic mean base flow	Water temperature (°C)		Total number of species	Percent coolwater and coldwater species
			Maximum	Mean		
Beaver Creek	16	0.63	26.1	19.6	12	0.3
Dilly Creek	11	0.69	22	16.2	11	14.0
West Branch tributary	8	1.15	22.6	16.3	7	35.1

channel characteristics, such as high gradients and coarse substrates, that are critical to the success of the native brook trout. Synoptic surveys can be used to identify these unique habitats and, hence, can guide protection and restoration efforts.

Acknowledgments. We wish to thank the Wisconsin Department of Natural Resources (WDNR) for funding a portion of this research. Lizhu Wang and John Lyons, of the WDNR, provided data on stream temperature and fish communities. We also thank Joseph Dorava, David Montgomery, Betty Palcsak, Aicardo Roa, and Daniel Sullivan for helpful comments on this manuscript.

REFERENCES

Becker, G. C., *Fishes of Wisconsin*, Univ. of Wisconsin Press, Madison, 1983.

Buchanan, T. J., and W. P. Somers, *Discharge measurements at gaging stations. Techniques of Water-Resources Investigations of the U. S. Geol. Surv.*, Bk. 3, Chap. A8, U.S. Geol. Surv., Alexandria, 1969.

Chapman, T. G., Comment on "Evaluation of automated techniques for base flow and recession analyses" by R. J. Nathan and T. A. McMahon, *Water Resour. Res.*, 27(7), 1783-1784, 1991.

Desai, M. C., *A Driftless Area watershed: synoptic analysis to evaluate agricultural management practices*, M.S. ind. study rpt., Univ. of Wisc.–Madison, 1996.

Gaffield, S. J., K. R. Bradbury and K. W. Potter, Hydrologic assessment of the Kickapoo Watershed, southwestern Wisconsin, *Wisc. Geol. & Nat. Hist. Surv., Open File Rpt. 1998-8*, 1998.

Johnson, W. C., *The impact of environmental change on fluvial systems: Kickapoo River*, Ph.D. thesis, Univ. of Wisc.–Madison, 1976.

Kent, C. A., *The Influence of Changes in Land Cover and Agricultural Land Management Practices on Baseflow in Southwest Wisconsin, 1969-1998*, Ph.D. thesis, Univ. of Wisc.–Madison, 1999.

Lyne, V., and M. Hollick, Stochastic time-variable rainfall-runoff modelling, *I.E. Aust. Natl. Conf. Publ. 79/10*, 89-93, Inst. of Eng., Aust., Canberra, 1979.

Nathan, R. J., and T. A. McMahon, Evaluation of automated techniques for base flow recession analyses, *Water Resour. Res.* 26(7), 1465-1473, 1990.

Potter, K. W., Hydrological impacts of changing land management practices in a moderately sized agricultural catchment, *Water Resour. Res.*, 27(5), 845-855, 1991.

Potter, K. W., A simple method for estimating base flow at ungaged locations. *Journ. Amer. Water Resour. Assoc.*, 37(1),177-184, 2001.

Sampson, N. R., *Farmland or wasteland: a time to choose*. Rodale Press, Emmaus, 1981.

Scott, W. E., and N. H. Hoveland, Report to the people of Wisconsin on cover destruction, habitat improvement and watershed problems on the state in 1950. *Wisc. Cons. Bull.*, 16(2), 3-78, 1951.

Trimble, S. W., and S. W. Lund, Soil conservation and the reduction of erosion and sedimentation in the Coon Creek basin, Wisconsin. *U.S. Geol. Surv. Prof. Pap., 1234*, Washington, 1982.

Young, H. L., Hydrogeology of the Cambrian-Ordovician aquifer system in the northern midwest, *U.S. Geol. Surv. Prof. Pap., 1405-B*, 1992.

Zaporozec, A., and R. D. Cotter, *Major ground-water units of Wisconsin*, Wisc. Geol. & Nat. Hist. Surv., Madison, 1985.

Kenneth W. Potter, 1261C Engineering Hall, 1415 Engineering Drive, Department of Civil & Environmental Engineering, University of Wisconsin, Madison, WI 53706-1607

Stephen J. Gaffield, Wisconsin Geological and Natural History Survey. 3817 Mineral Point Road, Madison, WI 53705-5100

Models of Habitat Use and Availability to Evaluate Anthropogenic Changes in Channel Geometry

James A. Gore

Department of Environmental Science and Public Health, Columbus State University, Columbus, Georgia

Five major hydraulic conditions most affect the distribution and ecological success of lotic biota: suspended load, bedload movement, and water column effects such as turbulence, velocity profile, and substratum interactions (near bed hydraulics). The interactions of these hydraulic conditions upon the morphology and behavior of the individual organisms govern the distribution of aquatic biota. Instream flow analyses assume that predictable changes in channel flow characteristics can, in turn, be used to predict the change in the density or distribution of lotic species or, at a minimum, the availability of useable habitat for those species. Historically, the Physical Habitat Simulation, PHABSIM, has been used to evaluate the flows necessary to maintain lotic biota under regulated flow conditions. However, PHABSIM has recently been employed to assess the success or potential value of instream restoration structures or the change in channel configuration as a result of land use change or riparian alteration. The model combines one- or two-dimensional simulations of instream hydraulics (linked to hydrological changes through back-step evaluations using Manning's or Chezy's equation) and biological habitat criteria to assess the change in available habitat for a target life stage or functional group of lotic species. This technique appears to be valuable in predicting changes in density and diversity among the macroinvertebrates and, with complex population models, can also predict recruitment success among some fish species. We have used the models to assess various restoration techniques (low-head weirs, deflectors, and artificial riffles) as well as bank stabilization activity. The model is sufficiently robust to predict habitat improvement, as well as changes in diversity and distribution of benthic species and improved maintenance and refuge habitat value for fish species. This paper will provide examples from projects in Tennessee and North Carolina along with generic recommendations for the application of this technique to other situations in which channel alteration is a focus of attention.

INTRODUCTION

The rehabilitation of lotic ecosystems involves an interdisciplinary approach that is focused upon recreating habitat characteristics and dynamics that will attract plant and animal colonists and support a sustainable ecosystem. Al-

Geomorphic Processes and Riverine Habitat
Water Science and Application Volume 4, pages 27-36

though stream restoration is most often a *response* to unplanned disturbance events, it is almost *as* often that some *a priori* planning can be made to minimize initial disturbance, thereby reducing the financial and personnel commitments to post-construction/disturbance restoration activity. Maintenance of a buffer zone, riparian corridor, or Streamside Management Zone (as denoted by the U.S. Forest Service) will reduce impacts from increased suspended loads, afford some amount of bank protection, control direct deposition of pollutants, and preserve a greater portion of stream habitat (through shade and food production) during human activity, which may take place adjacent to the river or stream (*Brookes et al.,* 1996a). If direct control of point sources of pollution is also provided, the restoration scientist can direct more effort towards the placement and construction of instream habitat structures, which enhance capture of organic detritus and *aufwuchs*, as well as, colonization by macroinvertebrate and fish species (*Gore,* 1985).

During the process of planning a stream restoration or rehabilitation project, it often becomes necessary to determine which types of structures will enhance existing habitat or colonization potential or which types will be the most stable under typical hydrographic conditions. Indeed, under certain regulatory conditions, it may be necessary to produce an *a priori* evaluation of the restoration or rehabilitation project in order to justify certain monetary investments. In these cases, the application of models of habitat preference and use by target species for management can be a valuable tool for resolving some of these issues.

HABITAT STRUCTURES

Traditionally, habitat structures have been designed and placed to enhance fish production. The most important physical and chemical criteria for fish production are acceptable water quality, food production areas, spawning/egg incubation areas, rearing areas for fry, feeding stations and cover [*Wesche*, 1985; *Nehring and Anderson*, 1993; *Bovee et al.,* 1994, *Brookes et al.*, 1996b]. In all cases, a plan for the design and placement of restoration devices is based upon the assumption that these habitat requirements can be controlled through design of structures that produce preferred physical and chemical conditions as related to flow conditions. In the same manner, restoration scientists are assuming that hydraulic conditions are one primary template that governs the distribution of lotic organisms. *Statzner et al.* [1988; for invertebrates] and *Heede and Rinne* [1990; for fish] have reviewed the data that support this assumption. Both reviews have shown that complex flow conditions such as Reynolds velocities and shear stress, which incorporate combined factors of depth, velocity, bed roughness, and viscosity of the water, are predictors of benthic densities and, perhaps, fish population response [*Bartholow et al.*, 1993].

Aquatic invertebrates are associated with a wide variety of substrates ranging from rooted vegetation and dead wood to periphyton to all sizes of inorganic particulates (silt, sand, gravel, and cobble). Although it is difficult to maintain, once in place, substrate composition is the most easily manipulated habitat characteristic in restoration projects. This newly emplaced substrate will become the "island" habitats for colonizing benthic macroinvertebrates (*Gore,* 1979, 1982). Thus, it will be necessary to consider such factors as degree of embeddedness of particles, size of particles, contour of the substrate, and heterogeneity of substrate types in the source and recipient areas so that the rate of establishment of permanent benthic communities can be maximized [*Gore,* 1985]. *Merritt and Cummins* [1996] have listed major habitat types for families and genera of aquatic insects in North America. This listing can be a useful guide if source areas of colonizers are unknown or unrestored. In most cases, highest diversities and production have been reported from channels with medium cobble (256 mm diameter) and gravel substrates [*Gore et al.*, 2001].

Spawning habitat has been measured extensively with special emphasis on salmonid fisheries. Generally, riffle areas with velocities between 15 and 90 cm s^{-1}, water depths greater than 15 cm, and cobbled and graveled substrates (0.6 to 7.6 cm diameter) are preferred by most coldwater fish [*Wesche*, 1985]. Demand for oxygen increases with embryonic development but must be at least 5.0 mg/l, measured interstitially. Because water percolation brings oxygen for incubation, the shape of the streambed is critical to supplying optimum oxygenation. *Vaux* [1962] found increased permeability and a convex streambed induced downwelling of water into egg nests within or on the substrate whereas concave streambeds caused a general upwelling. The downwelling provides oxygen as well as removal of metabolic wastes.

Cover is any structure that protects fish from mechanical damage by high current velocities and predation. Overhanging riparian bodies, submerged macrophytes, and submerged instream objects provide natural cover. Overhead cover is used by species showing photonegative behavior and thigmotaxis (contact with a solid body). Instream objects also create shear zones, which are favored feeding stations for a variety of fish species. Generally, fish establish territories around cover structures that are used for resting and feeding (*Hooper,* 1973).

The most common structures for fish habitat enhancement have been current deflectors, overpour structures (dams and weirs), bank cover, and boulder placements.

These instream structures also modify local hydraulic conditions to present preferred habitat to benthic invertebrates. Log-drop structures and other dams improve fish habitat by trapping sediment with minor impact on invertebrate densities whereas deflectors increase water velocity over riffles to remove accumulating sediments from cobble substrates [*Luedtke et al.*, 1976]. For macroinvertebrates, then, primary considerations include formation of sediment-free riffle areas and structures to control sediment deposition. Such structures as deflectors (gabion, single, or double wing), check dams, and large boulders have all been demonstrated to produce scoured areas of riffle habitat and pools [*Gore and Johnson*, 1981; *Wesche*, 1985; *Thompson*, 1985]. Construction details have been described in a number of publications [*Wesche*, 1985; *Newbury and Gaboury*, 1993; *Brookes et al.*, 1996b; *FISRWG*, 1998].

HABITAT MODELING

It may be the case that the value of certain structures must be demonstrated to permitting agencies *prior* to installation of those structures. Often, although many publications have demonstrated the utility of such structures, resource agencies will permit "exploratory" or "experimental" evaluations of restoration activities. Only after the demonstration of significant improvement to ecosystem structure or function can a full-scale restoration design be approved. Thus, there must be a mechanism to evaluate benefit without waiting for the recolonization cycles and biological adjustments to occur that usually demonstrate successful habitat enhancement.

Gore [1985] demonstrated a transect-hydraulic method for mapping changes in available habitat with changes in discharge before and after placement of structures and suggested that instream-flow computer simulations should be adequate to predict the benefit of restoration structures. In these cases, gravel bars created by scour and substrate particle displacement, as well as man-made riffles created optimum habitat conditions. More recently, *Shuler and Nehring* (1993) have demonstrated that the Physical Habitat Simulation (PHABSIM), the software package used in the Instream Flow Incremental Methodology, can be used to evaluate stream enhancement activities.

The Instream Flow Incremental Methodology (IFIM) is a series of concepts, techniques and computer programs to combine channel morphology, characteristics of flow and biological preferences of target organisms to predict gains or losses in physical habitat under new or modified flow regimes. IFIM is maintained by the U.S. Fish and Wildlife Service and is currently endorsed as one of the primary negotiating tools in disputes over flows needed to maintain ecological integrity; that is, to support target fish populations and/or community dynamics [*Gore and Nestler*, 1988]. IFIM and PHABSIM are often thought to be synonymous. In fact, IFIM is a generic decision-making model that employs systems analysis techniques. IFIM guides stream managers in the process of choosing appropriate targets, endpoints, and data requirements to achieve the management goal. At one level or another, IFIM requires a substantive knowledge of how aquatic habitat value changes as a function of incremental changes in discharge. Replicate habitat sampling, biological sampling for the development of habitat suitability curves, sediment and water-routing studies, as well as physical habitat, temperature, and water-quality simulations may be necessary to properly depict the condition of the catchment under new operating scenarios [*Sale*, 1985].

IFIM is based on the assumption that lotic biota have their distribution and certain phases of their life cycles controlled by the hydraulic conditions within the water column. The essence of IFIM is the computer simulation of the physical habitat, PHABSIM, which predicts changes in velocity, depth and channel resistance in each habitat element surveyed. These conditions are compared to the habitat preferences of target species, guilds or communities. The amount of useable habitat area (or volume) is predicted for each discharge of concern (Figure 1). *Bovee* [1986] and *Bovee et al.* [1998] have provided details of the IFIM procedures. *Gore and Nestler* [1988], *Nestler et al.* [1989], and *Gore and Mead* [2000] have provided reviews of this technique and appropriate application.

In IFIM, habitat suitability is treated as both *macrohabitat* and *microhabitat*. Macrohabitat suitability is predicted by measurement and/or simulation of changes in water quality, channel morphology, temperature, and discharge along the length of the targeted reach. These conditions may have an overriding impact upon decisions made at the microhabitat level.

Microhabitat suitability consists of individual species' preferences for these same criteria, reflected as depth, velocity, substrate or channel condition, and cover. Those individual preferences are incorporated into PHABSIM to obtain predictions of changes in available habitat at a selection of stream segments, "typical" of the reach being managed. In combination, microhabitat and macrohabitat provide the information necessary to adequately negotiate management alternatives (Figure 1).

The microhabitat evaluation within the IFIM methodology is completed through PHABSIM. Through a series of subroutine programs contained within PHABSIM, a prediction of the amount of available habitat (as weighted usable area, WUA) for a target organism over a range of discharges is created. HABTAT and its associate programs requires hydrologic information in the form of transect

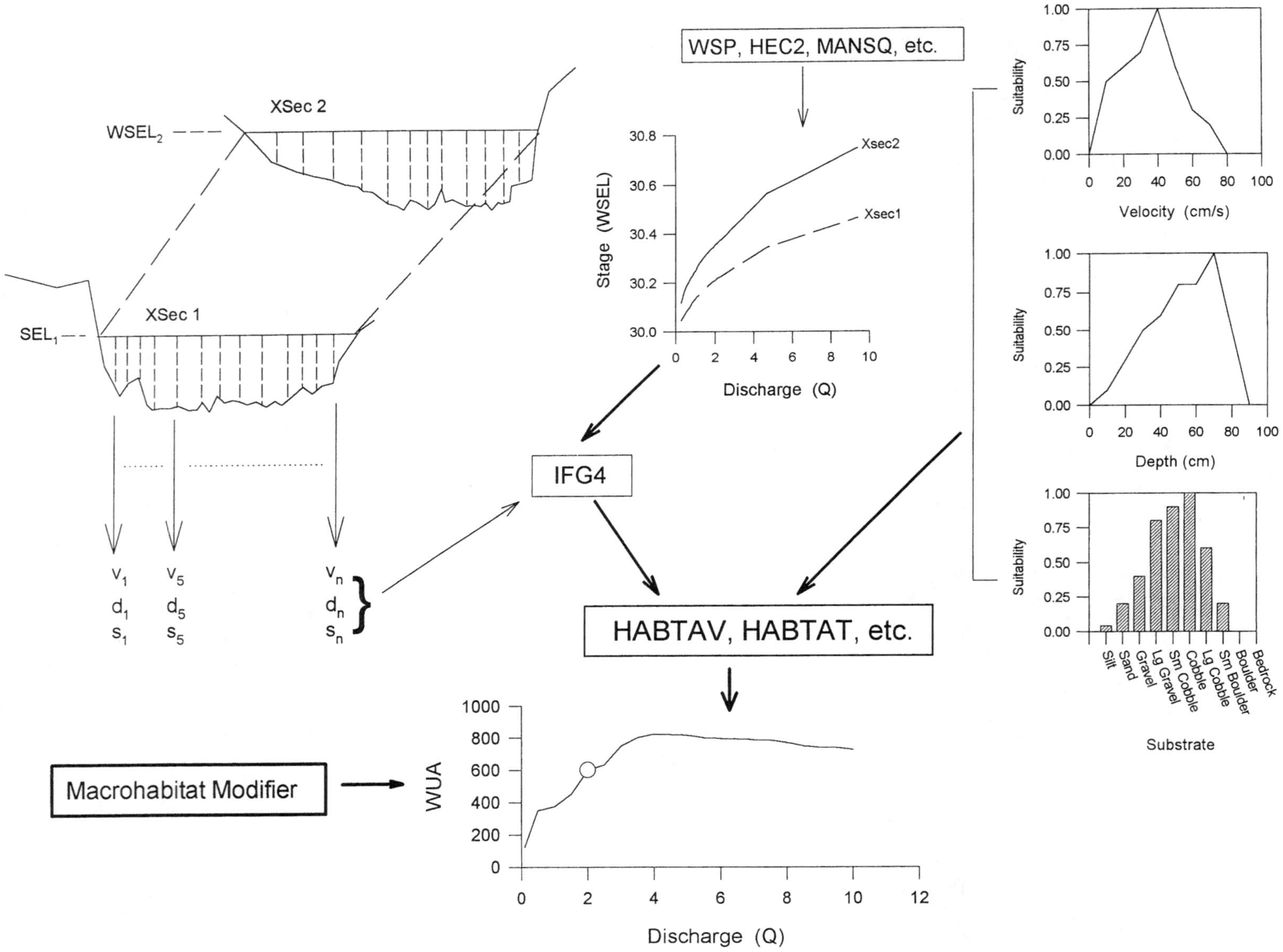

Figure 1. A schematic representation of the basic elements necessary to create predictions of habitat availability (WUA) using the Physical Habitat Simulation (PHABSIM). The macrohabitat modifier is derived from external evaluations or simulations.

(cell-by-cell) information on depth, velocity, cover value and/or substrate composition, and biological information in the form of preferences or suitabilities for these conditions by the target organism. Where possible, the hydraulic information for each transect should be measured. However, several desktop simulations can also simulate these data when field measurements are not available or are impossible to measure (in the case of very large rivers or those with rapidly varying, unsteady flow). In addition to simulations within PHABSIM (routines such as WSP, MANSQ, and IFG4), other hydraulic simulations are commonly used. These include steady-state models such as HEC-2 [*USACE*, 1982] and dynamic flow models such as RIV1H [*Bedford et al.*, 1983] and BIRM [*Johnson*, 1983]. Regardless of how the hydraulic information is provided to PHABSIM, stage-discharge relations are provided to the hydrologic simulation (usually IFG4) that predicts changes in velocity, cell-by-cell, with changes in water surface elevation. This prediction is accomplished through a series of back-step calculations through Manning's equation or, at the option of the user, Chezy's equation. This assumption assumes that substrate or channel geometry will remain stable over the range of discharges to be simulated. As an alternative, *Jowett* [1998] has suggested that on-site measurement of changes in hydraulic geometry provide estimates comparable to the back-step predictions contained within IFG4.

Thus, the PHABSIM model, in its current form, represents, at best, a quasi-two-dimensional model, because it distributes velocities and discharges laterally along each transect. Cell-by-cell evaluations of weighted usable area (WUA; the product of preference criteria for each of the hydraulic conditions simulated and the total surface area of each cell) are computed through HABTAT and related subroutines. The WUA/discharge relation can provide information on the potential gains and losses of habitat resulting from changes in discharge. If PHABSIM were to be applied to the evaluation of habitat structures, an evaluation of habitat availability with and without the channel modifications under the prevailing flow regime can demonstrate the improvement. That is, the decision-makers must decide what change in habitat will occur, given the hydrologic setting. This will be particularly important when decisions must be made during unusually wet or dry years. This is accomplished through the Habitat Time Series (HTS) component of IFIM [*Milhous et al.*, 1990]. Such conditions as median habitat value over 10 or 20 years of record, the percentage of available habitat if certain magnitudes of flood were attenuated or enhanced, and the duration of low habitat conditions are typical predictions of a HTS evaluation. Indeed, since the majority of large rivers and many low and medium-order rivers have flow regulation structures placed in them, the HTS becomes a valuable, if not mandatory, tool for managing flows to enhance or support the rehabilitation effort.

Traditionally, the IFIM protocol has been used to negotiate instream flow operations or standards; often as minimum-flow requirements to maintain biota in cases where reservoir manipulation or irrigation diversion and abstraction are potential disturbances. However, some recent applications have also included evaluations of rapidly fluctuating flows, as peaking hydropower generation, since high flow events can also be demonstrated to create habitat limiting conditions (*Gore et al.,* 1989). *Gore* [1985] suggested that IFIM could be easily modified to predict the changes in flow conditions after placement of restoration structures in surface-mined rivers. By re-surveying the channel after placement of structures, Gore showed that maintenance habitat for maximum macroinvertebrate community diversity had increased five-fold or more along selected reaches of a reclaimed coal strip-mined channel. This result suggests that by modifying the channel configuration in the hydraulic subroutine IFG4 so that structures are incorporated and new water surfaces estimated, a prediction of the benefit of placement of restoration structures could be obtained.

In a severely head-cut stream in Tennessee, three-log weirs were constructed at intervals of 30 to 40 m in order to control sedimentation from agricultural runoff and to assess the effectiveness of these structures as habitat improvements for lotic biota [*Gore and Hamilton*, 1996]. This section was surveyed according to recommended techniques and evaluated for available habitat for benthic macroinvertebrates at a series of typical discharges. After placement of the weirs, the reach was re-surveyed and available habitat was recomputed using PHABSIM. Habitat suitability curves used in the analysis did not represent single macroinvertebrate species but, using the techniques proposed by *Gore* [1989], were curves which represented conditions that promote the highest community diversity; highest sample diversities occurring at 65 cm s^{-1}, depths of 27 cm, and substrates of gravel and small cobble [see also *Gore et al.*, 2001, for more recent versions of those models]. These data were derived from samples of benthic macroinvertebrates collected from unimpacted streams in adjacent catchments.

Dramatic increases in available habitat for benthic macroinvertebrates occurred in the exposed gravel apron immediately downstream of each weir (Figure 2). Over the normal range of discharges during an annual hydrograph, PHABSIM analysis showed that the effect of the structures was to maintain high-quality habitat during periods of

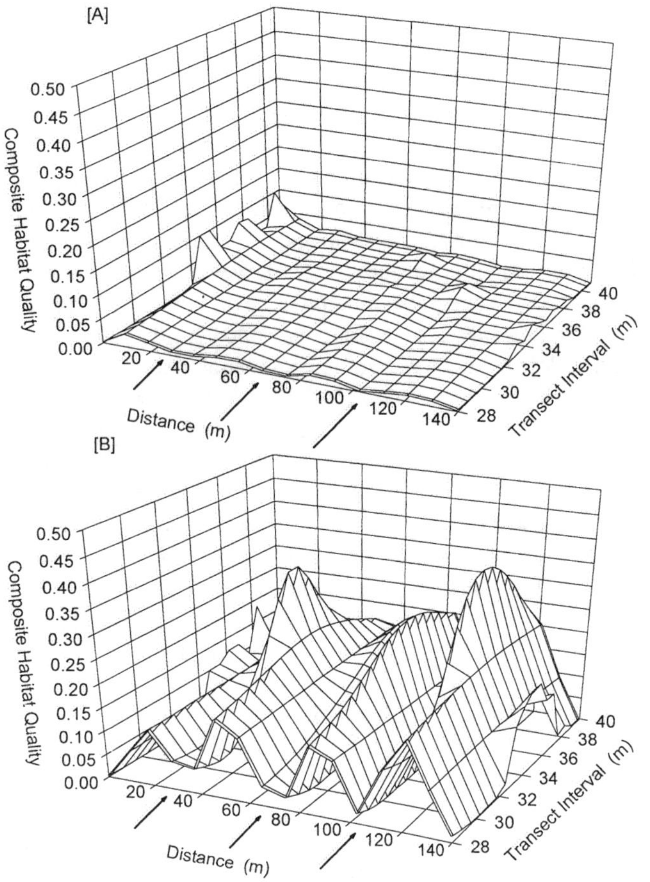

Figure 2. Evaluation of cell-by-cell habitat quality (product of velocity, depth, and substrate/cover preferences) for benthic macroinvertebrates prior to installation of restoration structures [A] on Brushy Branch, Tennessee. The arrows indicate location of proposed three-log weirs. A re-evaluation of habitat quality on Brushy Branch [B] after placement of wiers (at arrows) and stabilization of embankments and new channel geometry.

minimal flow by impounding water and discharging it at higher velocities across the gravel and cobble aprons (Figure 3).

A more sophisticated approach to predicting habitat enhancement by restoration structures has been suggested by *Gore and Nestler* [1988]. For dynamic flow analyses, both RIV1H [*Bedford et al.*, 1983] and BIRM (**B**ranched **Im**plicit **R**iver **M**odel) [*Johnson*, 1983] have been used to generate time-varying stage-discharge information to represent conditions during hydropeaking operation. RIV1H has the advantage of easily incorporating structures into the simulation. Using RIV1H, *Curtis et al.* [1982] demonstrated that placement of a single reregulation weir across the Cumberland River downstream of a peak-demand hydropower facility could be expected to increase habitat availability by 30% for brown trout and as much as 90% for rainbow trout as a result of the dampening of the rising limb of peaking-generation waves and the reduction of water velocity in sheltering pools. *Gore and Hamilton* [1996] demonstrated that three-log weirs in low-order streams had the same affect on habitat enhancement that reregulation structures did; that is, of habitat improvement during low-flow periods.

The linkage between water-quality models (CE-QUAL), dynamic flow (RIV1H), and habitat models (PHABSIM) has the potential of also answering many restoration questions even when modeling unregulated flow conditions. A monthly or daily hydrograph can be substituted and predictions of change in habitat availability with and without structures can be made. This sort of predictive ability is promising and continued research is underway although more studies on the nature of habitat quality are needed.

Similar studies on the impacts of placement of artificial riffles (Figure 4) have demonstrated that PHABSIM can be used to demonstrate improvement in quantity and quality, as well as the location of optimal habitat within the riffle (Figure 5) over a range of discharges [*Gore et al.*, 1998]. Indeed, *Gore et al.* [1998] demonstrated that most macroinvertebrates, like the more mobile fish species, were able to track the change in habitat conditions and relocate, apparently by active migration or drift-and-recolonization, to suitable habitat conditions as discharge changed over the artificial riffle. Thus, where possible, flows can be manipulated to optimize the ability of biota to recolonize and establish resident populations in a restored stream or river sections.

The greatest drawback to application of IFIM techniques to stream restoration is the lack of a clearly demonstrated relation between usable habitat area and population numbers or biomass support. Although *Bovee et al.* [1998] have suggested that the relationship between usable habitat area and carrying capacity of the stream reach is implicit in the development of PHABSIM, this has been difficult to demonstrate in practice. There has been any number of successes [*Cheslak and Jacobson*, 1990; *Nehring and Anderson*, 1993; *Bovee et al.* 1994] and purported failures [*Scott and Shirvell*, 1987] in predicting changes in fish populations using PHABSIM. However, *Statzner et al.* [1988] demonstrated that for more non-mobile organisms, such as benthic macroinvertebrates, the relationship between physical habitat and density is predictable. However, it may not be necessary to incorporate models of complex hydraulic conditions. Statzner *et al.* [1998] have shown that simple velocity, depth, and substrate models, created independently, will provide reasonable estimates of change in density or community structure with change in hydrologic conditions. The problem of mobility of most fish species is reflected in studies of peak-demand hydropower operations in which it is easy to estimate increases or losses in available habitat as the waves created by the peaking-power generation pass through a stream reach but difficult to demonstrate the response of the fish to that temporary gain or loss of habitat [*Nestler et al.*, 1988]. *Gore et al.* [1989)] found that observations of selected fish species living in rivers influenced by peaking hydropower generation suggest that habitat quality must be measured in two different ways: (1) as maintenance habitat during normal flow periods and (2) as refuge habitat during high flow periods. This method is probably a more accurate means of evaluating habitat quality but does not indicate the length of time or the ability of organisms to occupy the newly created or enhanced habitat.

PHABSIM currently uses a one-dimensional (1-D) hydraulic model to describe depth and velocity changes in a stream. This system is based on some simple energy and mass balance equations both between and across measured transects. A major criticism of this system has been the lack of accuracy when measuring a stream's frequently complex flow patterns such as eddies and divergences

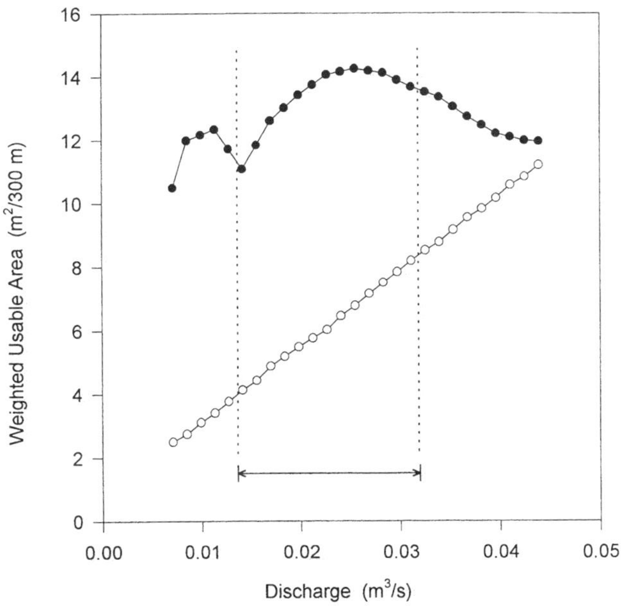

Figure 3. An overall evaluation of the "value" of the placement of three three-log weirs as restoration structures on Brushy Branch, Tennessee. Open circles represent habitat quality in the stream prior to restoration while solid circles represent habitat quality after placement of the structures. Double-headed arrow and broken lines represent the range of flows that exist at least 80% of the time.

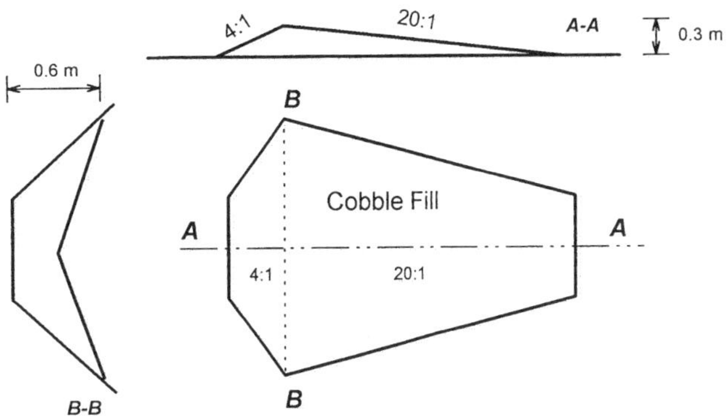

Figure 4. Design of an artificial riffle for placement in low order streams. The B-B distance is bank-to-bank. Small and medium cobble make up the fill of the riffle with slight taper from the center crest towards each embankment. Design is modified from *Newbury and Gaboury* [1993].

[*Waddle*, 1998]. Two-dimensional (2-D) models use finite element or finite difference techniques (rather than back stepping through Manning's equation) to describe fluid motion. Thus, it becomes easier to describe such phenomena as partially wetted side channels, point bars, islands, and intermittent inundation; these conditions being ignored or avoided in current IFIM analysis. Two-dimensional models rely upon explicit areal representations of a stream rather than data derived from individual transects. This makes 2-D model quite compatible with the latest technologies in Geographical Information Systems (GIS) abilities. When fully developed, 2-D models should be able to allow the evaluation of such conditions as velocity shear zones and side channels that are critical habitats for various life stages of both fish and invertebrate species.

Over the past decade, a number of investigators have been developing 2-D models and attempting to couple them with calculations of habitat availability [*Leclerc et al.*, 1995; *Bovee*, 1996]. The initial results are encouraging and seem to provide a more accurate representation of habitat conditions and the location of unusual habitat phenomena such as eddies. It remains for ecohydrologists to begin the development of new habitat criteria that reflect preference for these conditions and response to changes in eddies or wetting of side channels, for example.

Although certainly not the only templates that determine the successful occupancy of a certain section of a stream reach, hydraulic variables appear to be a major influence on the distribution of most lotic species. Hydraulic habitat models, then, can be useful tools for evaluating the benefit of certain restoration or rehabilitation activities. IFIM is a flexible, general method for evaluation of habitat conditions under a wide variety of hydrological conditions. Only

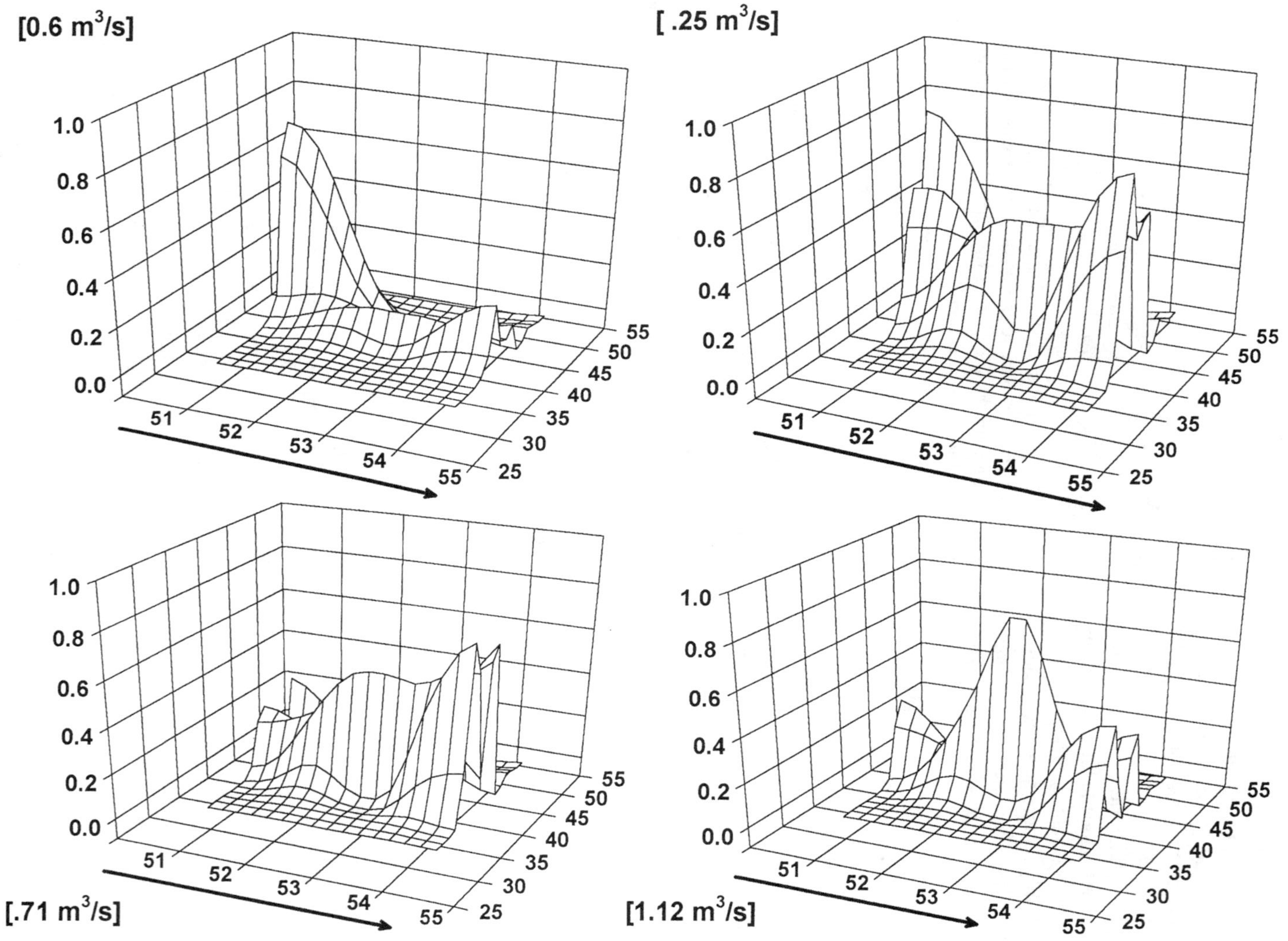

Figure 5. A cell-by-cell evaluation of habitat quality over a range of discharges across the artificial riffle described in Figure 4, placed in Holly Fork, Tennessee. Arrow indicates the direction of flow across the riffle and during the evaluation. Composite suitability is on the vertical axis in each plot, transect distance is along the axis parallel to the arrow, and transect interval is along the remaining axis. As discharge increases, optimal habitat conditions move from the forward crest on the artificial riffle towards the tail of the riffle and back towards the crest, while spreading laterally.

recently has this technique been explored for application to restoration or rehabilitation. The initial findings, as reported here, suggest that this is a beneficial avenue for further research with a goal of improving our ability to restore or rehabilitate lotic ecosystems and sustain their communities for long periods of time.

Acknowledgments. I thank Mr. Ken Bovee, U.S. Geological Survey, Dr. David Montgomery, Dept. of Geological Sciences, University of Washington, and two anonymous referees for their valuable contributions and suggestions that greatly improved the quality of this manuscript.

REFERENCES

Bartholow, J. M., J. L. Laake, C. B. Stalnaker, and S. C. Williamson, A salmonid population model with emphasis on habitat limitations, *Rivers* 4: 265-279, 1993.

Bedford, K. W., R. M. Sykes, and C. Libicki, Dynamic advective water quality model for rivers, *J. Environm. Eng. Div.*, Amer. Soc. Civil Eng., 109: 489, 1983.

Bovee, K. D., Development and evaluation of habitat suitability criteria for use in the instream flow incremental methodology, Instream Flow Info. Paper No. 21, U.S. Fish. Wildl. Serv., Bio. Rpt. 86(7), 1986.

Bovee, K. D., Perspectives on two-dimensional river habitat models: the PHABSIM experience, in *Ecohydraulics 2000, Proceedings of the 2nd International Symposium on Habitat Hydraulics* edited by M. Leclerc, H. Capra, S. Valentin, and Y. Côté , pp. B149-BB162, INRS-Eau, Québec, 1996.

Bovee, K. D., B. L. Lamb, J. M. Batholow, C. B. Stalnaker, J. Taylor, and J. Henriksen, Stream habitat analysis using the instream flow incremental methodology, U.S. Geol. Survey, Biol. Res. Div., Tech. Rpt. USGS/BRD-1998-0004, 1998.

Bovee, K. D., T. J. Newcomb, and T. G. Coon, Relations between habitat variability and population dynamics of bass in the Huron River, Michigan, U. S. Fish Wildl. Serv., Nat. Biol. Survey, Biol. Rpt. 21, Washington, DC, 1994.

Brookes, A., J. Baker, and C. Redmond, Floodplain restoration and riparian zone management, in *River Channel Restoration: Guiding Principles for Sustainable Projects.* edited by A. Brookes and F.D. Shields Jr., pp. 201-229, John Wiley & Sons, Ltd., London, 1996.

Brookes, A., S. S. Knight, and F. D. Shields, Jr., Habitat enhancement, pp. 103-126 in *River Channel Restoration: Guiding Principles for Sustainable Projects* edited by A. Brookes and F.D. Shields Jr., pp. 103-126, John Wiley & Sons, Ltd., London, 1996.

Cheslak, E. F., and A. S. Jacobson, Integrating the instream flow incremental methodology with a population response model, *Rivers* 1: 264-288, 1990.

Curtis, L. T., J. M. Nestler, and J. L. Martin, Comparative effects on trout habitat of hydropower modification with and without reregulation in the Cumberland River below Wolf Creek Dam, Kentucky, Misc. *Paper EL-87-2*, U.S. Army Engineer Waterways Experiment Station, Vicksburg, Miss., 1982.

Federal Interagency Stream Restoration Working Group (FISRWG), *Stream Corridor Restoration.* Washington, DC, 1998.

Gore, J. A., Patterns of initial benthic recolonization of a reclaimed coal strip-mined river channel, *Can. J. Zool.* 57: 2429-2439, 1979.

Gore, J. A., Benthic invertebrate colonization: source distance effects on community composition, *Hydrobiologia* 94: 183-193, 1982.

Gore, J. A., Mechanisms of colonization and habitat enhancement for benthic macroinvertebrates in restored river channels, in *The Restoration of Rivers and Streams*, edited by J.A. Gore, pp. 81-101, Butterworth, Boston, 1985.

Gore, J. A., Models for predicting benthic macroinvertebrate habitat suitability under regulated flows, in *Alternatives in Regulated River Management* edited by J.A. Gore and G.E. Petts, pp. 253-265, CRC Press, Inc., Boca Raton, FL 1989.

Gore, J. A., D. J. Crawford, and D. S. Addison, An analysis of artificial riffles and enhancement of benthic community diversity by Physical Habitat Simulation (PHABSIM) and direct observation, Regulated *Rivers* 14: 69-77, 1998.

Gore, J. A., and S. W. Hamilton, A comparison of flow-related habitat evaluations downstream of low-head weirs on small and large fluvial ecosystems, *Regulated Rivers* 12: 459-469, 1996.

Gore, J. A., and L. S. Johnson, Strip-mined river restoration, *Water Spectrum* 13: 31-38, 1981.

Gore, J. A., J. B. Layzer, and J. Mead, Macroinvertebrate instream flow studies after 20 years: a role in stream management and restoration, *Regulated Rivers* (In Press), 2001.

Gore, J. A., J. M. Nestler, and J. B. Layzer, Instream flow predictions and management options for biota affected by peaking hydropower releases, *Regulated Rivers* 3: 35-48, 1989.

Gore, J. A., and J. Mead, The Benefits and Dangers of Ecohydrological Models to Water Resource Management Decisions, in: *Ecohydrology: A new Paradigm,* edited by G. Janauer and M. Zalewski, United Nations/UNESCO, Geneva and Cambridge University Press, 2000.

Gore, J. A., and J. M. Nestler, Instream flow studies in perspective, *Regulated Rivers* 2: 93-101, 1988.

Gore, J. A., J. M. Nestler, and J. B. Layzer, Instream flow predictions and management options for biota affected by peaking-power hydroelectric operations, *Regulated Rivers* 3: 35-48, 1989.

Heede, B. H., and J. N. Rinne, Hydrodynamic and fluvial geomorphological processes: implications for fisheries management and research, *N. Amer. J. Fish. Mgmt.* 10: 249-268, 1990.

Hooper, D. R. *Evaluation of the Effects of Flows on Trout Stream Ecology*, Pacific Gas and Electric, Emeryville, Calif., 1973.

Johnson, B. H., User's guide for Branched Implicit River Model (BIRM) with application to the lower Mississippi

River, U.S. Army Engineer Waterways Experiment Station, Vicksburg, Miss., 1983.

Jowett, I. G, Hydraulic geometry of New Zealand rivers and its use as a preliminary method of habitat assessment, *Regulated Rivers* 14: 451-466, 1998.

Leclerc, M., A. Boudrealut, J. Bechara, and G. Corfa, Two-dimensional hydrodynamic modeling: a neglected tool in the instream flow incremental methodology, *Trans. Amer. Fish. Soc.* 124: 645-662, 1995.

Luedtke, R. J., M. A. Brusven, and F. J. Watts, Benthic insect community changes in relation to instream alterations of a sediment-polluted stream, *Melanderia* 23: 21-39, 1976.

Merritt, R. W. and K. W. Cummins (eds.), *An Introduction to the Aquatic Insects of North America*, 3rd. Ed., Kendall/Hunt, Dubuque, Iowa, 1996.

Milhous, R. T., J. M. Bartholow, M. A. Updike, and A. R. Moos, Reference *manual for generation and analysis of habitat time series - Version II.* U.S. Fish and Wildlife Service, Biol. Rpt. 90(16), 1990.

Nehring, R. B., and R. M. Anderson, Determination of population-limiting critical salmonid habitats in Colorado streams using the Physical Habitat Simulation system, *Rivers* 4: 1-19, 1993.

Nestler, J. M., J. A. Gore, L. T. Curtis, and J. L. Martin, Predicted effects of hydropower uprate on trout habitat in the Cumberland River, downstream of Wolf Creek Dam, Kentucky, Misc. Rpt. EL-88-10, U.S. Army Engineer Waterways Experiment Station, Vicksburg, Miss., 1988.

Nestler, J. M., R. T. Milhous, and J. B. Layzer, Instream habitat modeling techniques, in: *Alternatives in Regulated River Management,* edited by J. A. Gore and G. E. Petts, pp. 295-315, CRC Press, Inc., Boca Raton, Fla., 1989.

Newbury, R. W., and M. N. Gaboury, *Stream Analysis and Fish Habitat Design. A Field Manual.* Newbury Hydraulics Ltd., Gibsons, B.C., and The Manitoba Habitat Heritage Corporation, Winnipeg, 1993.

Sale, M. J, Aquatic ecosystem response to flow modification: an overview of the issues, in *Proceedings of the Symposium on Small Hydropower and Fisheries*, edited by F. W. Olson, R. G. White, and R. H. Hamre, pp. 22-31, American Fisheries Society, Bethesda, M., 1985.

Scott, D. S., and C. S. Shirvell, A critique of the instream flow incremental methodology and observations on flow determinations in New Zealand, in: *Regulated Streams,* edited by J. F. Craig and J. B. Kemper, pp. 27-43, Plenum Press, NY, 1987.

Shuler, S. W., and R. B. Nehring, Using the physical habitat simulation model to evaluate a stream habitat enhancement project, *Rivers* 4: 175-193, 1993.

Statzner, B., J. A. Gore, and V. H. Resh, Hydraulic stream ecology: observed patterns and potential application, *J.N. Amer. Benthol. Soc.* 7: 307-360, 1988.

Statzner, B., J. A. Gore, and V. H. Resh, Monte Carlo simulation of benthic macroinvertebrate populations: Estimates using random, stratified, and gradient sampling. *J.N. Am. Benthol. Soc.* 17: 324-337, 1998.

Thompson, C. S., Stream relocation on surface mined land, in: *Fish and Wildlife Relationships to Mining.* Edited by L. B. Starnes, pp. 39-49, Water Quality Section, American Fisheries Society, Milwaukee, Wis, 1985.

U.S. Army Corps of Engineers (USACE), *HEC-2, Water Surface Profiles*, USACE, Hydrologic Engineering Center, Davis, Calif., 1982.

Vaux, W. B., Interchange of stream and intergravel water in a salmon spawning riffle, Spec. Sci. Rep. Fish. 405, U.S. Fish and Wildlife Service, Washington, D. C., 1962.

Waddle, T., Development of 2-Dimensional habitat models, in: *Hydroecological modeling: research, practice, legislation, and decision-making*, edited by S. Blazkova, C. Stalnaker, and O. Novicky , pp. 19-22, T.G. Masaryk Water Research Institute, Prague, Czech Republic, and MESC, Fort Collins, Colo., 1998.

Wesche, T. A, Stream channel modifications and reclamation structures to enhance fish habitat, in: edited by J. A. Gore, pp. 103-163, Butterworth, Boston, 1985.

Mailing address: Dr. James A. Gore, Dept. of Environmental Science and Public Health, Columbus State University, Columbus, GA 31909 USA

Holocene and Recent Geomorphic Processes, Land Use, and Salmonid Habitat in two North Puget Sound River Basins

Timothy J. Beechie[1], Brian D. Collins[2], George R. Pess[1]

The quantity, quality, and distribution of salmonid habitats in the Skagit and Stillaguamish River basins have changed dramatically in response to post-glacial landscape evolution and volcanism over the last 16,000 years, and the more recent history of land use (approximately 150 years). After retreat of the Cordilleran ice sheet about 16,000 years ago, streams incised rapidly into valley-filling glacial sediments, lowering valley floors and creating terraces. Mainstems and floodplain sloughs on valley floors provided the majority of habitat, but moderate-gradient tributaries on terraces provided additional habitat for some salmonids. Channels in bedrock terrain were too steep to support anadromous salmonids and remain so today. Voluminous lahars from Glacier Peak approximately 5,500 years before present created an extensive low-gradient delta on the Skagit River, which then developed abundant habitats in wetlands and distributary channels. Since non-Native American settlers arrived in the mid-1800s, removal of beaver ponds, diking, ditching, and dredging of streams on the floodplains and deltas has isolated or obliterated approximately 50% of the coho salmon winter rearing habitat in both basins. These losses are associated mainly with agricultural practices, which occupy the same landforms as the majority of historical coho salmon habitat. Forestry activities are concentrated on the steeper slopes of the glacial sediments and bedrock terrain, and contribute to habitat losses by increasing sediment supplies and reducing wood abundance. Understanding the interplay of Holocene landscape evolution, geomorphic processes, land use, and salmonid habitat provides a context for developing habitat restoration programs.

INTRODUCTION

Changes in stream habitat conditions over the past 150 years have played a significant role in reducing salmonid populations in the Pacific Northwest [*Bisson et al.*, 1992]. In the northern Puget Sound region of Washington State, these changes have resulted from direct manipulations of streams and rivers such as levee construction, and filling or ditching of channels [*Beechie et al.*, 1994], as well as from changes to geomorphic processes including landsliding and recruitment of wood from riparian forests [*Beechie*, 1998]. However, there has been little attempt to understand how these changes compare to natural habitat changes prior to recent land uses. An accounting of natural changes to the landscape and habitat conditions since retreat of the continental ice sheet around 16,000 years before present (ybp) provides a long-term context for recent habitat alterations caused by land uses during the past 150 years, and illustrates how recent disturbances differ from those prior to

[1]Watershed Program, Northwest Fisheries Science Center, Seattle, Washington
[2]Department of Earth and Space Sciences, University of Washington, Seattle, Washington

Geomorphic Processes and Riverine Habitat
Water Science and Application Volume 4, pages 37-54

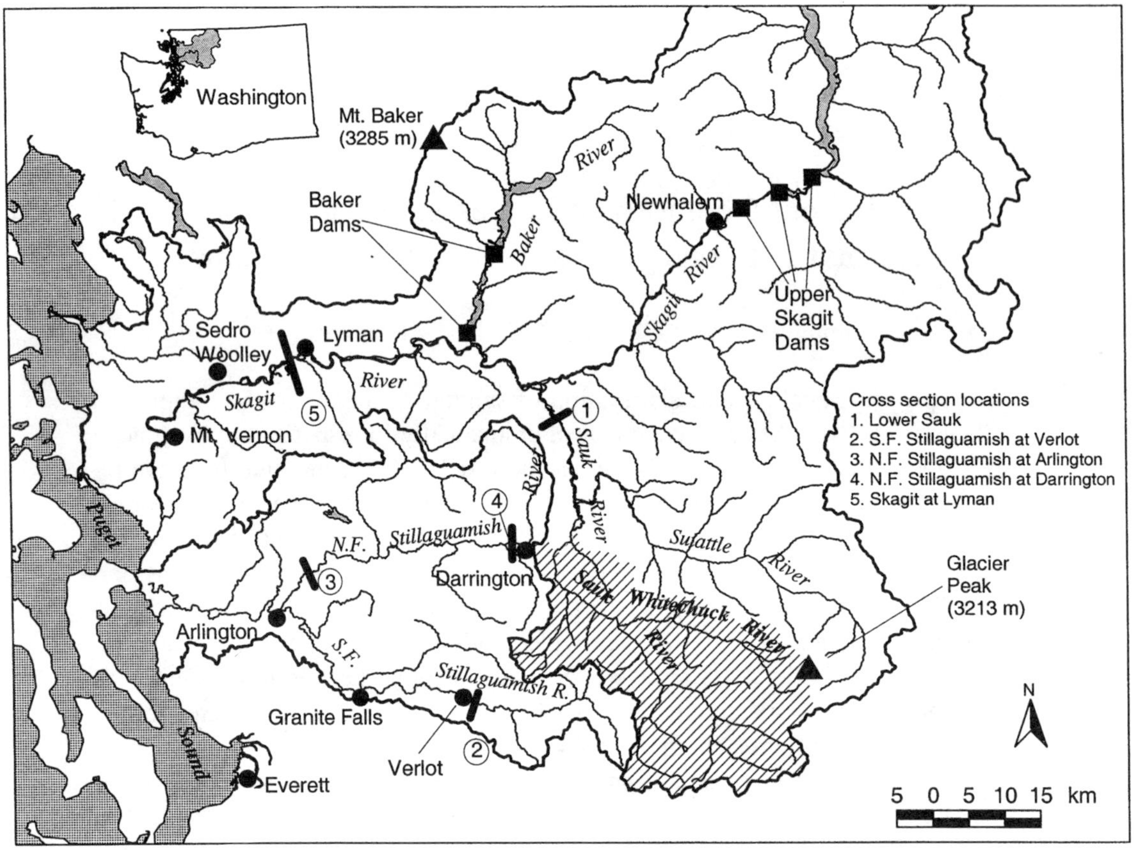

Figure 1. Study area map indicating watershed locations and boundaries, major rivers, and valley cross section locations. Hatched area indicates portion of basin that flowed into the North Fork Stillaguamish prior to 12,500 ybp, then was diverted into the Skagit drainage by lahars from Glacier Peak after approximately 12,500 ybp.

settlement by non-Native Americans. Understanding the Holocene history of geomorphic processes and fish habitat may be useful for interpreting the origins of genetic and phenotypic differences among salmon stocks [e.g., *McPhail and Lindsey*, 1986], whereas understanding how recent causes of habitat loss vary by land use and geomorphic setting is important for devising successful approaches to habitat protection or restoration [e.g., *Cairns*, 1990; *Beechie and Bolton*, 1999].

In this paper we synthesize previous studies to describe how post-glacial evolution of the landscape and recent land uses altered fish habitats in two watersheds of northern Puget Sound. We first recount the post-glacial geomorphic history of the area, describing long-term processes that altered valley floor elevation relative to sea level, which in turn altered stream channel lengths and slopes in both river basins. We then describe how short-term events such as valley-burying lahars (mudflows of volcanic origin) or forest fires may have affected habitat-forming processes and habitat conditions, and how land uses over the past 150 years have altered geomorphic processes and fish habitat in the study area. We also compare magnitudes of losses among different habitat types by estimating changes in potential coho salmon production. Finally, we describe how understanding the relationships among landforms, geomorphic processes, and fish habitat can help in developing strategies for salmonid habitat protection and restoration.

STUDY AREA

The 10,040 km^2 study area encompasses the 1,770 km^2 Stillaguamish River basin and the 8,270 km^2 Skagit River basin (Figure 1). Elevations in the study area range from sea level to 3,285 m on Mount Baker, and numerous peaks in the Skagit basin exceed 2,500 m. Elevations in the Stillaguamish basin are lower, with few peaks exceeding 2,000 m. Average annual rainfall in the study area ranges

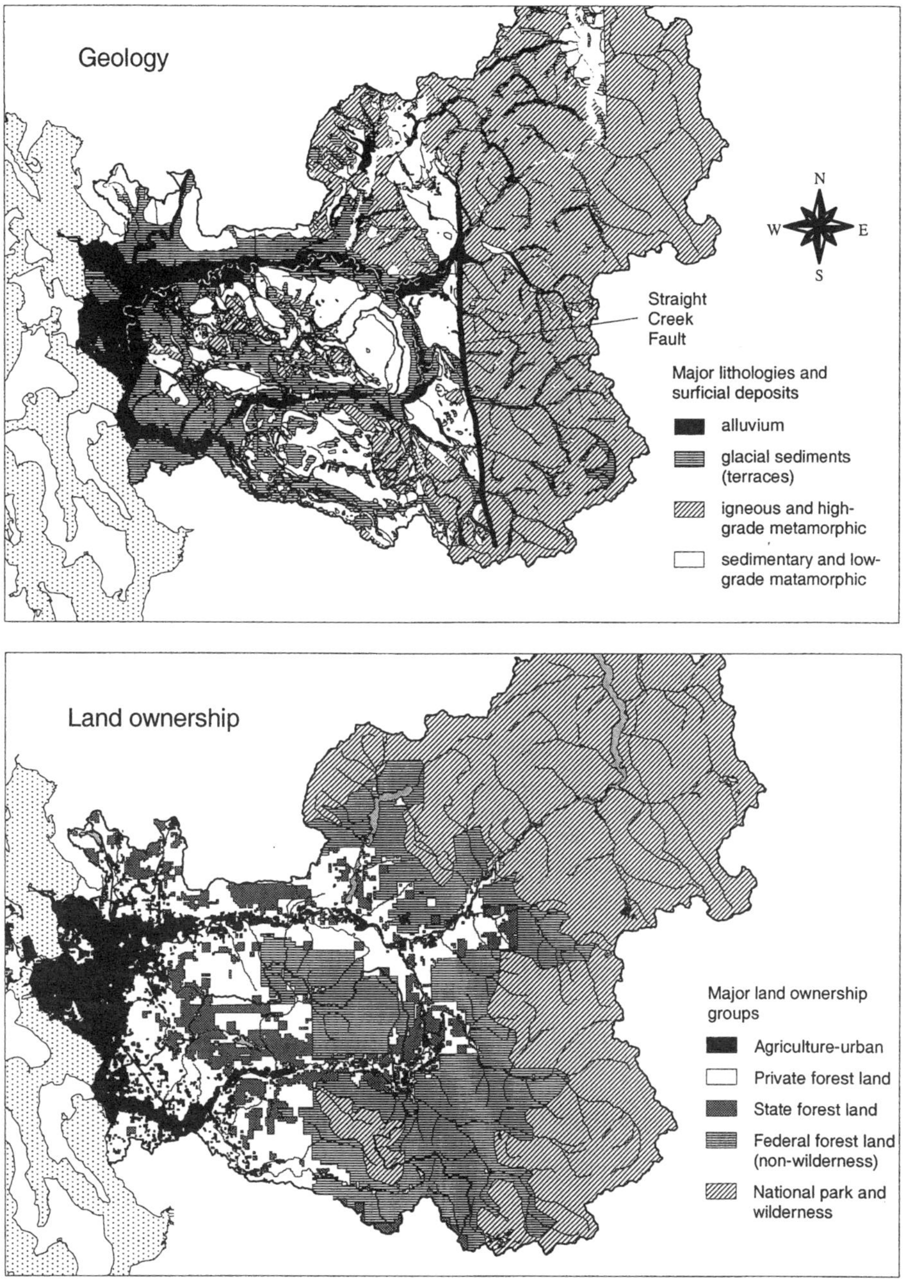

Figure 2. Simplified geology and land ownership in the study area basins. Geologic map based on 1:100,000 scale GIS themes obtained from Washington Division of Geology and Earth Resources, Olympia, Washington.

from about 80 cm in the lowlands to over 460 cm in the vicinity of Glacier Peak, and generally increases with elevation.

The core of the northern Cascade Range in the study area is composed mainly of high-grade metamorphic and igneous rocks of continental origin, which are located east of the Straight Creek Fault [*Brown et al.*, 1987] (Figure 2). The mountains and foothills west of the Straight Creek fault are composed generally of low-grade metamorphic and sedimentary rocks of marine origin, and consist of diverse rock units brought together by the northward movement of previously accreted terrains [*Brown et al.*, 1987]. Two Quaternary volcanoes, Glacier Peak and Mount Baker, are located at the eastern and northern boundaries of the study area, respectively. A lobe of the Cordilleran ice sheet repeatedly advanced into and retreated from Puget Sound during the late Pleistocene (40,000 ybp to 16,000 years ybp) [*Crandell*, 1965; *Booth*, 1987; *Porter and Swanson*, 1998], depositing a series of lacustrine clays, glacial till, and outwash sands and gravels that filled major river valleys to an elevation of at least 600 m (Figure 2) [*Heller*, 1979; *Brown et al.*, 1987].

Headwater streams are typically steep (channel slope > 0.2) and relatively small (bankfull width < 5 m), originating on mountain slopes underlain by various lithologies. Channel slopes decrease dramatically as streams traverse terraces carved into valley-filling glacial deposits (slopes typically between 0.01 and 0.08), and channel slopes are typically < 0.01 on floodplains.

Forest types in the study area vary widely depending on elevation and physical setting. Floodplain forests of the Skagit, Stillaguamish, and Sauk Rivers were historically densely populated with red alder (*Alnus rubra*), but large Sitka spruce (*Picea sitchensis*), western redcedar (*Thuja plicata*), and black cottonwood (*Populus trichocarpa*) grew within the alder forest [*Ayres*, 1899]. Upland forests to an elevation of about 600 m were dominated by Douglas-fir (*Pseudotsuga menziesii*), with the remainder in western redcedar, western hemlock, and Sitka spruce [*Gannett*, 1899; *Ayres*, 1899]. This area is currently known as the Western Hemlock Zone, based on climax forest type [*Franklin and Dyrness*, 1973]. In this zone, Douglas-fir dominated forest stands for the first 200 years after stand-replacing fire, and gradually succeeded to climax stands of western hemlock (*Tsuga heterophylla*) beyond 200 years [*Munger*, 1940]. In some wetter areas and along streams Sitka spruce or western redcedar were the most common species. Silver fir (*Abies amabilis*) and western hemlock dominate forests from about 600 m to 1,200 m (the Silver Fir Zone), and higher elevations are in the Alpine Fir (*A. lasiocarpa*) Zone [*Ayres*, 1899; *Franklin and Dyrness*, 1973].

Settlement of the study area by non-Native Americans began in the 1860s, and clearing and diking of lowlands for agriculture dominated land uses for the next few decades [*Interstate Publishing Company*, 1906]. Removal of kilometers-long raft jams by 1879 allowed rapid expansion of logging and agriculture up the river valleys, and agricultural use of upriver floodplains was extensive by 1909 [*Interstate Publishing Company*, 1906; *Mangum et al.*, 1911]. Agricultural and rural residential uses today cover 9% of the study area and are located predominantly in the lower floodplains and deltas (downstream of Lyman in the Skagit basin and downstream of Arlington in the Stillaguamish basin). Commercial forests owned by private companies, the State of Washington, or the U.S. Forest Service cover 41% of the basin and are located in the uplands. Urban development covers less than 1% of the study area. Forty-four percent of the area lies within the federally-owned North Cascades National Park, Mt. Baker and Ross Lake National Recreation Areas, Glacier Peak Wilderness, and Boulder River Wilderness. The remaining 1,040 km^2 of the Skagit basin is in the Province of British Columbia.

Anadromous salmonid species (i.e., salmonids that live the adult portion of their lives in the marine environment) in the study area include chinook salmon (*Oncorhynchus tshawytscha*), pink salmon (*O. gorbuscha*), chum salmon (*O. keta*), coho salmon (*O. kisutch*), sockeye salmon (*O. nerka*), steelhead trout (*O. mykiss*), cutthroat trout (*O. Clarkii*), and dolly varden char (*Salvelinus malmo*). There is significant overlap in the ranges of these species, and life history patterns vary considerably [*Groot and Margolis*, 1991]. Coho salmon, steelhead trout, cutthroat trout, and dolly varden char generally spawn in smaller and steeper streams (although at different times of year), and juveniles spend their first year or two in freshwater (*Williams et al.*, 1975). Chinook salmon typically spawn in the largest rivers, and most juveniles migrate to sea soon after they emerge from the gravel (*Williams et al.*, 1975). Pink and chum salmon tend to spawn in low gradient streams of widely varying sizes (but at different times and in different microhabitats), and juveniles migrate to sea immediately after emergence from the gravel (*Williams et al.*, 1975). Sockeye salmon spawn primarily in Baker Lake and upper Baker River, and juveniles rear in the lake for one year before moving to sea (*Williams et al.*, 1975). Estuaries provide an important rearing area during smoltification (the transition from freshwater to saltwater) for all species.

The ranges of coho salmon, steelhead trout, cutthroat trout, and dolly varden char in freshwater are limited mainly by natural barriers to upstream migration such as bedrock falls. Other species have shorter migrations and are limited in range by availability of preferred habitats. Upstream migration to the Baker River system has been blocked by the installation of two hydroelectric dams, but anadromous fish production is maintained through trapping and hauling operations, in addition to the maintenance of sockeye spawning beaches and smolt bypass trapping. Migration into the upper South Fork of the Stillaguamish River was naturally blocked at Granite Falls until 1954 when a fish ladder was constructed to permit anadromous fish access into the upper basin.

Adult coho salmon enter the Skagit and Stillaguamish Rivers in late summer and early fall. Spawning is concentrated in smaller tributaries, and occurs primarily between

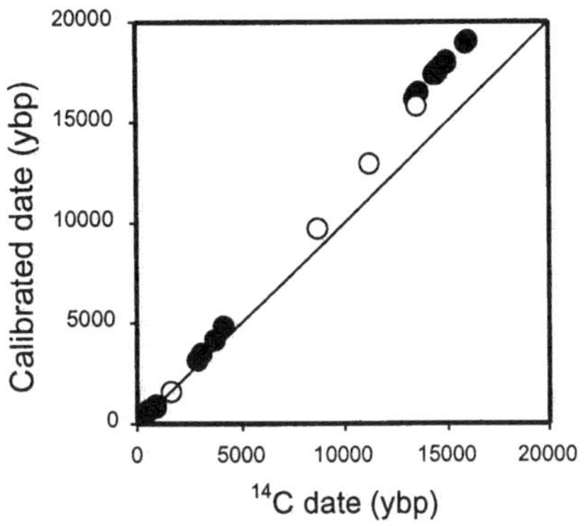

Figure 3. Calibrated dates of major events described in this paper plotted against their respective ^{14}C dates. Filled circles represent paired calibrated and ^{14}C dates reported by *Beale* [1991] and *Porter and Swanson* [1998]. Open circles indicate our calibrations of ^{14}C dates reported by *Benda et al.* [1992]. Calibrations were made using CALIB 4.0 [*Stuiver and Reimer*, 1993].

November and January. Fry emerge from the gravel in March and April and soon establish their summer rearing territories. In general, juvenile coho salmon typically remain in their natal streams, although some juveniles may be gradually displaced downstream as the summer progresses [*Chapman*, 1962; *Sandercock*, 1991]. With the first fall floods (usually in late September or October in the study area) juveniles migrate to winter rearing areas, including beaver ponds, off-channel ponds, and protected side channels [*Scarlett and Cederholm*, 1984; *Peterson and Reid*, 1984]. Coho leave their winter rearing areas in March and April and migrate to saltwater soon after.

DATA SOURCES AND METHODS

Our assessment of the distribution and function of habitats from 16,000 ybp to 150 ybp is less detailed than our assessment of habitats after 150 ybp because there are no direct means for characterizing aquatic habitats prior to 150 ybp. Therefore, we separate this paper into two sections with differing analysis methods and levels of resolution. The first section describes landscape and habitat change from 16,000 ybp to 500 ybp. For this time period we rely largely on studies of isostatic uplift, sea-level change, valley incision, and volcanic eruptions to explain changes in valley morphology and stream habitats. The second section describes pre-settlement geomorphic processes and habitat conditions (500 ybp to 150 ybp) and land use impacts to habitats over the past 150 years. We infer pre-settlement rates of habitat-forming processes such as fire or mass wasting based on their current behavior or field evidence of past occurrences. We use more detailed information on abundance of different habitat types to quantify habitat change due to land uses from 150 ybp to present. For all time periods we assume that anadromous salmonid species have habitat preferences similar to those at present. A brief summary of data sources and methods for each time period follows.

Holocene History of the Landscape and Fish Habitat (16,000 ybp to 500 ybp)

To describe landscape and habitat changes over the past 16,000 years, we compiled and analyzed data from numerous studies of ice retreat, sea-level change, isostatic uplift, and lahars. In the process of assembling these data we encountered some difficulty in reconciling ^{14}C dates and calibrated dates from the various sources. Differences between calibrated dates and ^{14}C dates are relatively small during the past 5,000 years, whereas differences between calibrated dates and ^{14}C dates near 16,000 ybp may be as much as 2,900 years (Figure 3). While most recent studies reported both calibrated dates and ^{14}C dates, a few earlier studies reported only ^{14}C dates. For studies reporting only ^{14}C dates, we calibrated ^{14}C dates using CALIB 4.0 [*Stuiver and Reimer*, 1993]. Our calibrations were based only on information included in the published sources, and the original sources did not include all of the inputs requested by CALIB 4.0. This missing information resulted in a relatively small systematic error in the calibration (evident in Figure 3). Therefore, the timeline presented here should be considered approximate.

During the Vashon Glaciation of the Puget Lowland (18,500 ybp to 16,000 ybp), the land surface was significantly lower than today due to the weight of the ice sheet [*Thorson*, 1980]. Isostatic depression was probably at least 200 m in northern Puget Sound during the glaciation [*Thorson*, 1980; *Booth*, 1987], and for the purposes of this paper we assume that depression in the study area was 200 m at ice retreat (Figure 4). Rebound of the land surface after ice retreat was initially more than 100 mm/yr, but then slowed to 20 mm/yr before 11,000 ybp [*Dethier et al.*, 1995]. We calculated uplift using a rate of 100 mm/yr from 16,000 ybp to 12,500 ybp, and 20 mm/yr after 12,500 ybp based on *Dethier et al.* [1995].

Estimates of sea level at various locations around the world approximately 18,000 ybp range from 102 to 163 m lower than today, whereas sea level in 13,000 ybp was approximately 60 m lower than today [*Matthews*, 1990]. From these data we calculated that sea level at ice retreat (16,000 ybp) was about 90 m lower than today, and that sea level rose about 12 mm/yr between 18,000 ybp and 13,000 ybp. A study of six Puget Sound salt marshes

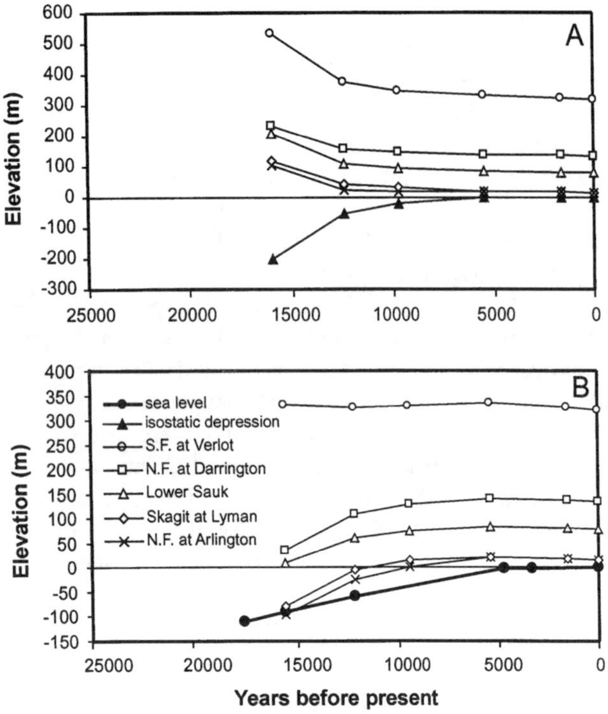

Figure 4. (A) Present-day elevations of ^{14}C-dated terraces at five cross sections (open symbols, locations shown in Figure 1), and isostatic rebound over the past 16,000 years. (B) Subtraction of isostatic depression from present-day terrace elevations yields corrected valley floor elevations relative to present-day sea level (open symbols), for comparison to documented changes in sea level over the past 16,000 years (filled circles).

showed that by 5,000 ybp sea level was approximately 3 m lower than today [*Beale*, 1991]. These data indicate that sea level rise averaged 7 mm/yr between 13,000 ybp and 5,000 ybp (rising from –60 m to –3 m), and has been less than 1 mm/yr since 5,000 ybp (rising from –3 m to 0 m).

To estimate changes in river valley elevations over the last 16,000 years, we first assumed that the highest terrace at each of five valley cross sections was the elevation of the valley floor at ice retreat (i.e., the valley floor elevation prior to erosion of the glacial sediments). (Cross section locations are shown in Figure 1.) We then measured elevations of other ^{14}C-dated valley floor surfaces on USGS topographic maps. Dated valley floor deposits were either lahar deposits from Glacier Peak [*Beget,* 1982; *Dragovich et al.*, 2000] or river gravel deposits on terrace surfaces [*Benda et al.*, 1992].

Once we had compiled the data for isostatic uplift, sea level, and river incision, we combined these data to estimate changes in valley floor elevations relative to sea level. We subtracted isostatic depression from the terrace elevations to estimate post-glacial valley elevation at each cross section. By comparing these data to changes in sea level, we estimated changes in location of river mouths in each study basin. We also described changes in drainage patterns, tributary length, and habitat quality resulting from volcanic eruptions and lahars, based on previous studies.

Recent Changes to Geomorphic Processes and Fish Habitat (500 ybp to present)

We describe geomorphic processes and natural disturbance regimes between 500 ybp and 150 ybp based primarily on previous studies of fire regimes, mass wasting, and recruitment of wood to streams in the study area. Effects of geomorphic processes on habitat characteristics are also summarized based on the scientific literature. We describe resultant pre-settlement habitat conditions using data from reference sites [*Beechie et al.*, 1994; *Pess*, unpublished data], as well as historical reconstructions of habitat conditions at approximately 150 ybp [e.g., *Collins and Montgomery*, this volume].

To describe how habitats have been altered by land uses, we rely primarily on data from *Beechie et al.* [1994] for the Skagit basin, and on *Pess and Collins* [unpublished data] for the Stillaguamish basin. Both studies employ the simple stratification of habitat types listed in Table 1. Methods for estimating historic and current habitat areas differ by habitat type, so we briefly summarize these methods here.

We examined reduction of pool areas in tributary habitats (streams < 10 m wide) by comparing pool areas in managed streams to pool areas in unlogged reference streams. Data for reference sites and for present-day conditions in the Skagit River basin were from *Beechie et al.* [1994]. Data for present-day conditions in the Stillaguamish River basin were from *Pess* [unpublished data]. For mainstem rivers we were unable to directly estimate changes in habitat conditions, but we describe likely changes based on recent studies in Puget Sound.

Historical areas of slough habitats (both side channels and distributaries) were estimated from historical maps, notes, and photos, as well as field evidence of their prior locations [*Beechie et al.*, 1994; *Collins*, unpublished data; *Pess*, unpublished data]. Present day areas were measured from aerial photographs and in the field. Lake areas were measured directly from historical and current maps [*Beechie et al.*, 1994].

Pre-settlement beaver pond areas in tributaries were estimated based on natural frequencies of beaver ponds in other parts of the U.S. and average pond area from present-day ponds in the study area. Natural beaver pond frequencies in the U.S. range from 2 ponds/km to 16 ponds/km

Table 1. Description of habitat types used to quantify loss in coho salmon rearing habitat [adapted from *Beechie et al.*, 1994].

Habitat type	Description
Mainstem	Channels with summer wetted width greater than 6 m (bankfull width generally greater than 10 m).
Tributary	Channels with summer wetted widths less than 6 m (bankfull widths generally less than 10 m).
Side-channel slough	Channels branching from a mainstem river and rejoining the river downstream. They typically have more than 90% of their surface area in pools in the summer, and maintain pool or pond-like characteristics during floods.
Distributary slough	Channels branching from the mainstem and entering estuaries. They typically have more than 90% of their surface area in pools in the summer and maintain pool-like conditions during winter base flow.
Ponds	Beaver ponds and other natural impoundments with surface area less than 5 ha.
Lakes	Natural impoundments or reservoirs with surface area greater than 5 ha.

[*Naiman et al.*, 1988]. We used the conservative lower limit of this range (2 ponds/km) for the historical pond frequency, and an average pond area of 1660 m^2 [*M.M. Pollock*, Northwest Fisheries Science Center, Seattle, Washington, unpublished data]. Present-day pond areas were measured from aerial photographs and in the field [*Beechie et al.*, 1994; *Pess*, unpublished data].

We calculated smolt production as the product of total habitat area (for each habitat type), density of juvenile coho occupying that habitat type, and survival to smolt stage [*Beechie et al.*, 1994]. Densities of rearing juvenile coho and survival rates to smolt stage (the freshwater-saltwater transition) are shown in Table 2. Juvenile coho populations are limited by availability of winter rearing habitat at present [*Beechie et al.*, 1994; *Pess*, unpublished data], so we focus our analyses of habitat losses on winter rearing habitats. ("Winter limited" means that summer habitats are more abundant and produce more fish than winter habitat areas can accommodate. Therefore, availability of winter habitat limits smolt production.) We do not consider changes to habitat quality (e.g., stream temperature or contaminants) because we are currently unable to quantify their effects on salmonid populations.

HOLOCENE HISTORY OF THE LANDSCAPE AND FISH HABITAT

Since retreat of the continental ice sheet about 16,000 ybp [*Porter and Swanson*, 1998], three long-term processes have driven landscape evolution in the two basins: isostatic uplift, erosion of glacial sediments from the valleys, and changes in sea level. Fish habitats were also temporarily altered by valley-burying lahars during the 12,500-ybp and 5,500-ybp eruptive periods of Glacier Peak (Figure 5). We separate our description of landscape evolution and fish habitats into three time periods separated by the Glacier Peak eruptions: 16,000—12,500 ybp, 12,500—5,500 ybp, and 5,500—500 ybp.

Glacial Retreat to 12,500 ybp

At ice retreat (approximately 16,000 ybp), land elevations and sea level were both much lower than today [*Thorson*, 1980; *Matthews*, 1990], and both began rising as the ice moved northward from Puget Sound [*Dethier et al.*, 1995; *Matthews*, 1990]. Accounting for isostatic depression and river incision at ice retreat we calculated that the 16,000-ybp Skagit and Stillaguamish River valley floors were 80 m and 90 m lower than today, respectively (Figure

Table 2. Densities of juvenile coho salmon (fish/m^2), density independent survival to smoltification, and resultant smolt production estimates for each of the six habitat types (smolts/m^2) [adapted from *Beechie et al.*, 1994]. Smolt production estimates for mainstem and lake habitats are in smolts/km (sm/ka) and smolts/ha (sm/ha), respectively.

Habitat Type	Density at a stage (fish/m^2)	Survival to smolt	Potential smolt production (smolts/m^2)
Side Channel and Distributary Slough			
Summer	1.3	0.25	0.319
Winter	2.5	0.31	0.775
Tributary			
Pool (summer, all pools)	1.7	0.25	0.425
Pool (winter, only backwater pools)	3.5	0.31	1.085
Riffle (Summer)	0.7	0.25	0.17
Riffle (Winter)	0.0		0.0
Mainstem			600 sm/km
Pond			
Summer	1.5	0.25	0.375
Winter	3.8	0.31	1.163
Lake			25 sm/ha

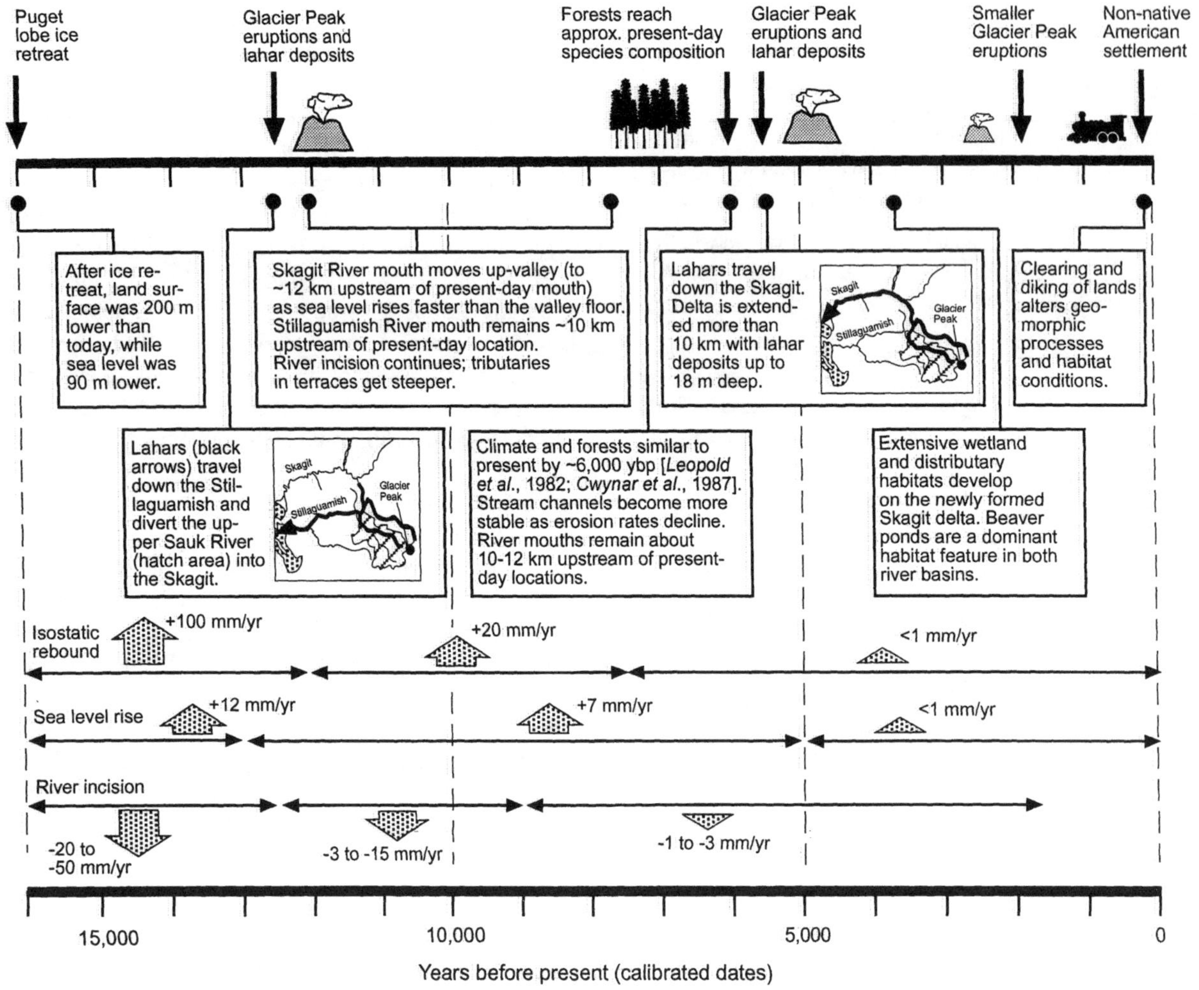

Figure 5. Timeline of major historical events (top) and rates of long–term processes (bottom). Narrative description of main effects of processes and events in center panels. See text for further explanation and sources of information.

4). Nevertheless, these valley floors were not below the 16,000-ybp sea level, as sea level was also 90 m lower than today. The 16,000-ybp valley floor of the Stillaguamish River at Arlington was at sea level, indicating that the 16,000-ybp river mouth was about 25 km upstream of the present-day river mouth. The Skagit River at Lyman was only about 10 m above sea level at ice retreat, placing the river mouth 4 km downstream of Lyman, or about 35 km upstream of its present-day location.

Between 16,000 ybp and 12,500 ybp the lower valleys rose relative to sea level, suggesting that both river mouths moved down-valley. The Stillaguamish valley floor at Arlington rose about 70 m relative to sea level during this time, whereas the Skagit valley floor at Lyman rose 75 m. Assuming that the Stillaguamish valley slope downstream of Arlington was similar to that between Arlington and Darrington, we estimated that the 12,500-ybp Stillaguamish River mouth was about 10 km upstream of its present-day location. A similar calculation for the Skagit suggests that its 12,500-ybp river mouth was approximately 5 km upstream of its present-day location.

As rivers incised into the glacial sediments between 16,000 ybp and 12,500 ybp, they created a series of terraces along the main valleys. River incision decreased the base level for tributaries originating on the valley slopes, and tributary slopes would have increased in response to the decreasing base level (Figure 6). Because anadromous salmonids generally prefer habitats in channels with slope < 0.03 [e.g., *Montgomery et al.*, 1999], the extent of usable stream length probably decreased between 16,000 ybp and 12,500 ybp as some tributaries in glacial sediments became too steep to provide suitable habitat. The distribution of present-day channel slopes in the study area (Figure 7) supports this interpretation, as nearly 75% of streams in the

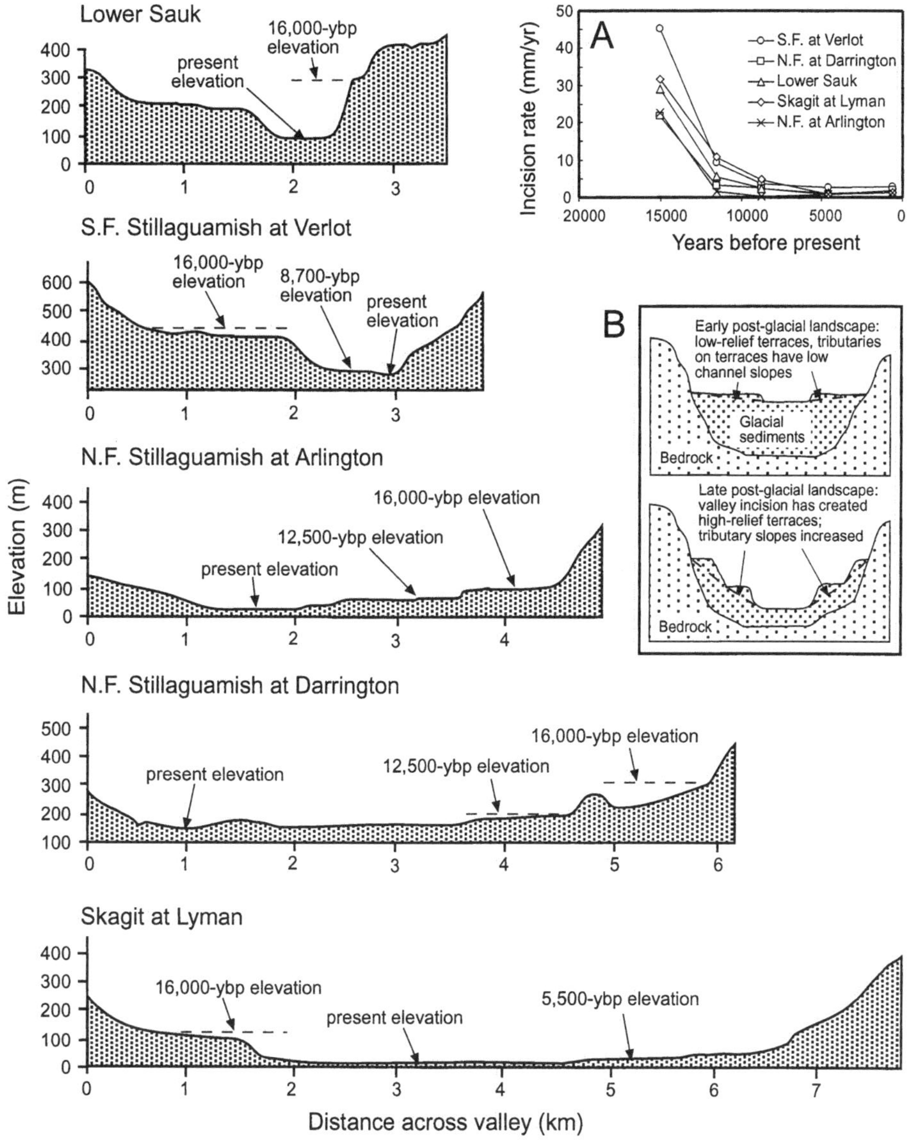

Figure 6. Valley cross sections showing ^{14}C-dated terrace surfaces (cross section locations shown in Figure 1). Inset (A) shows the decline in rate of valley floor incision at each cross section. Inset (B) illustrates how valley floor incision lowered tributary base levels, resulting in steeper tributary slopes in the terraces.

terraces are steeper than 0.03 [based on 30 m digital elevation models, *Lunetta et al.*, 1997].

The end of this first time period is marked by the 12,500-ybp eruptive episode of Glacier Peak. Lahar deposits from this episode are 2 m deep only 20 km upstream of the present-day Stillaguamish River mouth, and the lahar likely reached Puget Sound [*Beget*, 1982; *Mastin and Waitt*, 1995]. A branch of this lahar also traveled down the Sauk River valley nearly to its confluence with the Skagit River [*Dragovich et al.*, 2000]. This lahar changed the course of the Sauk River at Darrington, forcing it to flow northward to the Skagit [*Beget*, 1982]. Approximately 830 km^2 of drainage area (over 140 km of stream length accessible to anadromous salmonids) was shifted from the Stillaguamish River basin to the Skagit River basin.

The 12,500-ybp lahars likely reduced the quality of salmonid habitats in both river basins. Considering the 1980 Mount St. Helens eruption a modern analogue to the Glacier Peak eruptions (i.e., assuming that dome collapses, pyroclastic flows, and lahars of Glacier Peak produced deposits similar to those on Mount St. Helens), we infer that erosion rates increased dramatically on the slopes of the

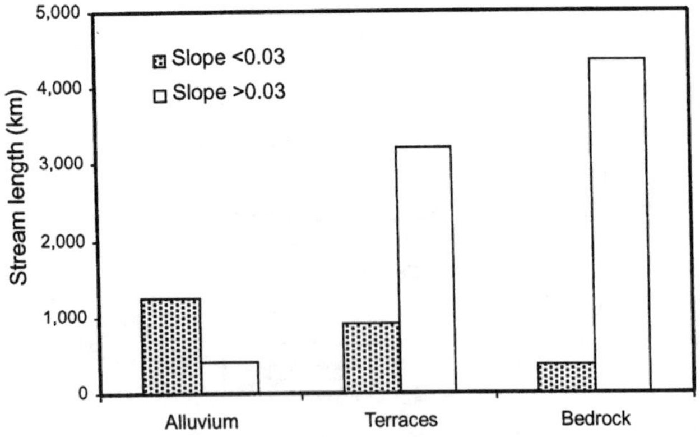

Figure 7. Distribution of channel slopes in areas mapped as alluvium, terraces (deep glacial sediments), and various bedrock lithologies in Figure 2, based on stream slopes generated from 30 m digital elevation model [*Lunetta et al.*, 1997].

volcano and lahar surfaces, leading to increased sediment loads in stream channels [*Major et al.,* 2000]. Extreme post-eruption sediment loads create a variety of habitat changes including lateral and vertical channel instability, greater exposure to high stream temperatures in summer, and reduced availability of rearing pools and cover [*Martin et al.*, 1986]. These changes then cause a significant reduction in use by salmonids [e.g., *Leider*, 1989]. Because the Glacier Peak eruptive episodes also included thickly bedded pyroclastic flow and dome collapse deposits proximal to the volcano [*Beget*, 1982], high sediment loads may have continued for decades. Sediment yields from 1980 Mount St. Helens debris avalanche remained high two decades after the eruption [*Major et al.*, 2000], and forests apparently require many decades to reestablish on the deposit [*Dale*, 1991]. Channels in distal mudflow-affected valleys may have stabilized within a few years to a decade [e.g., *Meyer and Martinson*, 1989]. Forests probably colonized such lahars soon after they stabilized, just as red alder colonized Mount St. Helens lahars within a few years [*Heilman*, 1990]. However, advanced successional stages of forest development likely spanned many decades to a few centuries [e.g., *Munger*, 1940].

12,500 ybp to 5,500 ybp

After 12,500 ybp, isostatic uplift and river incision appear to have decreased significantly, but uplift continued to outpace incision (Figure 4). As a result, valley floors rose more slowly, and valley floor elevations relative to sea level were typically stable or decreasing. Incision of the lower Skagit during the latter half of this time period approximately matched the rate of uplift, but sea level continued to rise. Thus, the mouth of the Skagit River would have moved up-valley as the lower valley was flooded by the rising Puget Sound. Assuming a valley slope similar to that of today, we estimate that by 5,500 ybp the mouth of the Skagit River had moved up-valley to approximately 12 km upstream of its present-day location. This position is consistent with the pre-lahar location estimated by *Dragovich et al.* [2000]. By contrast, the Stillaguamish River valley floor rose at about the same rate as sea level, suggesting that the mouth of the Stillaguamish River remained about 10 km upstream of its present location throughout this period.

Between 12,500 ybp and 5,500 ybp, base levels of tributaries continued to drop as the main rivers eroded into glacial sediments. Consequently, tributary slopes would have increased, suggesting that the total length of low-gradient tributary streams continued to decrease. Such changes may have favored steelhead trout and coho salmon (which utilize small streams of moderate slope), whereas species such as chum salmon were confined to habitats in low-elevation floodplains.

Around 5,500 ybp Glacier Peak entered its second eruptive period since ice retreat [*Beget*, 1982]. At least one lahar during this period reached the mouth of the Skagit River [*Dragovich et al.*, 2000]. It deposited 3 m of dacitic debris 35 km upstream of the present-day Skagit River mouth [*Beget*, 1982; *Mastin and Waitt*, 1995], indicating a 5,500-ybp valley floor elevation only a few meters above the present river elevation. Deposits up to 18 m thick near the present-day delta shoreline [*Dragovich et al.*, 2000] indicate that much of the delta downstream of Sedro Woolley was created during this eruptive episode. We estimated extension of the mainstem river during this episode at approximately 15 km based on delta aggradation equivalent to the depth of the deposit mapped by *Dragovich et al.* [2000]. This would put the post-lahar mouth of the Skagit River near its present location. Once again these lahars probably reduced the quality of salmonid rearing habitats for at least several decades in the mainstem Skagit, Sauk, and Stillaguamish Rivers.

5,500 ybp to 500 ybp

Since 5,500 ybp sea level rose about 3 meters [*Beale*, 1991], suggesting that river mouths may have moved up-valley. This also raises the possibility that 5,500-ybp deltas were larger than at present if there were no uplift or deposition on the deltas during the past 5,500 years. However, the salt marshes studied were located on shores away from major river delta deposits, and aggradation of deltas during the same time period may have effectively prevented movement of river mouths up-valley if aggradation were also at least 3 m. Assuming an average sediment supply rate of 70 $m^3/km^2/yr$ [*Paulson*, 1997] over the last 5,500 years, we estimate that the Skagit River exported more than 3 x 10^9 m^3 of sediment to the 450 km^2 delta (including sub-tidal areas). If all of the sediment were deposited in the delta, aggradation could have been as much as 6 m and could easily have prevented up-valley movement of the river mouth over the past 5,500 years.

Since 5,500 ybp, rates of valley floor incision decreased (Figure 6). Overall, proportions of the landscape mapped as bedrock terrain, glacial terraces, and floodplains have probably been more or less constant over the last 1,700 years. Therefore, the distribution of channel slopes has probably been relatively constant as well. Subsequent lahar deposits from Glacier Peak did not extend beyond the Suiattle and Whitechuck river valleys [*Beget*, 1982], and Mount Baker has not produced significant avalanches or lahars entering the study area since 16,000 ybp [*Gardner et al.*, 1995]. Thus, neither of the lower valleys appears to have experienced a major volcanic disturbance in the last 5,500 years.

RECENT HISTORY OF THE LANDSCAPE AND FISH HABITAT

Rates of change in valley elevation or sea level over the past 500 years have not appreciably affected the spatial extent of fish habitat. Instead, land development by non-Native Americans since 1850 has removed a large proportion of historical habitats from the landscape, and fundamentally altered some of the geomorphic processes that form and sustain remaining salmonid habitats. These changes have resulted in extensive losses of habitat on floodplains and deltas, as well as in beaver ponds and tributaries in the terraces [*Beechie et al.*, 1994; *Pess*, unpublished data]. In addition, land uses currently constrain geomorphic and biological processes that allowed habitats to recover after natural disturbances (e.g., levees block river migration and flooding, riparian forest removal prevents wood recruitment, etc.).

The four main land uses that alter rates of geomorphic processes and subsequently affect the quantity or quality of fish habitat in the study area are agriculture, forestry, rural residential development, and hydropower dams. (Urban areas occupy less than 1% of the study area, so we group urban areas under the category rural residential.) Agricultural and rural residential areas are almost exclusively within the floodplains and deltas, and overlap more than 59% of the historical range of salmon in the Skagit basin and 38% of the range in the Stillaguamish basin (Figure 8). Forest management in the study area can be grouped into two main categories: (a) intensive commercial forest management with timber harvest rotations of less than 60 years (state and private forests), and (b) multi-use management including longer rotation forestry, wilderness areas, and national parks (federal forests and parks). Intensive commercial forestry activities are concentrated on the terraces, and overlap the majority of the remaining historical range of salmon in both basins. Federal forests under multi-use management are concentrated in the bedrock terrain, and contain less than 5% of the salmon habitat in the study area.

Spatial Distribution of Habitats on the Landscape

By 500 ybp, the landscape had reached its present configuration of steep hillslopes at higher elevations, valley-filling glacial deposits eroded into terraces, and relatively narrow floodplains. This configuration controlled the distribution of channel slopes in the two basins (Figure 7), which in turn controlled some aspects of habitat formation such as the basic morphology of channel beds [*Montgomery and Buffington*, 1997], pool spacing [*Montgomery et al.*, 1995], and size of gravels available for spawning [*Beechie and Sibley*, 1997].

Channels in the bedrock terrain contained a very small percentage of total pre-settlement anadromous salmonid habitat. Examination of channel slopes in the study area revealed that most channels in the bedrock terrain have slopes > 0.2 (78% of the stream length) and only 4% have

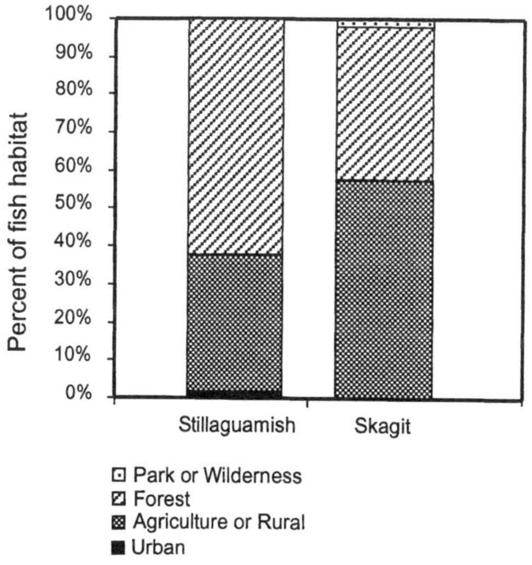

Figure 8. Proportions of anadromous salmonid habitat located in each land use category in the Skagit and Stillaguamish River basins.

slope < 0.03. Channels > 0.03 are generally too steep to support anadromous salmonids, in part because migration barriers (falls and impassable cascades) are commonly located at the lower end of streams in this terrain, and in part because these channels provide little spawning or rearing habitat [*Montgomery et al.*, 1999]. Streams steeper than 0.02 and with bankfull width less than 15 m have very little spawning gravel available, primarily because average basal shear stress is high enough that median particle size of the bed surface tends to be cobble or larger [*Beechie and Sibley*, 1997]. Thus, particle sizes in steeper channels typically exceed those useable by salmonids that occupy these small streams (i.e., gravels < 64 mm diameter used by coho salmon, steelhead trout, or cutthroat trout). In addition, pool area tends to be much lower in steeper channels [*Beechie and Sibley*, 1997].

Channels on terraces are less steep than those in bedrock areas, with 70% of channels < 0.2 and 22% of channels < 0.03. Field observations in both study basins show that small streams on terraces (bankfull widths < 15 m) are typically used by coho salmon, steelhead trout, and cutthroat trout. These channels also typically have narrow floodplains, creating preferred locations for beaver ponds in the study area. Historically, beaver ponds occupied a minimum of 8% of tributary channel length in both basins. In between beaver ponds, channels would have been predominantly forced pool-riffle channels due to large amounts of woody debris (> 0.4 pieces per meter of channel length) that provided pool-forming structure in otherwise straight reaches [*Montgomery et al.*, 1995; *Beechie and Sibley*, 1997]. Step-pool reaches (slope typically > 0.03) may have been occupied more by steelhead and cutthroat trout than by coho salmon because they contain less coho salmon spawning and rearing habitat [*Bisson et al.*, 1988; *Beechie and Sibley*, 1997].

Based on present-day topography and historical reconstructions of habitat areas, we estimate that much of the 150-ybp habitat area in both study basins was located on the deltas and floodplains. Most channels (62%) on areas mapped as alluvium in Figure 2 have slopes < 0.01, and at least 75% have a slope < 0.03. The large mainstem rivers in the study area (Skagit, Sauk, and North Fork Stillaguamish Rivers) typically migrate laterally or avulse across the alluvium, resulting in meandering or anastomosing channel patterns. The anastomosing channel pattern produces many side-channels (channels branching off the mainstem and re-entering the mainstem downstream), which account for a significant proportion of the pre-settlement channel length. For example, at least 44% of channel length in the Skagit and Sauk River floodplains were side channels prior to non-Native American settlement. Distributary channels (channels branching off the mainstem and entering Puget Sound) were numerous on the deltas, and wetland complexes covered more than half of the total delta area [*Collins and Montgomery*, this volume]. In combination, the floodplains and deltas contained more than half of the total salmon habitat area in both basins, yet they constituted less than 10% (1,010 km^2) of the study area.

Post-settlement Alteration of Geomorphic Processes and Fish Habitat

Changes in sediment supply and wood recruitment. Prior to settlement by non-Native Americans, the natural disturbance regime of fires and storms periodically altered rates of mass wasting and wood recruitment to streams in the study area, which subsequently altered channel morphology and the physical structure of tributary habitats [*Beechie*, 1998]. The recurrence interval of stand-replacing fires was about 200 years in the Western Hemlock Zone (< 600 m elevation), and about 380 years at higher elevations [*Beechie*, 1998]. Based on standard techniques for relating return intervals to stand age distributions, *Beechie* [1998] estimated that these disturbance regimes maintained an average of 12% of forest stands < 20 years old in the Western Hemlock Zone, and 7% of stands < 20 years old at higher elevations. Sediment supply rates from these immature stands (i.e., stands < 20 years old) average four times that of rates from mature forests [*Paulson*, 1997], leading to long-term average sediment supply rates 1.2 to 1.3 times the rate one would expect in the absence of fires [*Beechie*, 1998].

Based on sediment budgets for 10 sub-watersheds representing the range of lithologies and landforms in the study area (covering 672 km^2), *Paulson* [1997] found that landslide rates from roads are roughly 45 times the rate from mature forests, and landslide rates from clearcut areas are roughly 4 times that of mature forests. Total bed load supplies resulting from these altered landslide rates average 2 times the estimated natural rate in commercial forests, and 1.5 times the natural rate in multi-use federal forests [based on data in *Paulson*, 1997].

Natural fires also killed riparian forests along small streams prior to 150 ybp, which may have caused short-term pulses of wood recruitment followed by several decades of low recruitment. After a fire, forests of the Western Hemlock Zone took several decades to grow to 1 m diameter [*Munger*, 1940], which would be large enough to provide pool-forming wood to streams. Based on a model of forest growth, wood recruitment, and pool formation for northwestern Washington forests, *Beechie et al.* [2000]

calculated the percentage of riparian forests that were too small to provide pool-forming wood to streams. They estimated that under the natural fire regime, more than 60% of riparian forests consisted of trees large enough to provide pool-forming wood to tributaries less than 15 m wide (bankfull). Logging of riparian forests after 150 ybp reduced the supply of pool-forming wood to channels. Forest management regimes of the past century have reduced the proportion of riparian forests that produce pool-forming wood to less than 50% in very small channels (< 4 m wide), and to less than 25% in channels 4 to 15 m wide [*Beechie et al.*, 2000].

Habitat changes resulting from altered sediment supply and wood recruitment. Decreased wood abundance in channels reduces pool abundance and pool surface area within the study area [*Beechie and Sibley*, 1997] and throughout the Pacific Northwest [*Bilby and Ward*, 1991; *Montgomery et al.* 1995]. Large increases in sediment supply also reduce the depth, frequency, and total area of pools in the study area [*Nelson*, 1998] and throughout the western U.S. [e.g., *Lisle*, 1982; *Madej and Ozaki*, 1996]. Increased supply of bed load causes a modest reduction in the average depth of pools when sediment supply increases to more than 90 $m^3/km^2/yr$ [*Nelson*, 1998], and dramatically reduces the number and area of pools in extreme cases (i.e., sediment supply > 1,000 $m^3/km^2/yr$) [*Collins and Beechie,* unpublished data].

In both basins, almost all tributaries in agricultural lands are low-slope channels, and pool areas are much less than in the reference sites or in streams with other adjacent land uses (Table 3). Removal of wood alone is unlikely to cause such decreases in pool areas because pool areas in low-slope channels are relatively unresponsive to changes in wood abundance [*Beechie and Sibley*; 1997]. Much of this difference in pool areas is likely due to the repeated channel dredging that occurs in many agricultural streams of the study area, or to increased supply of fine sediments from pastures and croplands.

Decreased pool area in terrace tributaries was largely the result of forestry activities over the last 150 years. Increased mass wasting from forest road construction and logging has contributed to loss of pool area in tributaries, but pool losses are more directly attributable to decreased wood abundance [*Nelson*, 1998]. Reduced wood recruitment in forest lands has caused the greatest reductions in pool areas in moderate-slope streams (0.02 – 0.04) where other mechanisms of pool formation do not compensate for the loss of wood-forced pools [*Beechie and Sibley*, 1997].

Reductions of pool area in low-slope channels in rural areas are similar to those in forestry areas (Table 3), and reflect the lesser sensitivity of pool area to wood abundance in low-slope channels [*Beechie and Sibley*, 1997]. In moderate-slope channels in rural lands, percent pool areas are 36% and 46% in the Skagit and Stillaguamish basins, respectively, compared to 54% in the unlogged sites. Thus, reductions in pool area are slightly less in rural lands than in forested lands in both basins, apparently due to the limited protection of riparian forests afforded by the steep terrain adjacent to many streams. There appears to have been no change in pool areas in steeper streams due to urban or residential development.

Pool frequencies in the main stem Skagit, Sauk, and North Fork Stillaguamish Rivers have probably also decreased, based on decreased abundance of wood large enough to create log jams and pools in the similarly modified Snohomish River [*Collins and Montgomery*, this volume]. Limited data from six reference sites and 11 altered mainstem channels in the study area (bankfull widths > 24 m) also suggest that total pool area has decreased by more than 35% [*Pess*, unpublished data]. In large channels such as these, key pieces more than 1 m diameter create stable jams, which in turn create more and deeper pools [e.g., *Abbe and Montgomery*, 1996; *Collins and Montgomery*, this volume]. Clearing of this wood debris has dramatically decreased abundance of pool-forming wood in the main rivers [*Collins and Montgomery*, this volume], and logging of riparian forests has reduced recruitment of trees large enough to form pools where channels are wider than 15 m [*Lunetta et al.*, 1997].

Table 3. Average percent pool by channel slope class and land use for the two river basins in the study area. Percent pool is pool area expressed as a percentage of total wetted area in summer. N/A indicates that sample size (number of reaches) was less than ten, and an average value was not calculated. Averages for unlogged lands are from *Beechie et al.* [1994].

	Channel Slope		
	<2%	2%-4%	>4%
Unlogged lands	64	54	35
Skagit basin			
Agriculture (n=25)	47	N/A	N/A
Forestry (n=172)	61	29	27
Rural (n=95)	61	36	36
Stillaguamish basin			
Agriculture (n=20)	36	N/A	N/A
Forestry (n=76)	44	29	33
Rural (n=32)	42	46	N/A

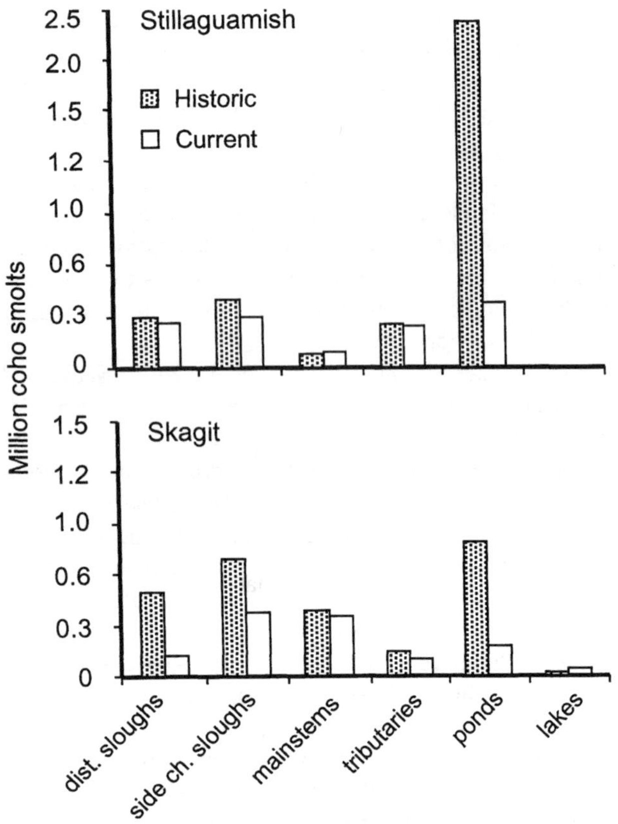

Figure 9. Historic and current winter rearing habitat capacities by habitat type for the Stillaguamish (upper panel) and Skagit (lower panel) River basins.

Estimated changes in coho salmon smolt production due to habitat alteration in tributaries (i.e., reduced pool area) are relatively small compared to losses in ponds and sloughs (Figure 9). In the Skagit basin, tributaries have lost more than 30% of their potential coho smolt production, whereas tributaries in the Stillaguamish basin have lost less than 5% of their potential coho smolt production. We cannot make similar estimates of the change for mainstem channels because there are insufficient data in the literature to quantify coho salmon use of habitats in large rivers [*Beechie et al.*, 1994].

Changes in channel migration and formation of floodplain habitats. Recent studies indicate that river migration, avulsion, and recruitment of floodplain trees to channels were fundamentally important mechanisms of habitat formation in Pacific Northwest watersheds [*Peterson and Reid*, 1984; *Featherston et al.*, 1995; *Abbe and Montgomery*, 1996]. Channel movement across floodplains forms a variety of off-channel habitat types in abandoned channels, including off-channel ponds, oxbow lakes, and side channels [*Peterson and Reid*, 1984]. Recruitment of trees from the floodplain initiates formation of log jams, which can then force channel movement and the formation of forested islands, side channels, and deep pools in the main channel [*Featherston et al.*, 1995; *Abbe and Montgomery,* 1996]. Therefore, floodplains contained a diverse array of habitat types, including small and large channels, ponds, sloughs, abundant woody debris, and pool-riffle channels. This dynamic habitat system encompassed nearly half of the low-slope channel length in the study area. Similar processes also occurred on deltas, resulting in extensive freshwater and estuarine habitats, particularly on the Skagit delta [*Collins and Montgomery*, this volume].

Historically, most coho salmon habitat was on the deltas and floodplains, and these landforms have been nearly completely occupied by agricultural uses over the past 150 years. The dominant mechanism of habitat loss on the deltas and floodplains has been the diking and draining of sloughs, wetlands, and beaver ponds [*Beechie et al.*, 1994; *Collins and Montgomery*, this volume]. In combination, these practices account for more than 90% of coho habitat losses in both river basins (Figure 10). Many channels that remain within leveed areas have been converted to ditches, and now serve as a drainage network for farmlands. These channels are inaccessible from their upstream ends during floods, so juvenile salmon can no longer access them as freshwater refugia. Some channels have partial access through tidegates at their lower (saltwater) ends, and minimal use is made of these channels as salmonids transition from freshwater to saltwater. Isolation of 40% of the distributary slough area in the Stillaguamish delta and 75% of distributary slough area in the Skagit delta has eliminated much of the winter habitat capacity from the deltas of these two basins [*Beechie et al.*, 1994; *Pess,* unpublished data]. In the Stillaguamish River basin 28% of floodplain side-channel habitats have been isolated or obliterated [*Pess,* unpublished data] and 45% have been isolated or obliterated in the Skagit River basin [*Beechie et al.*, 1994].

Blocked access to tributary habitats. Culverts and other stream crossing structures block an estimated 6% and 1% of coho rearing capacity in the Skagit and Stillaguamish basins, respectively (Figure 10). Habitats have also been inundated by the reservoirs of two dams built in the Baker River canyon. Coho winter rearing habitats inundated by the dams constituted about 5% of the total habitat loss [*Beechie et al.*, 1994]. However, increased rearing capacity in the reservoirs was about 4%, resulting in a negligible change of total winter rearing capacity caused by the dams. Three dams in the upper Skagit River do not block passage, as there is no record of anadromous fish historically

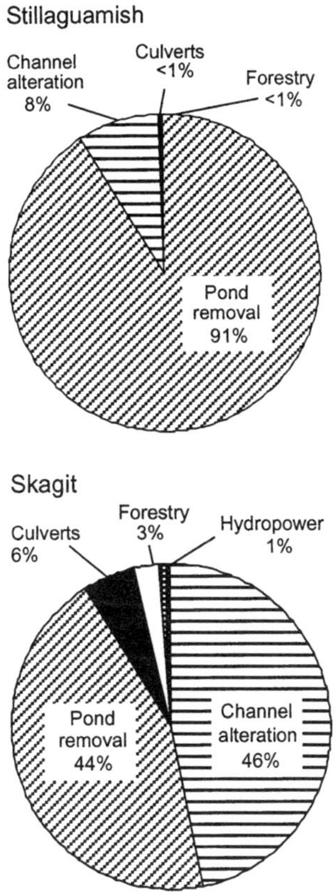

Figure 10. Distribution of losses in average smolt production (from winter rearing habitat) by type of impact in the Stillaguamish (upper panel) and Skagit (lower panel) River basins. [Beaver pond data from *M.M. Pollock*, Northwest Fisheries Science Center, Seattle, Washington, unpublished data.]

passing through the gorge above Newhalem. There are no hydropower dams in the Stillaguamish basin.

Cumulative changes to salmonid habitats

Historically, coho salmon smolt production from the Stillaguamish basin was summer limited, whereas summer and winter smolt production capacities from the Skagit basin may have been nearly equal [*Beechie et al*., 1994; *Pess*, unpublished data]. Overall losses in coho salmon rearing areas have been greatest for winter rearing habitat, resulting in winter limited smolt production from both basins at present. In both basins estimated declines in winter rearing habitat capacity are greater than 50% (Figure 11), suggesting that pre-settlement coho smolt production was more than twice that of today.

Historic winter rearing capacity appears to have been greatest in ponds in the Stillaguamish River basin, and more evenly split among ponds and sloughs in the Skagit River basin. The greater historic areas of distributary and side-channel slough habitat in the Skagit River basin was a result of the much larger area of alluvial deposits in the Skagit basin, which allowed formation of more extensive slough channels. Dredging, diking, and ditching of channels were responsible for more than 46% of estimated winter habitat loss in the Skagit River basin, and more than 8% of estimated losses in the Stillaguamish (Figure 10). Removal of beaver ponds and trapping of beaver account for 44% and 91% of the coho salmon rearing habitat losses in the Skagit and Stillaguamish basins, respectively. Estimated losses from reduced pool areas in forestry areas account for a small proportion of habitat losses in both basins. Because riffle areas also provide rearing habitat for coho salmon (but less than pools), reductions in potential coho smolt production from loss of pools are relatively small compared to complete losses of habitat area resulting from channel alteration, culverts that block fish passage to upstream habitats, and removal of beaver ponds.

SUMMARY AND CONCLUSIONS

Historical assessments of landform development, geomorphic processes, and salmonid habitats put recent declines of salmonid habitats into a context of long-term landscape evolution. The landscape and fish habitats of the study area have been changing gradually throughout the last 16,000 years as a result of isostatic uplift, valley incision, and sea level rise. These changes slowly altered the suite of habitats available to anadromous fishes, with low-slope habitats declining in extent while moderate-slope habitats increased. These shifts probably favored species adapted to slightly steeper streams (such as steelhead and

Figure 11. Historic and current total winter rearing habitat capacities for the Skagit and the Stillaguamish River basins.

cutthroat trout or coho salmon) while species such as chum salmon experienced decreasing availability of habitat. (However, we note that chum salmon are more abundant than other species today, probably due to lower harvest levels and a life history strategy that avoids significant rearing time in the freshwater environment.)

Reconstruction of geomorphic processes and fish habitats prior to non-Native American settlement illustrates that the majority of pre-settlement salmon habitat (ca. 1860) in the Skagit and Stillaguamish basins was on floodplains and deltas. Over the past 150 years, the most intensive land uses and the most severe alterations to salmonid habitats also have been concentrated on alluvial landforms. Because most of the habitat potential is in floodplains and deltas, restoring the rearing capacity of salmon habitats in the study area will now require reversal of much of the channel alteration and isolation that caused these losses. However, simple solutions (such as levee removal) are in conflict with present land uses, and habitat restoration will likely be technically and politically arduous.

These assessments document how different land uses affect different habitat-forming processes, and therefore indicate where land managers must change land management practices in order to restore habitats and the geomorphic processes that form them. For example, forestry activities are the primary land use affecting mass wasting processes because they are the only land use concentrated in the steep slopes of the bedrock terrain. In order to prevent or reverse the effects of forestry on landsliding and pool abundance in streams, timber harvest and road construction must be altered to reduce impacts on landslide-prone areas. By contrast, agricultural practices have almost no effect on mass wasting processes because they are concentrated on alluvial deposits, but they have been accompanied by widespread isolation of habitat areas by levee construction. Therefore, agricultural practices have little need to address mass wasting processes in habitat restoration, but must address the isolation and obliteration of habitats that existed on the floodplains and deltas.

A fundamental difference between habitat changes by natural disturbances and those of recent land uses is that land use changes interrupt natural process that would recover habitats. For example, a valley-burying lahar may have caused extensive habitat change, but river migration, woody debris, and re-establishment of floodplain forests all could contribute to habitat recovery. By contrast, floodplain habitats isolated by levees and simplified mainstem channels cannot recover by natural processes because rivers are presently disconnected from their floodplains and prevented from migrating. Therefore, habitat conditions cannot improve without management interventions.

Finally, the historical context for habitat-forming processes provides reasonable expectations of restored processes and habitat conditions. Without knowledge of the extent of habitats prior to intensive land uses, we have little understanding of what to expect for recovery of processes that form and sustain salmonid habitats. For these two basins, we have estimates of historical and current habitat capacity for coho salmon in different habitat types, which indicate that losses of sloughs on the Skagit delta and beaver ponds in both basins are the largest habitat losses that have occurred over the past 150 years. Therefore, restoration strategies targeting recovery of coho salmon populations must include restoration of overwintering habitats such as sloughs and ponds. Additionally, estimates of natural rates for processes such as landsliding and delivery of sediment to streams help us understand where sediment supply rates were relatively higher or lower under pre-settlement conditions, as well as where current rates of these processes deviate from natural rates. These estimates can be used to identify where habitat-forming processes are in need of restoration, and can help identify the expected outcomes of various restoration activities.

Acknowledgments. We thank Peter Kiffney, Martin Liermann, David Montgomery, David Meyer, Faith Fitzpatrick, and Betty Palcsak for their helpful reviews of the manuscript.

REFERENCES

Abbe, T.B. and D.R. Montgomery, Large woody debris jams, channel hydraulics and habitat formation in large rivers, *Regulated Rivers: Research and Management* 12:201-221, 1996.

Ayres, H.B., Washington Forest Reserve, in *Nineteenth Annual Report of the United States Geological Survey, 1897-1898, Part V – Forest Reserves*, edited by H. Gannett, pp. 283-313, Government Printing Office, Washington, D. C., 1899.

Beale, H., *Relative Rise in Sea-level During the Past 5,000 Years at Six Salt Marshes in Northern Puget Sound, Washington*, Masters Thesis, Western Washington University, Bellingham, 1991.

Beechie, T.J., *Rates and Pathways of Recovery for Sediment Supply and Woody Debris Recruitment in Northwestern Washington Streams, and Implications for Salmonid Habitat Restoration*, Doctoral Dissertation, University of Washington, Seattle, 1998.

Beechie, T., E. Beamer, and L. Wasserman, Estimating coho salmon rearing habitat and smolt production losses in a large river basin, and implications for restoration, *North American Journal of Fisheries Management* 14:797-811, 1994.

Beechie, T.J. and S. Bolton, An approach to restoring salmonid habitat-forming processes in Pacific Northwest watersheds, *Fisheries* 24(4): 6-15, 1999.

Beechie, T.J. and T.H. Sibley, Relationships between channel characteristics, woody debris, and fish habitat in northwestern Washington streams, *Transactions of the American Fisheries Society* 126:217-229, 1997.

Beechie, T.J., G. Pess, P. Kennard, R.E. Bilby, and S. Bolton, Modeling recovery rates and pathways for woody debris recruitment in northwestern Washington streams, *North American Journal of Fisheries Management* 20:436-452, 2000.

Beget, J.E., Postglacial volcanic deposits at Glacier Peak, Washington, and potential hazards from future eruptions, USGS Open-file Report 82-830, 1982.

Benda, L., T.J. Beechie, R.C. Wissmar, and A. Johnson, Morphology and evolution of salmonid habitats in a recently deglaciated river basin, Washington State, USA, *Canadian Journal of Fisheries and Aquatic Sciences* 49:1246-1256, 1992.

Bilby, R.E. and J.W. Ward, Characteristics and function of large woody debris in streams draining old-growth, clear-cut, and 2nd-growth forests in southwestern Washington, *Canadian Journal of Fisheries and Aquatic Sciences* 48:2499-2508, 1991.

Bisson, P.A., T.P. Quinn, G.H. Reeves, and S.V. Gregory, Best management practices, cumulative effects, and long-term trends in fish abundance in Pacific Northwest river systems, in *Watershed Management: Balancing Sustainability and Environmental Change*, edited by R.J. Naiman, pp. 189-232, Springer-Verlag, New York, 1992.

Bisson, P.A., K. Sullivan, and J.L. Nielsen, Channel hydraulics, habitat use, and body form of juvenile coho salmon, steelhead, and cutthroat trout in streams, *Transactions of the American Fisheries Society* 117: 262-273, 1988.

Booth, D. B., Glaciofluvial infilling and scour of the Puget Lowland, Washington, during ice-sheet glaciation, *Geology* 22:695-698, 1987.

Brown, E.H., D.L. Blackwell, B.W. Christenson, F.I. Frasse, R.A. Haugerud, J.T. Jones, P.A. Leiggi, M.L. Morrision, P.M. Rady, G.J. Reller, J.H. Sevigny, D.S. Silverberg, M.T. Smith, J.N. Sondergaard, and C.B. Ziegler, *Geologic map of the northwest Cascades, Washington*, Geological Society of America, Map and Chart Series MC-61, 1987.

Cairns, J., Jr., Lack of theoretical basis for predicting rate and pathways of recovery, *Environmental Management* 14:517-526, 1990.

Chapman, D.W., Aggressive behavior in juvenile coho salmon as a cause of emigration, *Journal of Fisheries Research Board of Canada*, 19:1047-1080, 1962.

Collins, B.D. and D.R. Montgomery, Importance of archival and process studies to characterizing pre-settlement riverine geomorphic processes and habitat in the Puget Lowland, in *Geomorphic Processes and Riverine Habitat*, edited by J.M. Dorava, B.B. Palcsak, F.A. Fitzpatrick, and D.R. Montgomery, American Geophysical Union, Washington, D.C., this volume.

Crandell, D.R., Glacial history of western Washington and Oregon in *Quaternary of the United States*, edited by H.E. Wright, Jr. and D.G. Frey, pp. 341-354, Princeton University Press, Princeton, New Jersey, 1965.

Cwynar, L.C., Fire and the forest history of the North Cascade range, *Ecology* 68:791-802, 1987.

Dale, V.H., Revegetation of Mount St. Helens debris avalanche 10 years post eruption, *National Geographic Research and Exploration* 7:328-341, 1991.

Dethier, D.P., F. Pessl, Jr., R.F. Keuler, M.A. Balzarini, and D.R. Pevear, Late Wisconsinan glaciomarine deposition and isostatic rebound, northern Puget Lowland, Washington, *Geological Society of America Bulletin* 107:1288-1303, 1995.

Dragovich, J.D., D.T. McKay, Jr., D.P. Dethier, and J.E. Beget, Holocene Glacier Peak lahar deposits in the lower Skagit River valley, Washington, *Washington Geology* 28:19-21, 2000.

Featherston, K.L., R.J. Naiman, and R.E. Bilby, Large woody debris, physical process, and riparian forest development in montane river networks of the Pacific Northwest, *Geomorphology* 13:133-144, 1995.

Franklin, J. F. and C. T. Dyrness, Natural Vegetation of Oregon and Washington, USDA Forest Service General Technical Report PNW-8, Pacific Northwest Forest and Range Experiment Station, Portland, Oregon, 417 pp, 1973.

Gannett, H., *Nineteenth annual report of the United States Geological Survey, 1897-1898. Part V -- Forest Reserves*, Government Printing Office, Washington, D.C., 1899.

Gardner, C.A., K.M. Scott, C.D. Miller, B Meyers, W. Hildreth, and P.T. Pringle, Potential volcanic hazards from future eruptions of Mount Baker, Washington, U.S. Geological Survey Open-file Report 95-498, 1995.

Groot, C. and L. Margolis, *Pacific Salmon Life Histories*, UBC Press, University of British Columbia, Vancouver, 564 pp., 1991.

Heilman, P.E., Growth and N status of *Populus* in mixture with red alder on recent volcanic mudflow from Mount St. Helens, *Canadian Journal of Forest Research* 20:84-90, 1990.

Heller, P.L. Map showing surficial geology of parts of the lower Skagit and Baker valleys, North Cascades, Washington, U.S. Geological Survey Open-file Report 79-964, 1979.

Interstate Publishing Company, *An Illustrated History of Skagit and Snohomish Counties; Their People, Their Commerce and Their Resources, with an Outline of the Early History of the State of Washington*, Interstate Publishing Company, Chicago, 1117 pp., 1906.

Leider, S. A., Increased straying by adult steelhead trout, *Salmo gairdneri*, following the 1980 eruption of Mount St. Helens, *Environmental Biology of Fishes* 24:219-229, 1989.

Leopold, E.B., R. Nickman, J.I. Hedges, and J.R. Ertel, Pollen and lignin records of late Quaternary vegetation, Lake Washington, *Science* 218:1305-1307, 1982.

Lisle, T.E., Effects of aggradation and degradation on riffle-pool morphology in natural gravel channels, northwestern California, *Water Resources Research* 18:1643-1651, 1982.

Lunetta, R.S., B.L. Cosentino, D.R. Montgomery, E.M. Beamer, and T.J. Beechie, GIS-based evaluation of salmon habitat in the Pacific Northwest, *Photogrammetric Engineering and Remote Sensing* 63:1219-1229, 1997.

Madej, M.A. and V. Ozaki, Channel response to sediment wave

propagation and movement, Redwood Creek, California, USA, *Earth Surface Processes and Landforms* 21:911-927, 1996.

Major, J.J., T.C. Pierson, R.L. Dinehart, and J.E. Costa, Sediment yield following severe volcanic disturbance – a two decade perspective from Mount St. Helens, *Geology* 28:819-822, 2000.

Mangum, A. W. and Party, *Reconnaissance Soil Survey of the Eastern Part of Puget Sound*, U. S. Soils Bureau, 1911.

Martin, D.J., L.J. Wasserman, and V.H. Dale, Influence of riparian vegetation on posteruption survival of coho salmon fingerlings in west-side streams of Mount St. Helens, Washington, *North American Journal of Fisheries Management* 6:1-8, 1986.

Mastin, L.G. and R.B. Waitt, Is Glacier Peak a dangerous volcano? U.S. Geological Survey Open-file Report 95-413, 1995.

Matthews, R.K. Quaternary sea-level change, in. *Studies in Geophysics: Sea-level Change,* pp. 88-103, National Academy Press, Washington, D.C., 1990.

McPhail, J.D. and C.C. Lindsey, Zoogeography of the freshwater fishes of Cascadia (the Columbia system and Rivers north to the Stikine), in *The Zoogeography of North American Freshwater Fishes*, edited by C.H. Hocutt and E.O. Wiley, pp. 615-637, John Wiley and Sons, New York, 1986.

Meyer, D.F. and H.A. Martinson, Rates and processes of channel development and recovery following the 1980 eruption of Mount St. Helens, Washington, *Hydrological Sciences Journal* 34:115-127, 1989.

Montgomery, D.R. and J.M. Buffington, Channel-reach morphology in mountain drainage basins, *Geological Society of America Bulletin* 109:596-611, 1997.

Montgomery, D.R., E.M. Beamer, G.R. Pess, and T.P. Quinn, Channel type and salmonid spawning distribution and abundance, *Canadian Journal of Fisheries and Aquatic Sciences* 56:377-387, 1999.

Montgomery, D.R., J.M. Buffington, R.D. Smith, K.M. Schmidt, and G. Pess, Pool spacing in forest channels, *Water Resources Research* 31:1097-1105, 1995.

Munger, T.T., The cycle from Douglas-fir to hemlock, *Ecology* 21:451-459, 1940.

Naiman, R.J., C.A. Johnston, and J.C. Kelly, Alteration of North American streams by beaver, *BioScience* 38:753-762, 1988.

Nelson, K., *The influence of sediment supply and large woody debris on pool characteristics and habitat diversity*, Masters Thesis, University of Washington, Seattle, 1998.

Paulson, K., *Estimating changes in sediment supply due to forest practices: a sediment budget approach applied to the Skagit River basin*, Washington, Masters Thesis, University of Washington, Seattle, 1997.

Peterson, N.P. and L.M. Reid, Wall-base channels: their evolution, distribution, and use by juvenile coho salmon in the Clearwater River, Washington, in *Proceedings of the Olympic wild fish conference*, edited by J.M. Walton and D.B. Houston, pp. 215-225, Fisheries Technology Program, Peninsula College, Port Angeles, Washington, 1984.

Porter, S.C. and T.W. Swanson, Radiocarbon age constraints on rates of advance and retreat of the Puget Lobe of the Cordilleran ice sheet during the last glaciation, *Quaternary Research* 50: 205-213, 1998.

Sandercock, F.K, Life history of coho salmon, in *Pacific Salmon Life Histories*, edited by C. Groot and L. Margolis, Pages 397-445, University of British Columbia Press, Vancouver, British Columbia, 1991.

Scarlett, W.J. and C.J. Cederholm, Juvenile coho salmon fall-winter utilization of two small tributaries of the Clearwater River, Jefferson County, Washington, in *Proceedings of the Olympic Wild Fish Conference*, edited by J.M. Walton and D.B. Houston, pp. 215-225, Fisheries Technology Program, Peninsula College, Port Angeles, 1984.

Stuiver, M. and P.J. Reimer, Extended ^{14}C data base and revised CALIB 3.0 ^{14}C age calibration program, *Radiocarbon* 35:215-230, 1993.

Thorson, R.M, Ice sheet glaciation of the Puget Lowland, Washington, during the Vashon stade (late Pleistocene), *Quaternary Research* 13:303-321, 1980.

Williams, R.W., R.M. Laramie, and J.J. Ames, *A Catalog of Washington Streams and Salmon Utilization, Volume 1: Puget Sound*, Washington Department of Fisheries, Olympia, 1975.

Timothy J. Beechie, Watershed Program, Northwest Fisheries Science Center, 2725 Montlake Blvd. E., Seattle, WA 98112

Brian D. Collins, Department of Geological Sciences, Box 351310, University of Washington, Seattle, WA 98105

George R. Pess, Watershed Program, Northwest Fisheries Science Center, 2725 Montlake Blvd. E., Seattle, WA 98112

Section 2

Pristine Undisturbed Systems

Channel Changes and Habitat Diversity in a Warm-Water, Gravel-Bed Stream

Rose McKenney

Geoscience Department, Minot State University, Minot, North Dakota

Physical habitat characteristics strongly influence community diversity in streams; geomorphic theories and techniques are often used to quantify the extent of aquatic habitats and estimate changes in these habitats. Although the effect of flood magnitude and frequency on channel change is widely studied, the assumption that habitat change is proportional to geomorphic change has not been verified for many stream systems. Detailed, short-term observations of channel change, habitat distributions, and habitat diversities in warm-water, gravel-bed Ozark streams indicate that channel change and habitat diversity have fundamentally different relations with peak flood magnitude. Channel morphology was substantially altered by erosion and deposition that occurred during a large, overbank (50- to 100-year recurrence interval) flood. However, when the linear geomorphic change relationship was coupled with flood frequency, it produced a relationship in which maximum channel change occurs at moderate flood magnitudes. Habitat diversity, in contrast, has a logarithmic relationship with flood magnitude. As a result, small (less than 5-year recurrence interval) and moderate, bankfull to slightly overbank floods (5- to 10-year recurrence interval) result in substantially different habitat distributions and diversities. These differences in habitat diversity occur primarily because glide area increases following small floods, whereas riffle and pool area increases following moderate and large floods. As a result, substantial fluctuations of habitat diversity occur in warm-water, gravel-bed Ozark streams on a decadal time frame.

INTRODUCTION

Research that strives to determine the effects of abiotic factors on aquatic populations requires quantification of physical habitat variability in freshwater streams. This need is particularly pressing in river systems that experience frequent channel change and hydrologic variability. Quantification of the effects of hydrologic change on habitat generally relies on hydraulic geometry relations [*Hogan and Church*, 1989], one-, or two-dimensional modeling [*Bovee*, 1978; *Orth*, 1987; *Leclerc et al.*, 1995; *Panfil and Jacobson*, 1999, for example]. Although low-flow hydrologic variability is an important source of habitat disturbance in river systems, it is frequently superimposed on systems that experience channel change. Effects of these channel changes are rarely incorporated into evaluations of habitat distribution [*Heritage et al.*, 1996]. Additional management problems arise in systems that have large numbers of species which are managed to preserve biodiversity and to increase or maintain the numbers of native and/or introduced game fish. Conflicting habitat requirements of organisms often complicate these management goals [*Jowett and Richardson*, 1995].

Geomorphic Processes and Riverine Habitat
Water Science and Application Volume 4, pages 57-71

Table 1. Hydrologic characteristics from U. S. Geological Survey streamflow gages on Jacks Fork.

U. S. Geological Survey Stream Gages (Fig. 1)	Period of record	Drainage area (km^2)	Date of recorded peak discharge	Peak discharge of record (m^3/s)	Unit peak discharge of record ($m^3/s/km^2$)	95 Percent discharge [a] (m^3/s)	5 Percent discharge [b] (m^3/s)	Estimated bankfull discharge (m^3/s)
At Alley Spring (07065495)	March 1993 to present[c]	772 [c]	November 14, 1993[c]	1,380 [c]	1.79	2.3	30.1	350
At Eminence (07066000)	October 1921 to present[c]	1,030 [c]	November 15, 1993[c]	1,660 [c]	1.61	3.1	39.3	400

[a] Discharge is exceeded 95% of the time. Value is calculated from daily mean discharges of the period of record.

[b] Discharge is exceeded 5% of the time. Value is calculated from daily mean discharges of the period of record.

[c] Hauck et al., 1999.

Although many studies have linked channel morphology to habitat and organism distribution [*McClendon and Rabeni*, 1987; *Lobb and Orth*, 1991; *Walters and Wilson*, 1996; *Payne and Lapointe*, 1997, for example], quantification of the effects of channel change on physical habitat distribution remains difficult. Many techniques and theories that are commonly used to evaluate physical habitat in freshwater streams have been devised for geomorphic study of stream systems. Before these techniques are used extensively in physical habitat and biologic studies, their efficacy in determining physical habitat distribution and diversity should be evaluated.

Physical habitat diversity, while related to channel morphology, is often not completely described by assessments of channel change. Physical habitat distribution and diversity are generally measured at low flows, because many activities of aquatic organisms' life history occur during low-flow periods [*Bovee and Milhous*, 1978]. Additionally, documented habitat use during flood conditions indicates that during floods, fish move to areas outside of the low-flow channel with characteristics similar to their preferred habitat [*Jowett and Richardson*, 1994]. Cross-sectional channel change, in contrast, is generally determined for the bankfull channel [*Emmett*, 1974; *Madej and Ozaki*, 1996, for example], and in Ozark streams bankfull discharges are rarely met or exceeded (less than 5% of the time, Table 1, for example). In addition, lateral-channel migration often causes large amounts of cross-sectional channel change as measured by area eroded and area deposited, but in areas where the channel morphology is similar before and after migration, there may be little to no change in physical habitat distribution.

This study compares relations among channel change, habitat distribution and diversity, and flood frequency and magnitude in a warm-water, gravel-bed Ozark stream from 1991 to 1998. Channel change is defined in this paper as erosion and deposition within the surveyed cross section. This change includes vertical as well as lateral erosion and deposition. This comparison is used to determine if the relationship between channel change and flood magnitude is a reasonable indicator of habitat distribution and diversity in similar Ozark streams.

Study Area

This study was conducted at the Burnt Cabin reach of the Jacks Fork within the Ozark Plateaus (Figure 1, Table 1). Streams of the Ozark Plateaus are incised into uplifted Paleozoic sedimentary rocks, which form gently rolling uplands. Valley bottoms are generally flat; valley walls rise nearly vertically out of the valley bottom. Bedrock lithologies are dominantly cherty carbonates which supply the streams with chert-rich gravel-sized sediment.

Although the bedrock walls are generally eroding very slowly, the alluvial bottomland is being extensively reworked in many Ozark streams [*Saucier*, 1983; *Jacobson and Pugh*, 1997]. Reworking generally occurs during moderate (approximately bankfull) and large, overbank floods which cause substantial lateral channel migration. In addition, large floods (recurrence interval >50 years) have caused overbank deposition of gravel and intense floodplain scour [*Ritter*, 1988; *McKenney and Jacobson*, 1996].

At the Burnt Cabin reach, the Jacks Fork drains 780 km^2 and flows through a wide valley with a substantial alluvial bottomland (Figures 1, 2). Two U. S. Geological Survey gages are located within the basin--one at Eminence (07066000, installed 1921) and another at Alley Spring (07065495, installed March 1993). Low flow in the Jacks

Fork basin, approximately 3 m^3/s at Eminence, is maintained by spring flow from an extensive cavern system in the dolomitic rocks [*Beckman and Hinchey*, 1944]. During intense storms, the cavern system is by-passed and water flows overland, resulting in fast-rising floods. The peak discharge of record, 1,660 m^3/s (50-year recurrence interval), was recorded on November 15, 1993 at the U. S. Geological Survey gage at Eminence (Figures 1, 3, Table 1).

An abundance of fish and aquatic invertebrate species live within the streams of the Ozark Plateaus. Over 100 fish species live in Ozark streams--56 fish species and subspecies are primarily restricted to the Ozarks [*Pflieger*, 1989]. Twenty species and subspecies of crayfish are also largely restricted to the Ozarks [*Pflieger*, 1989]. The large number of fish and invertebrate species in the Ozarks, coupled with extensive urbanization in some areas has led to increased interest in the effects of flood magnitude on habitat distribution and diversity. In diverse biologic communities such as those found in Ozark streams, a variety of habitats are essential so evaluations of habitat must include the diversity of habitats available, rather than the extent of a single type of habitat.

METHODS

Channel cross-section and hydrologic data were collected at the Burnt Cabin reach on the Jacks Fork (Figure 2) to evaluate the effect of flood magnitude and frequency on channel change and habitat diversity. The 19 cross-sections in the reach extend over 1 km of channel length and encompass a pool-riffle-pool-riffle-pool sequence in the stream. The study reach was located in a basin with two U. S. Geological Survey gages to compare flood frequency and magnitude to channel change and habitat diversity.

Channel Change

Net erosion and deposition were monitored using the network of channel cross sections. Cross sections were spaced less than five low-flow channel widths apart and were oriented perpendicular to bankfull flow directions. Locations were adjusted to make individual cross sections representative of habitats common in streams of the Ozark Plateaus [*Rabeni and Jacobson*, 1993a, 1993b]. Surveying was conducted using an electronic total station and digital data logger [*McKenney and Jacobson*, 1996]. Cross sections were resurveyed annually from 1991 to 1995 [*McKenney and Jacobson*, 1996] and again in 1997 and 1998 [*Jacobson*, unpublished data].

Erosional and depositional areas were determined for consecutive surveys using a geographic information system to calculate changes in the channel cross sections [*McKenney*, 1997]. Distance-elevation data for each survey

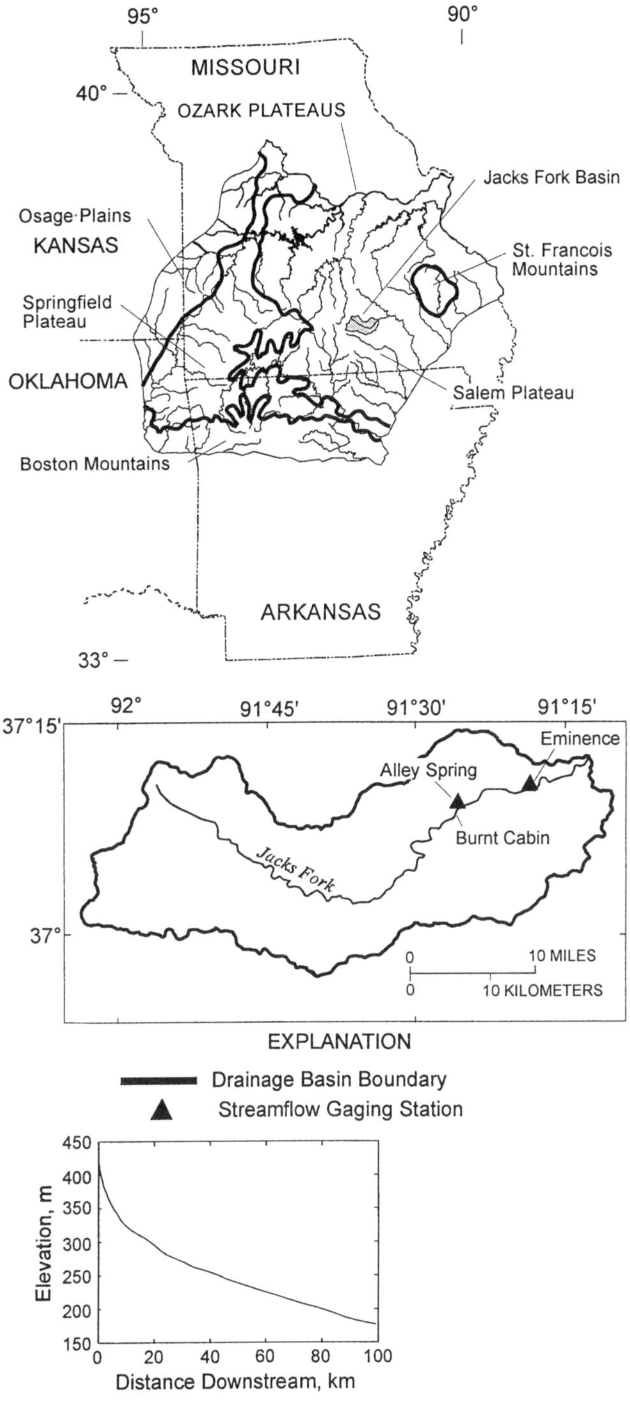

Figure 1. Location maps for the Jacks Fork, and the Burnt Cabin reach, and longitudinal profile of the Jacks Fork.

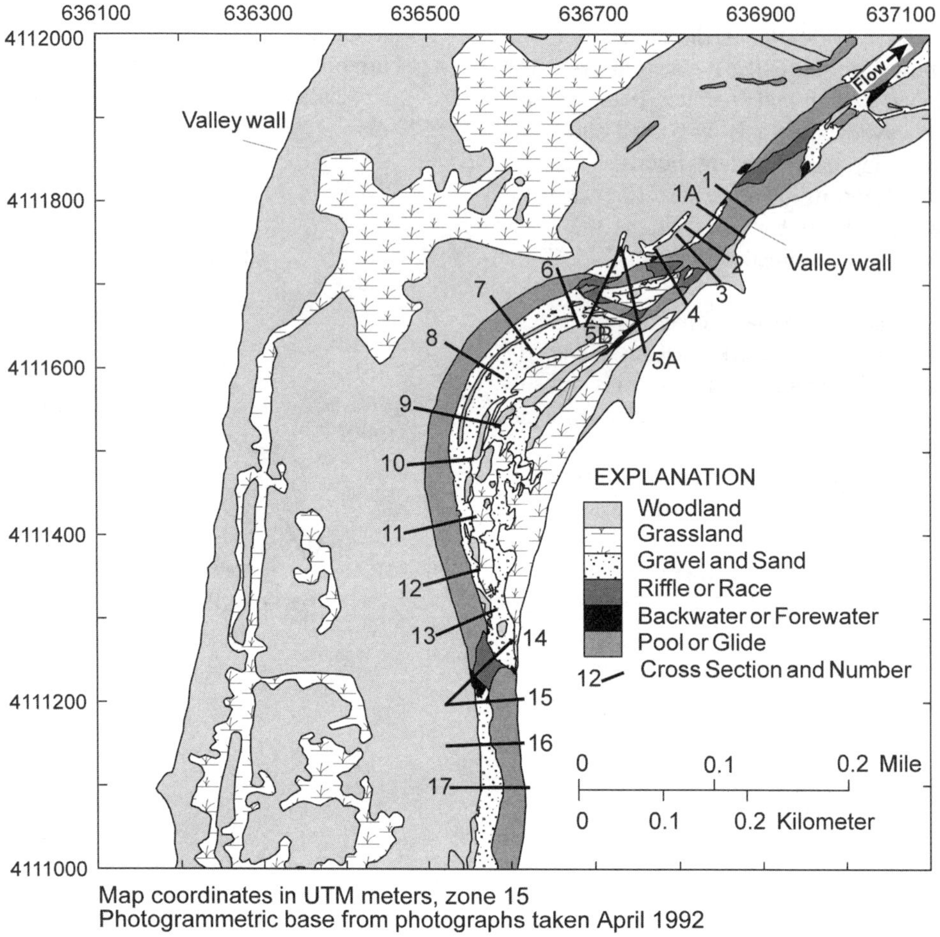

Figure 2. Burnt Cabin reach, Jacks Fork, Missouri (from McKenney and Jacobson, 1996).

of each cross-section were converted to lines and nodes in the geographic information system and polygons of 'channel' and 'sediment' were created (Figure 4). The areas which changed from 'channel' in the first survey to 'sediment' in the second survey were tabulated as area deposited. The areas which changed from 'sediment' in the first survey to 'channel' in the second survey were tabulated as area eroded. Net cross-sectional channel change was calculated by subtracting the area eroded from the area deposited for each cross section. Total cross-sectional channel change was calculated by summing erosion and deposition for each cross section. Mean cross-sectional change was calculated by summing total cross-sectional channel change values and dividing by the number of cross sections. Geomorphic change along the reach was calculated by multiplying each cross-sectional channel change value by one-half the average distance to the adjacent cross sections (Figure 5).

Habitat Diversity

Habitat diversity was also determined using cross-section data. Cross sections were divided into habitats using a modification of a hierarchical classification developed for Ozark streams [*Rabeni and Jacobson*, 1993b; *McKenney*, 1997]. This classification system is based on field observations of distinct, repeatable sub-riffle-pool units, which can be separated using water-surface gradient, morphology and bank material criteria [*Rabeni and Jacobson*, 1993b]. These criteria were modified as indicated here to consistently identify an individual area within a channel as one habitat type over a range of low-flow stages. These modifications result in a method which provides objective rule-based criteria to subdivide channel cross-sections into multiple habitat types. Habitat diversity was calculated from habitat distributions using Simpson's index [*Simpson*, 1949]. To standardize the amount of wetted channel, sur-

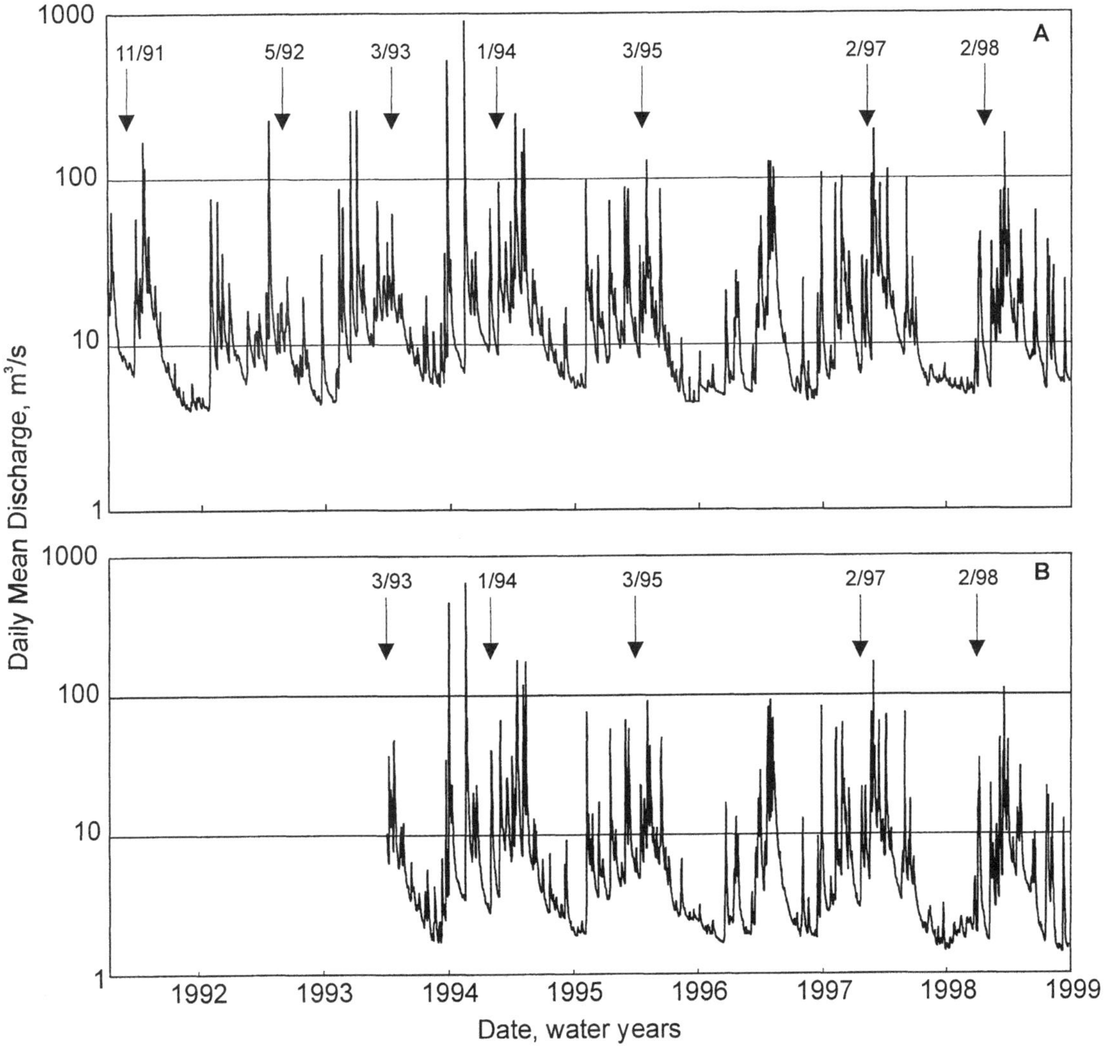

Figure 3. Daily mean discharge hydrographs from the Jacks Fork (A) gage at Eminence, Missouri (07066000) and (B) gage at Alley Spring (07059495) during the study period. Arrows indicate cross-section survey dates.

veyed water surface profiles were corrected to a single arbitrary low-flow stage.

Nine habitat types were found in this reach throughout the period of study. Glides, bluff pools and lateral pools characteristically have low velocities and low water-surface gradients. Glides have poorly-sorted sand- to cobble-sized bed material and trapezoidal cross-sections. Bluff pools are characterized by poorly sorted sand- to large boulder-sized bed material, v-shaped cross sections, and bedrock or coarse talus banks. Lateral pools have gravel- to cobble-sized bed material skewed cross sections, and alluvium banks. Riffles and races [named coarse runs in *Rabeni and Jacobson*, 1993a, b] have high velocities, sorted gravel- or cobble-sized bed material and high water-surface gradients. Riffles characteristically have low depths and trapezoidal cross sections. Races characteristically are deeper than riffles and have u-shaped cross sections. Edgewaters occupy the shallow channel margins where low-flow velocities are near zero. Backwaters and forewaters are shallow, have near zero velocity and are isolated from flowing water at low flow.

Cross-sections were initially divided into low- and high-gradient groups, using a water-surface gradient criterion of 0.0075. This process divided the cross sections into longitudinal groups of pools and glides (low-gradient group), and riffles and races (high-gradient group). These longitudinal groups were subdivided into more detailed habitat classes using depth, morphology, and bank material crite-

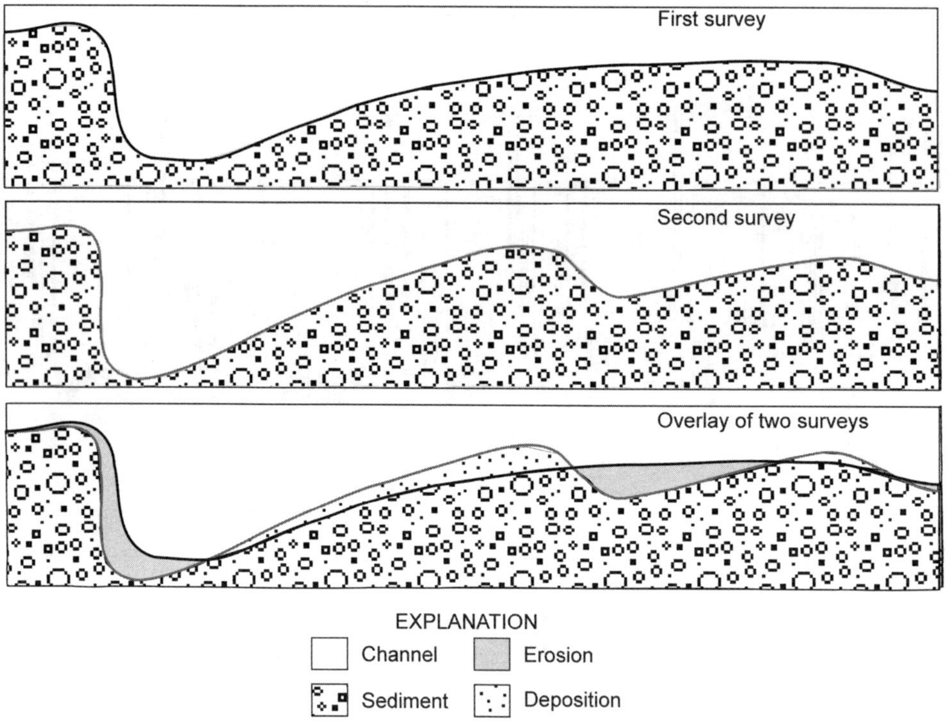

Figure 4. Schematic diagram of erosion and deposition calculations performed in a geographic information system.

ria. Parts of cross-sections that were submerged, but not affected by the main flow, were classified as forewaters and backwaters. Forewaters are connected to the main channel at their upstream end. Backwaters are connected to the main channel at their downstream end.

Within low-gradient cross-sections, edgewaters were separated from glides using a 10-cm depth criterion. This criterion was based on depths associated with increased velocity in glide and pool cross sections [*McKenney*, 1997]. Glides were separated from pools using a depth criterion of 3.5% of the reach-averaged wetted width. The *reach-averaged wetted width* is calculated by averaging the water widths of all the cross-sections in a study reach for any given stage. The criterion is based on the distribution of maximum depth to wetted width ratios for field identified glides and pools. Pools were classified as bluff pools if the adjacent bank was composed of bedrock or a coarse talus deposit, or as lateral pools if the adjacent bank was composed of alluvium.

Similar to low-gradient cross sections, high-gradient cross sections were separated into riffles and races using a depth criterion of 2.5% of the reach-averaged wetted width. This depth criterion was based on the distribution of maximum depth to reach average wetted width ratios for riffles and races identified in the field at all reaches.

The areal extent of each habitat type was calculated by multiplying the width of each habitat within the cross-section times the sum of one-half the distance to the nearest cross section upstream and one-half the distance to the nearest cross section downstream. Habitat diversity was calculated from areal habitat distributions for each survey by using Simpson's [1949] index $1/\lambda$, where

$$\lambda = \Sigma(n_i/N)^2$$

Each square meter of each habitat type was counted as an individual, n_i, and the total number of square meters was the total count, N. The minimum value of Simpson's index is 1, indicating the presence of a single habitat type. The maximum possible value of the index is the number of types of habitat, indicating equal amounts of each type of habitat are present. Although it is theoretically possible to have a diversity index equal to the number of habitats, forewaters, backwaters, and edgewaters generally occupy small areas in Ozark streams (< 5% of total reach length), resulting in lower maximum diversity values.

Flood Frequency and Magnitude

Flood frequency and magnitude were evaluated using available U. S. Geological Survey gages and published

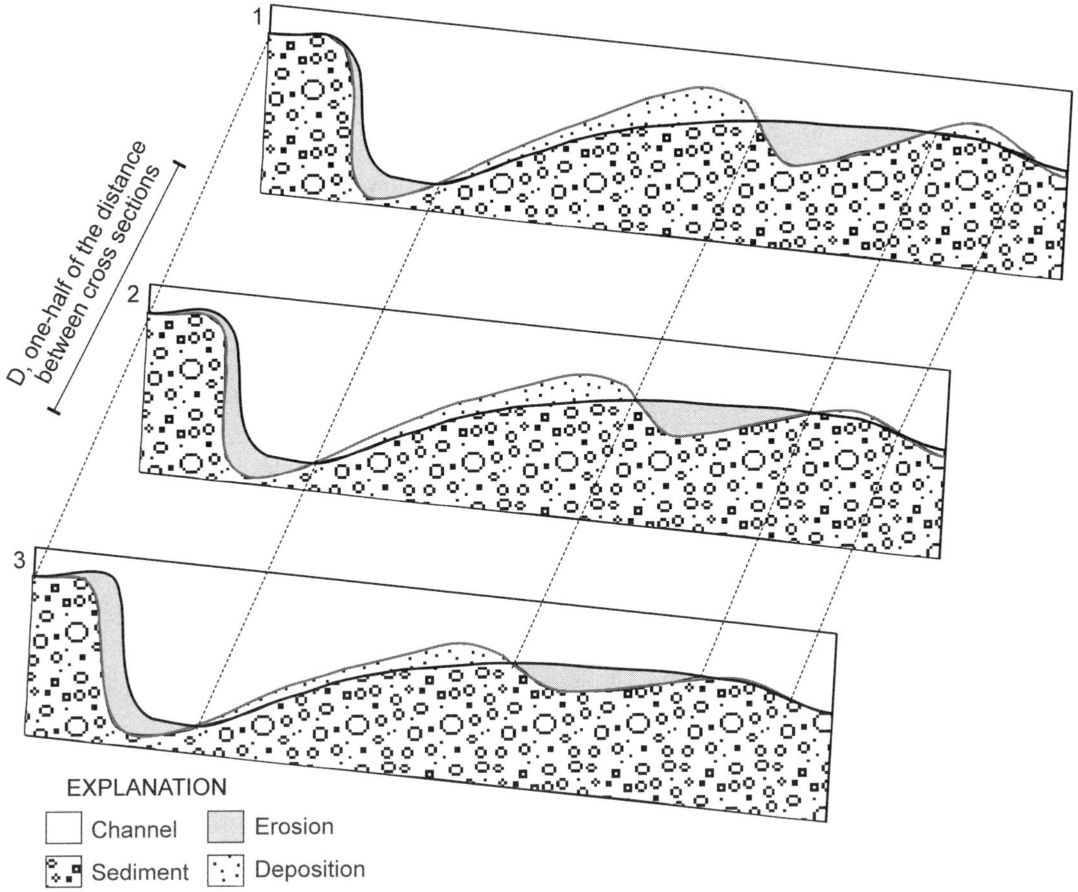

Figure 5. Schematic diagram of geomorphic change calculations. Geomorphic change is area eroded plus area deposited times D, one-half of the average distance between cross sections.

flood frequency-magnitude regression relations. The regression relationship (R^2=0.988),

$$q_a = q_e * 0.89$$

between the peak flow record at Eminence (q_e, gage 07066000) and the peak flow record at Alley Spring (q_a, gage 07065495) from March 1993 to September 1997 (Figure 6) was used to calculate the magnitude of peak discharges for floods upstream of Alley Spring that occurred prior to March 1993. Calculated and observed peak discharges of the Jacks Fork at Alley Spring were used to determine the approximate recurrence intervals of floods that occurred during this study (Tables 2, 3).

RESULTS

Data indicate that erosion and deposition increased at the Burnt Cabin reach with increasing flood magnitude (Figures 3, 7, Table 4). The amount of glide habitat decreased with increasing flood magnitude, whereas the amount of riffle habitat increased (Plate 1).

Flood Frequency, Magnitude and Channel Changes

Small floods with magnitudes of less than 400 m^3/s (< 5-year recurrence intervals) caused an average of 9.7 to 11.6 m^2 of total cross-sectional channel change in each cross section (Table 4). Moderate floods with magnitudes between 400 m^3/s and 700 m^3/s (5- to 10-year recurrence intervals) caused an average of 18.3 to 18.9 m^2 of total cross-sectional channel change. The large flood with a magnitude of 1,370 m^3/s (50-year recurrence interval) caused an average of 32.9 m^2 of total cross-sectional channel change. In addition to the increased total cross-sectional channel change, the variability in the amount of erosion and deposition in individual cross sections increased with increasing flood magnitude (Figure 7, Table 4).

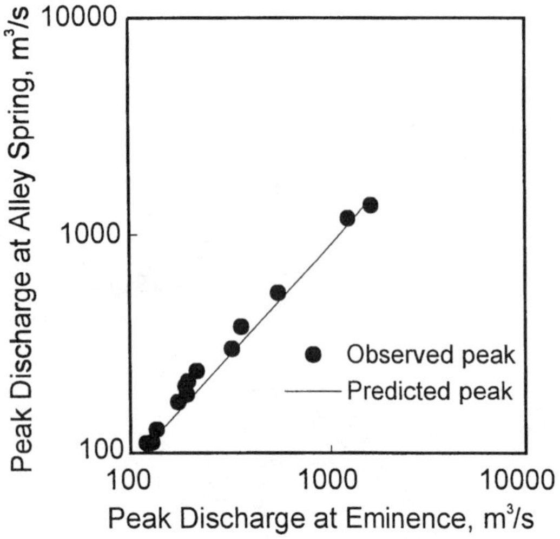

Figure 6. Regression relation between peak discharges at Eminence gage (07066000) and peak discharges at Alley Spring gage (07059495). R^2 value is 0.988.

Although floods of similar magnitude caused similar amounts of total cross-sectional channel change, the balance between erosion and deposition within the reach varied. For example, comparison of net cross-sectional channel change for the moderate flood in January 1993 and the moderate flood in May 1994 indicates greater amounts of erosion occurred during the January 1993 flood than occurred during the May 1994 flood (Figure 7, Table 3). This variation may be caused by slight differences in flood magnitude or may result from flood sequence. The flood in January 1993 was larger than the May 1994 flood, however, the May 1993 flood also occurred after the large flood in November 1993, which caused extensive erosion (Figures 7, 8). Large amounts of deposition during the May 1994 flood may have resulted from subsequent filling of areas eroded by the November 1993 flood.

The location of the erosion and deposition also changed with increasing magnitude [Figure 8, *McKenney and Jacobson*, 1996; *McKenney*, 1997]. Small floods deposited small amounts of sediment in the channel, especially in pools (Figure 8, BC 9, March 1995 through February 1998). Sediment transported during moderate floods, in contrast tended to be deposited on bar surfaces at or near the bankfull channel margin (Figure 8, BC 9, March 1993). In addition, during these floods pools were scoured and cutbanks eroded (Figure 8, March 1993). The large, overbank flood deposited large amounts of sediment on high bars and as coarse splays on the floodplain (Figure 8, January 1994), causing large amounts of geomorphic change outside of the low-flow channel.

Flood Frequency, Magnitude and Changes in Habitat Distribution

Similar to channel morphology, habitat distribution varies with flood magnitude. The area of glide habitat, for example, substantially increased following small floods and decreased after moderate and large floods (Plate 1). Riffle habitat area, in contrast, generally decreased after small floods and increased after moderate and large floods. Pool habitat generally increased after moderate and large floods (Plate 1). Unlike riffle habitat area, however, the increase in pool habitat area after large floods was small compared to the increase in pool habitat after moderate floods. In particular, lateral pool habitat did not substantially increase following the large flood. Although floods alter edgewater, backwater, and forewater habitat areas, the relationship between flood magnitude and these habitats is unclear. For example, backwater habitat area was greatest

Table 2. Estimated discharges of selected recurrence intervals for the Burnt Cabin reach based on regional regression relations.

Recurrence Interval, years	Discharge[a], m^3/s
2	270
5	504
10	691
25	953
50	1,140
100	1,340

a. Calculated using Alexander and Wilson's (1995) regression relation.

Table 3. Peak discharges and estimated recurrence intervals for the Jacks Fork at the Alley Spring gage during the study period.

Flood Date	Flood Discharge, m^3/s	Estimated Recurrence Interval of Flood, years
April 14, 1991	166	< 2
April 20, 1992	275	2
January 4, 1993	611	10
November 15, 1993	1,370	[a]100
May 7, 1994	539	5
April 29, 1996	234	<2
February 27, 1997	375	3

a. Recurrence interval at Eminence gage - 50-year (McKenney et al., 1995).

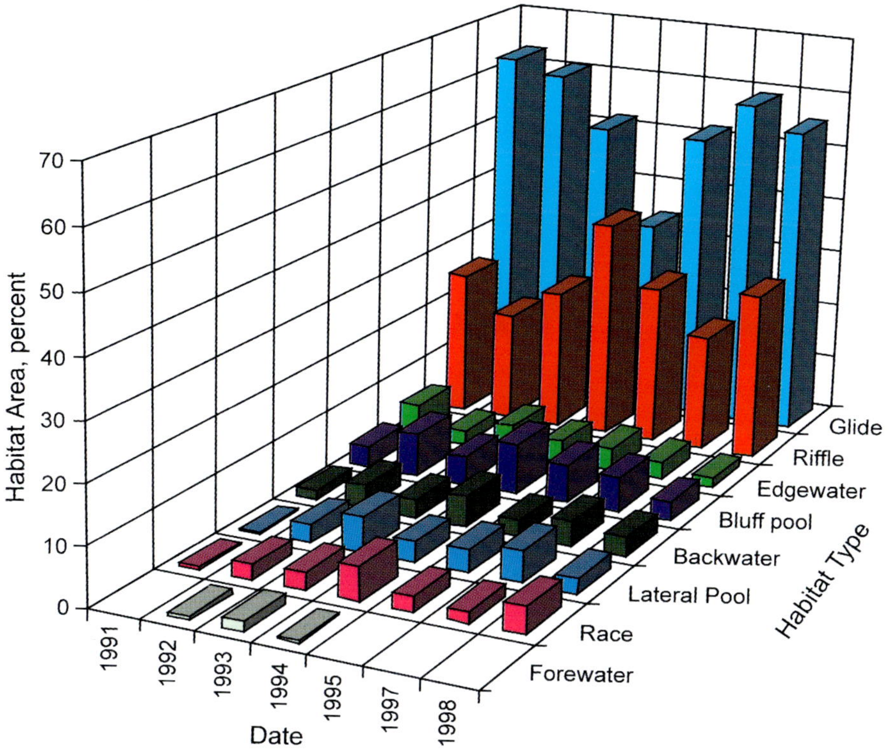

Plate 1. Habitat distribution of the Burnt Cabin reach November 1991 to February 1998. Habitats classified using modification of Rabeni and Jacobson (1993b).

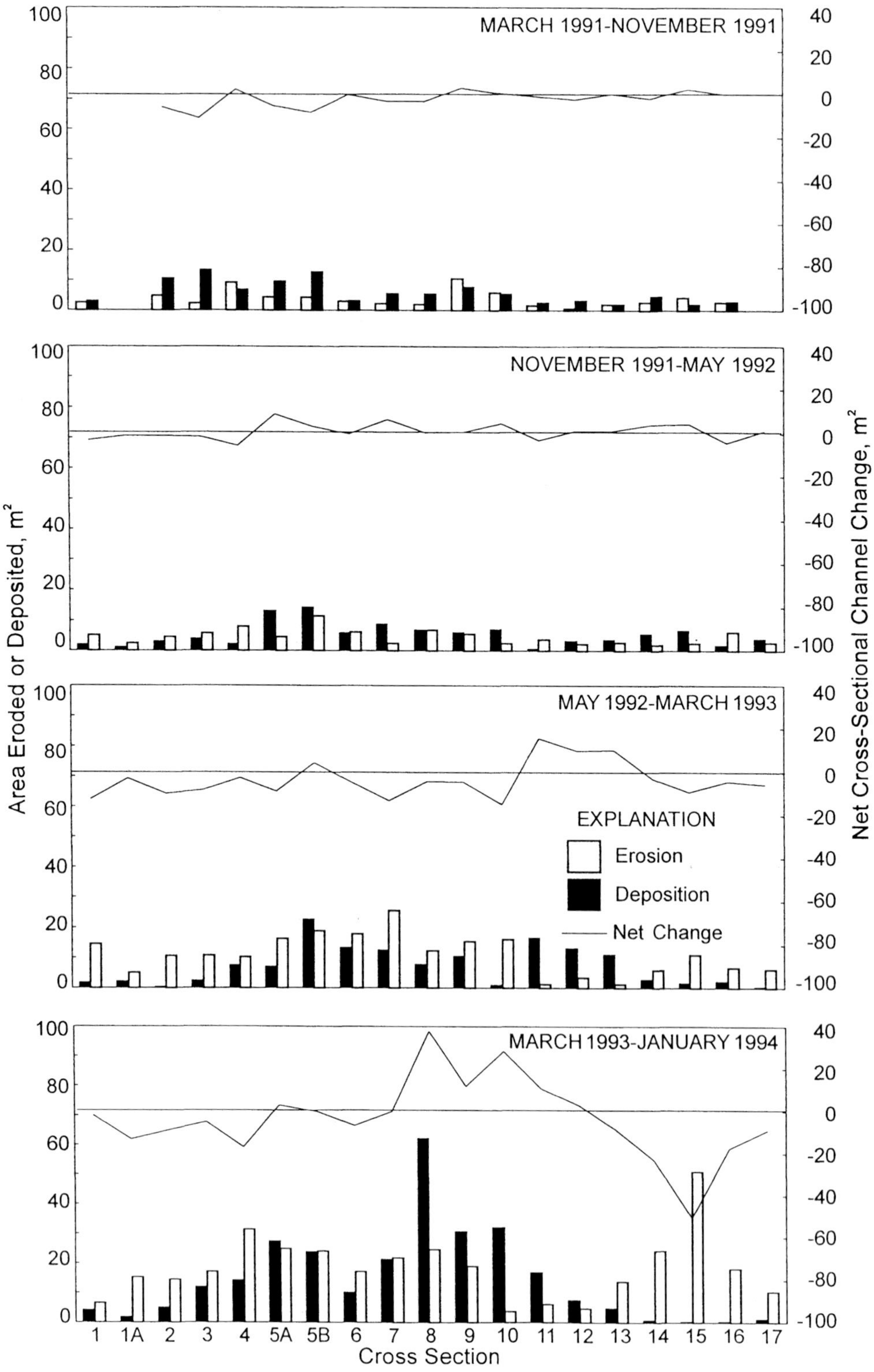

Figure 7. Area eroded, area deposited, and net cross-sectional channel change at each cross section for surveys from November 1991 to February 1998 at the Burnt Cabin reach, Jacks Fork, Missouri.

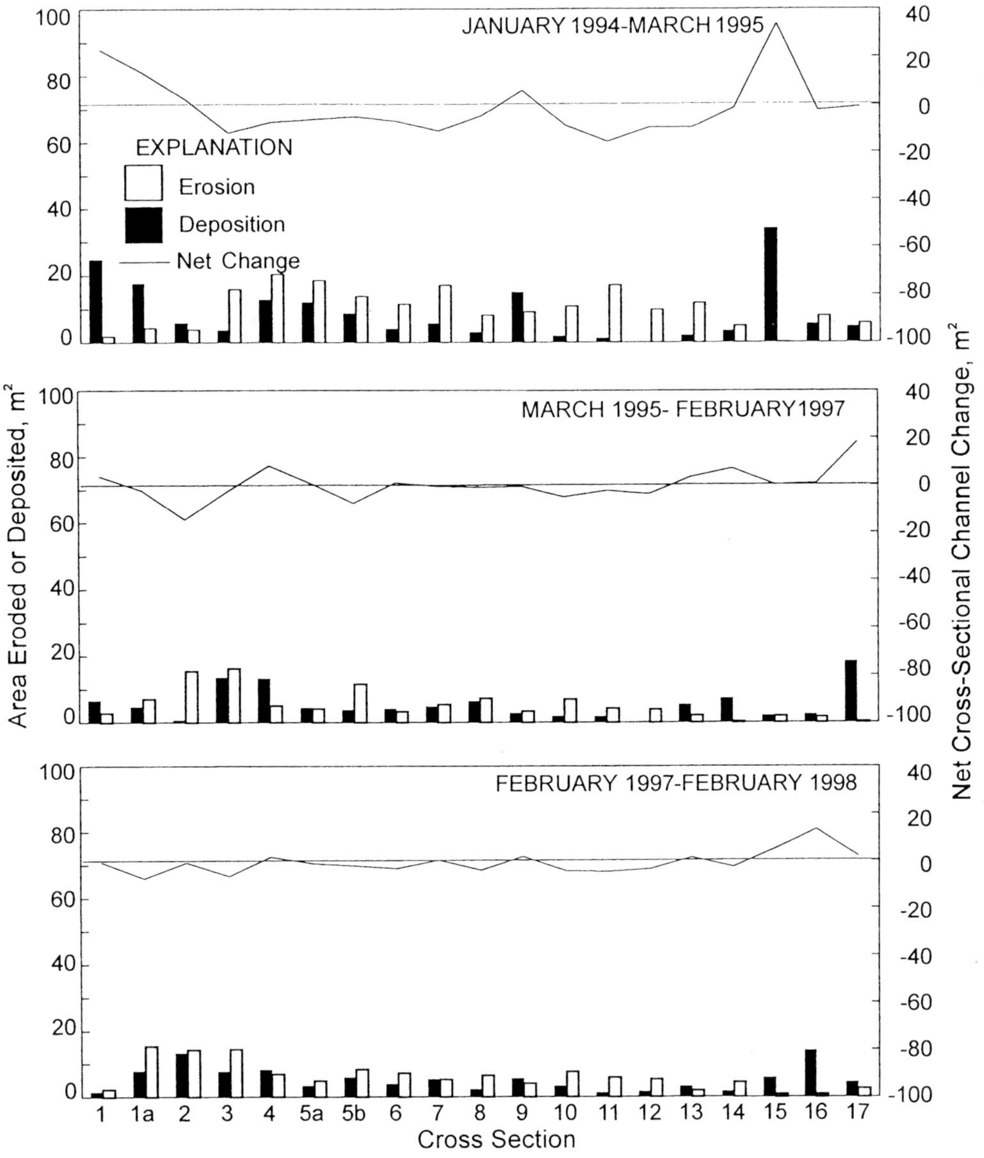

Figure 7. (continued)

Table 4. Mean and confidence intervals of total cross section change for each survey date at the Burnt Cabin reach.

Survey Dates	Peak Discharge, m^3/s	Mean of total cross section change, m^2	Upper 95% Confidence Interval	Lower 95% Confidence Interval
March 1991 to November 1991	166	9.7	12.3	7.0
November 1991 to May 1992	275	9.8	12.3	7.3
May 1992 to March 1993	611	18.3	23.1	13.5
March 1993 to January 1994	1,370	32.9	42.3	23.5
January 1994 to March 1995	539	18.9	22.9	14.9
March 1995 to February 1997	234	10.9	14.0	7.8
February 1997 to February 1998	375	11.6	14.8	8.4

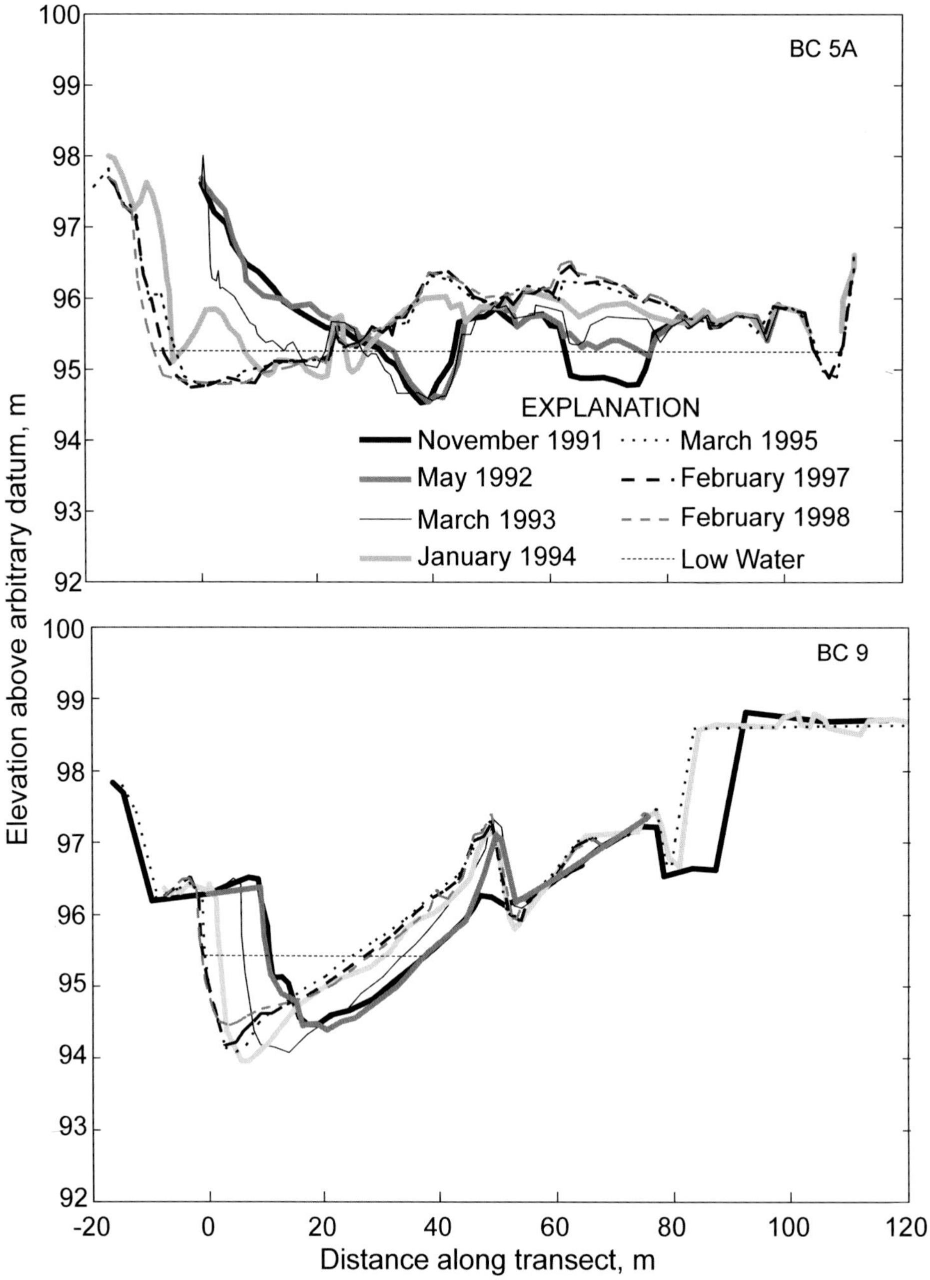

Figure 8. Example cross sections from the Burnt Cabin reach for surveys from November 1991 to February 1998. BC 5A is a high-gradient (riffle) cross section; BC 9 is a low-gradient (pool) cross section.

following the largest and two of the three smallest floods (Plate 1, Table 3). Edgewater habitat area, however, was approximately the same for moderate and large floods, but was the largest after the smallest flood.

Although both total channel change and habitat diversity increase with increasing flood magnitude, the relations observed in Ozark streams are fundamentally different. Geomorphic change increases linearly with peak discharge according to the statistical relationship ($R^2 = 0.98$, Figure 9):

$$GC = 22.4\ q + 3935$$

where GC is geomorphic change and q is peak discharge. Habitat diversity, however, increases logarithmically according to the statistical relationship ($R^2 = 0.96$, Figure 9):

$$1/\lambda = 1.5 \log q - 1.1$$

where $1/\lambda$ is Simpson's diversity index.

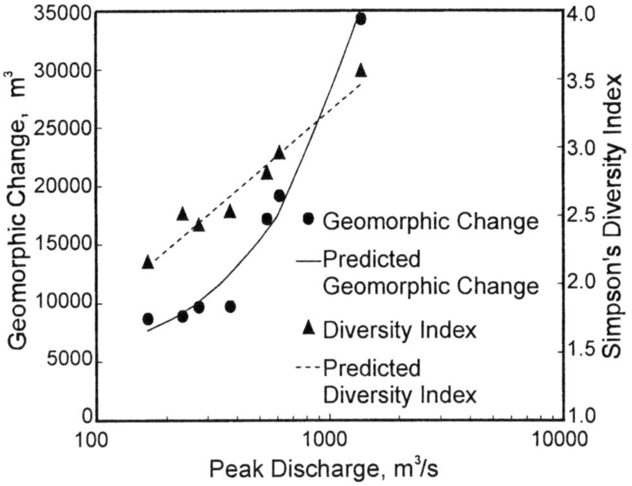

Figure 9. Relations between geomorphic change and flood magnitude, and habitat diversity and flood magnitude at the Burnt Cabin reach for the period March 1991 to March 1998. The statistical relation between geomorphic change and flood magnitude follows the equation GC = 22.4 q + 3935 ($R^2 = 0.98$) where GC is geomorphic change and q is peak discharge. Habitat diversity, however, increases logarithmically according to the equation $1/\lambda = 1.5 \log q - 1.1$ ($R^2 = 0.96$), where $1/\lambda$ is Simpson's diversity index.

DISCUSSION

The amount of erosion and deposition that occurs in Ozark streams increases linearly with flood magnitude (Figure 9). The increase in geomorphic change that occurs from large floods is approximately double the channel change caused by moderate floods (Figure 7, Table 4). Large floods, however, occur an order of magnitude less frequently than moderate floods in Ozark streams. Over time, therefore, the maximum amount of erosion and deposition results from moderate, bankfull to slightly overbank floods (recurrence interval 5-10 years). This relationship is similar to the relationship between flood frequency and geomorphic work (Figure 10a) described by *Wolman and Miller* [1960].

Habitat diversity also increases as a function of flood magnitude, however, the shape of the curve indicates that changes in habitat diversity do not increase as dramatically with flood magnitude as geomorphic change (Figure 9). In addition, these changes did not persist after the moderate flood, which followed the large overbank flood (Plate 1). Lower rates of change and lack of persistence, when coupled with the low frequency of large floods, suggests that most changes in habitat diversity result when small and moderate floods alternate within a stream (Figure 10b). In addition, because small floods dominantly alter low-flow channel morphology, channel changes caused by small floods are critical in determining habitat distribution.

The disparity between effects of flood magnitude on geomorphic change and on habitat diversity results from differences in depositional and erosional styles of small, moderate, and large floods in Ozark streams. Aggradation in pools during small floods decreases bed-elevation variations between pool troughs and riffle crests, favoring the formation of glides over riffles and pools. Decreased longitudinal and lateral bed-elevation variation results in low hydraulic and habitat diversity as measured at low flow. A similar decrease in longitudinal bed-elevation variations was observed in Redwood Creek, California following flood-associated sediment input [*Madej*, 1999].

Scour in pools during moderate floods increases the bed-elevation variation between the pool trough and riffle crest. Avulsion tends to form channel cutoffs that contain deep backwaters. In addition, moderate floods flow through dense vegetation bands on bars, forming ridge and swale topography [*McKenney et al.*, 1995]. Because these changes occur near the low-flow channel margin, this topography tends to form shallow ephemeral backwaters and forewaters at low flow. Increased longitudinal and lateral

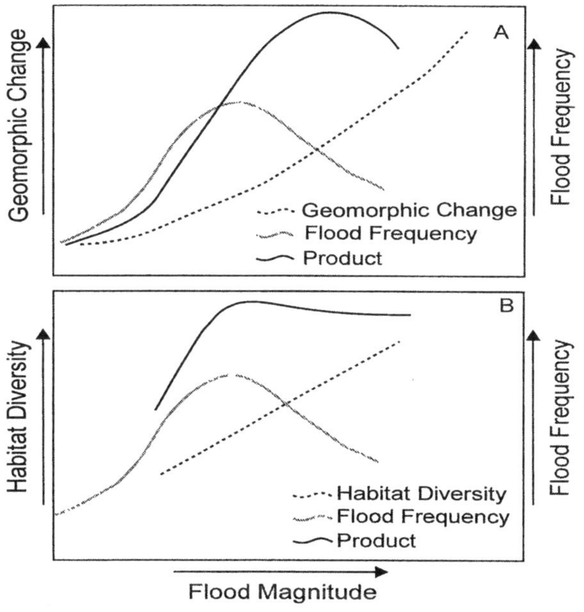

Figure 10. Relations between (A) geomorphic change and flood magnitude (after Wolman and Miller, 1960), and (B) habitat diversity and flood magnitude in Ozark streams. The amount of geomorphic change is an increasing function of flood magnitude. When this relation is coupled with flood frequency, the largest amount of geomorphic change occurs for moderate magnitude floods. Habitat diversity is also an increasing function of flood magnitude. However when this relation is coupled with flood frequency, the largest fluctuations in habitat diversity occur between small and moderate floods. As a result, habitat diversity continually fluctuates.

bed-elevation variations, channel cutoffs, and bar ridge and swale topography result in high hydraulic and habitat diversity as measured at low flow.

During large floods, large amounts of scour occur in pool cross-sections, and bank erosion is extensive. Large floods cause the largest bed-elevation variations between pool troughs and riffle crests. Lateral bed-elevation variations, although large during the high magnitude flood, occupy both the bankfull and overbank flow areas. Therefore, the bankfull channel tends to be uniformly deepened during large floods. This results in a morphologically simplified low-flow channel cross section. Large longitudinal bed-elevation variations are partially offset by simplification of the low-flow channel cross section. As a result, hydraulic and habitat diversity as measured at low flow do not increase as dramatically as geomorphic change.

CONCLUSIONS

Relatively minor geomorphic changes that result from alternating small and moderate magnitude floods cause large changes in habitat diversity. As a result, habitat distribution is continually fluctuating in many Ozark streams. This fluctuation of habitat distribution occurs because small floods that transport sediment occur every two years on average and moderate floods occur every five years on average. As a result, most habitat distributions in Ozark streams are probably altered frequently on a decadal time frame. These data suggest that relations describing the location of geomorphic change with increasing flood magnitude are needed to evaluate habitat diversity in stream systems that experience frequent channel changes.

Acknowledgments. This study was a part of the Ozark Stream Geomorphology Project (OSGP) administered by Robert B. Jacobson. The OSGP was funded by the Water Resources Division of the U. S. Geological Survey (Missouri District), the Missouri Department of Conservation, and the Global Climate Change Program of the Biological Resources Division of the U. S. Geological Survey (previously the National Biological Service and National Park Service Global Climate Change research program). Logistical support for this study was provided by the National Park Service Ozark National Scenic Riverways. Rob Wertheimer, Alex Primm, Maria Panfil, Geoff Redmond, Beth Lambert, Chris Gutmann, Aaron Pugh, Andrew Jacobson, Gina Michl, Sara Gran, Karen Bobbitt, Steve Lancaster, and Lori Wood assisted in field data collection. Thoughtful reviews by David R. Montgomery, Michael A. Urban, and Greg F. Koltun improved the quality of this paper.

REFERENCES

Alexander, T. W. and G. L. Wilson, Technique for estimating the 2-to 500-year flood discharges on unregulated streams in rural Missouri: U.S. Geol. Surv. *Water Res. Invest. Rep.* 96-4038-B, 33pp., 1995.

Beckman, H. C. and N. S. Hinchey, Large springs of Missouri: Missouri Department of Natural Resources *Second Series pub.* 29, 141pp., 1944.

Bovee, K. D., The incremental method of assessing habitat potential for coolwater species, with management implications: *Am. Fish. Soc. Special Pub.*, 11, pp. 340-346, 1978.

Bovee, K. D. and R. Milhous, Hydraulic simulation in instream flow studies: theory and techniques: Cooperative Instream Flow Service Group, Fish and Wildlife Service, *Instream Flow Information Pap.* 5, 1978.

Emmett, W. W., Channel changes: *Geol.*, 3, pp. 271-272, 1974.

Hauck, H. S., L. G. Huber and C. D. Nagel, Water resources data, Missouri, water year 1998: U.S. Geol. Surv. *Water-Data Rep.*, MO-98-1, 434 pp., 1999.

Heritage, G. L., A. W. van Niekerk and D. Weeks, A model of fish assemblage response to hydraulic and fluvial geomorphological change in a pool-rapid channel type: in M. Leclerc, H. Capra, S. Valentin, A. Boudreault, and Y. Cote (Eds.), *2nd International Symposium on Habitat Hydraulics*, INRS-Eau, Quebec, pp. 355-366, 1996.

Hogan, D. L. and M. Church, Hydraulic geometry in small, coastal streams: progress toward quantification of Salmonid habitat: *Can. J. of Fish. and Aqua. Sci.*, 46, pp. 844-852, 1989.

Jacobson R. B. and A. L. Pugh, Riparian vegetation and the spatial pattern of stream-channel instability, Little Piney Creek, Missouri: U.S. Geol. Surv. *Water Supply Pap.* 2494, 33pp., 1997.

Jowett, I. G. and J. Richardson, Comparison of habitat use by fish in normal and flooded river conditions. *New Zealand J. of Marine and Freshwater Res.*, 28, pp. 409-416, 1994.

Jowett, I. G. and J. Richardson, Habitat preferences of common, riverine New Zealand native fishes and implications for flow management: *New Zealand J. of Marine and Freshwater Res.*, 29, pp. 13-23, 1995.

Leclerc, M., A. Boudreault, J. A. Bechara and G. Corfa, Two-dimensional hydrodynamic modeling: A neglected tool in the instream flow incremental methodology: *Trans. of the Am. Fish. Soc.*, 124, pp. 645-662, 1995.

Lobb, M. D. I. and D. J. Orth, Habitat use by an assemblage of fish in a large warmwater stream: *Trans. of the Am. Fish. Soc.* 120, pp. 65-78, 1991.

Madej, M. A., Temporal and spatial variability in thalweg profiles of a gravel-bed river: *Earth Surf. Proc. and Land.*, 24, pp. 1153-1169, 1999.

Madej, M. A. and V. Ozaki, Channel response to sediment wave propagation and movement, Redwood Creek, California, USA: *Earth Surf. Proc. and Land.*, 21, pp. 911-927, 1996.

McClendon, D. D. and C. F. Rabeni, Physical and biological variables useful for predicting population characteristics of Smallmouth Bass and Rock Bass in an Ozark stream: *N. Am. J. of Fish. Man.*, 7, pp. 46-56, 1987.

McKenney, R., Formation and Maintenance of Hydraulic Habitat Units in the Streams of the Ozark Plateaus, Missouri and Arkansas: The Pennsylvania State University, *unpublished PhD*, 347pp., 1997.

McKenney, R. and R. B. Jacobson, Erosion and deposition at the riffle-pool scale in gravel-bed streams of the Ozark Plateaus, Missouri and Arkansas: U. S. Geol. Surv. *Open-File Rep.* 96-655A, 171pp., 1996.

McKenney, R., R. B. Jacobson and R. C. Wertheimer, Woody vegetation and channel morphogenesis in low-gradient, gravel-bed streams in the Ozarks Region, Missouri and Arkansas: *Geomorphology*, 13, pp. 175-198, 1995.

Orth, D. J., Ecological consideration in the development and application of instream flow-habitat models: *Regulated Rivers: Res. and Man.*, 1, pp. 171-181, 1987.

Panfil, M. S. and R. B. Jacobson, Hydraulic modeling of in-channel habitats in the Ozark Highlands of Missouri: Assessment of physical habitat sensitivity to environmental change (http://www.cerc.usgs.gov/model/), 1999.

Payne, B. A. and M. F. Lapointe, Channel morphology and lateral stability: effects on distribution of spawning and rearing habitat for Atlantic salmon in a wandering cobble-bed river: *Can. J. of Fish. and Aqua. Sci.*, 54, pp. 2627-2636, 1997.

Pflieger, W. L., Aquatic community classification system for Missouri: Missouri Department of Conservation *Aquatic Series* No. 19, 69pp., 1989.

Rabeni, C. F. and R. B. Jacobson, Geomorphic and hydraulic controls on the abundance and distribution of stream Centrarchids in Ozark Streams: *Polaskie Archiwum Hydrolbiologii*, 40, pp. 87-99, 1993a.

Rabeni, C. F. and R. B. Jacobson, The importance of fluvial hydraulics to fish habitat restoration in low gradient alluvial streams: *Freshwater Biology*, 29, pp. 211-220, 1993b.

Ritter, D. F., Floodplain erosion and deposition during the December 1982 floods in southeast Missouri: in V. R. Baker, R. C. Kochel, and P. C. Patton (Eds.), *Flood Geomorphology*, John Wiley and Sons, Inc., N. Y., pp. 243-260, 1988.

Saucier, R. T., Historic changes in Current River meander regime: American Society of Civil Engineers *Proceedings of Conference, Rivers '83*, pp.180-190, 1983.

Simpson, E. H., Measurement of diversity: *Nature*, 163, p. 688, 1949.

Walters, J. P. and J. R. Wilson, Intraspecific habitat segregation by smallmouth bass in the Buffalo River, Arkansas: *Trans. of the Am. Fish. Soc.*, 125, pp. 284-290, 1996.

Wolman, M. G. and J. P. Miller, Magnitude and frequency of forces in geomorphic processes: *J. of Geol.*, 68, pp. 54-74, 1960.

Rose McKenney, Geoscience Department, 500 University Ave., Minot State University, Minot, ND 58707

Biophysical Gradients Related to Channel Islands, Middle Snake River, Idaho

W. R. Osterkamp

Water Resources Division, U. S Geological Survey, Tucson, Arizona

W. C. Johnson

Department of Horticulture, Forestry, Landscape, and Parks, South Dakota State University, Brookings, South Dakota

M. D. Dixon

Department of Zoology, University of Wisconsin, Madison, Wisconsin

Biophysical gradients of fluvial systems refer to systematic landform and habitat change resulting from differences in hydrologic process. Where physical gradients change both horizontally and vertically, habitat is varied and species diversity may be large. Islands in the middle Snake River, Idaho, vary in structure, shape, size, and age, yielding complex down-valley habitat gradients that are responses to position (reach conditions), processes of formation, and flood magnitudes. Species richness and diversity typically respond to these physical changes in habitat, but owing to a natural paucity of sediment as well as to storage in reservoirs, the formation of riffles, gravel bars, and accreted banks is minimal in the upper reaches of the middle Snake River; tributary inputs of water and sediment cause an increase of horizontal habitat diversity in the lower reaches. Physical changes along the river largely related to geomorphic history result in vertical diversity of vegetation on islands above (relict) and at (regime) flood-plain level.

1. INTRODUCTION

The use and management of drainage basins involve the administration, regulation, and conservation of resources. Increasingly, watershed managers must combine scientific principles with knowledge of the basin, and thus decisions and procedures change with increases in information and sophistication of models that characterize watersheds. Conceptual models describing the channel section, stream reach, drainage network, stream order, or watershed itself, often help to understand physical site conditions but historically have proved inadequate when applied to changed practices elsewhere within a basin. Consequently, integrative models that consider all resources and interactive processes have steadily increased in popularity among watershed managers and now provide the conceptual basis from which many procedures of landscape and bottomland restoration are applied.

This paper extends recent conceptual models of bottomland form and process into one combining spatial and temporal gradients of riparian-zone habitat. Although riparian-zone habitat gradients are well known, recent application of integrated models to resource management has been limited. An example of how coupling of physical and biological features,

Geomorphic Processes and Riverine Habitat
Water Science and Application Volume 4, pages 73-83

as biophysical gradients, can be applied to basin processes is described here for the Snake River of southwestern Idaho. The example illustrates a close dependency of hydrologic and biotic resources of the basin on the spatial and temporal scales considered.

Many investigators have recorded large variation in features such as bottomland width, riparian-zone vegetation, and channel size, shape, and gradient, along relatively short reaches. For simplicity, however, the conceptualization of river systems often disregarded stream-reach diversity while stressing average conditions. Examples are power functions of downstream hydraulic geometry [*Leopold and Maddock*, 1953] and zones of supply, transfer, and deposition within a fluvial system [*Schumm*, 1977].

To characterize bottomland habitats, gradational physical change along a river was observed explicitly by *Vannote et al.* [1980], who viewed a river continuum as a biological analog to geomorphic concepts of minimal-energy expenditure. *Vannote et al.* [1980] noted the tendency for streams to widen and deepen down-valley while increasing in mean velocity and decreasing in gradient, but reach-specific variations in fluvial-energy conditions were the basis for relating habitat to local channel characteristics. In short, the river-continuum "concept proposes that understanding of the biological strategies and dynamics of river systems requires consideration of the gradient of physical factors formed by the drainage network" [*Vannote et al.*, 1980, p. 131]. This view was extended by *Montgomery* [1999], who suggested an alternative term, process domain, to describe the linkage between the physical dynamics of watersheds and ecosystems.

The one-dimensional river continuum describes adjusted (time-independent) stream channels. *Stanford et al.* [1996] applied this approach to all habitats of valley bottoms. Their biophysical continuum recognizes gradients (gradational changes through space and time) of all habitat types, thus all biota from headwaters to a stream mouth that result from and respond to down-valley change in geomorphic process. This three-dimensional model includes ecological variation dependent on elevation of a geomorphic surface above mean stream level. A fourth gradient, time, could be long-term response to a rare flood or perhaps climatic change.

Biophysical gradients of the middle Snake River are described here, as indicated by changes in channel and island characteristics. The 186-km river length receives water and sediment from the Rocky Mountains and the eastern Columbia Plateaus. The Snake River flows south from Wyoming, turns west into Idaho, and arcs south and then north before joining the Columbia River in Washington. The middle Snake River begins near Swan Falls and ends above Farewell Bend State Park, OR. Based on landforms and channel characteristics, five river reaches are distinguished (Figure 1):

A -- 10.0 km, from the east edge of Canyon County, ID, to Idaho State Highway 45,

B -- 68.5 km, from State Highway 45 to county highway 1 km south of Adrian, OR,

C – 10.6 km, from 1 km south of Adrian to the mouth of the Boise River,

D -- 48.9 km, from the Boise River nearly to Horse Island (latitude 44° 7' 30"),

E -- 39.4 km, from near Horse Island to Farewell Bend State Park, OR.

2. SETTING

Much of the semiarid Snake River Basin is underlain by flat-lying and permeable lava flows of basaltic composition. The low-relief Snake River Plain (Figure 1, inset) occupies the basin interior where, in many places, basalt is capped by terrace gravel, loess, and alluvium [*Othberg*, 1994]. Enclosing mountains are mostly of volcanic and igneous-intrusive rocks. Precipitation is largely snow that melts, infiltrates, and later emerges as ground-water contributions to streamflow [*Maupin*, 1995]. As a result, recorded peak discharges in the middle Snake River have been about four times as great as the mean discharges, an unusually low ratio [*U.S. Geological Survey*, various years].

Landforms document catastrophic outflow of Pleistocene-age Lake Bonneville into the Snake River Basin [*Gilbert*, 1890; *Malde*, 1968; and *O'Connor*, 1993]. Flood features preserved in the upper reaches show that Holocene flows of the Snake River have been ineffective in modifying some relict flood deposits. Recent riverine processes have been altered by diversions, ground-water extractions, and storage of water and sediment in reservoirs. Owing to the attenuated, meltwater-dominated streamflow, highly erosive floods have probably been rare in the middle Snake River Valley during Holocene time.

About 150 permanent channel islands occur in the middle Snake River. Islands (vegetated landforms bounded by river branches) of most rivers are adjusted to on-going fluvial processes and connote isolated flood plain [*Osterkamp*, 1998]; thus, they are "regime" islands and reflect recent biophysical variation. Based on observations of river stage, on flow durations computed from discharge records collected at gage sites, and on stage-discharge records from seven study islands in reaches B, D, and E, regime islands of the Snake River have accreted to a modern flood-plain level and are overtopped, on average, every 1 to 2 years; most probably formed during low flow by plant establishment on a riffle or channel bar. Relict islands, which are higher than flood-plain level, represent former flood plain or elevated deposition older than that of regime islands. The upper surfaces of many relict islands lack evidence of having been overtopped even by the largest floods of the past century. Because relict islands may be too high for their upper surfaces to be altered by floods, they exhibit (by aerial photography) less spatial

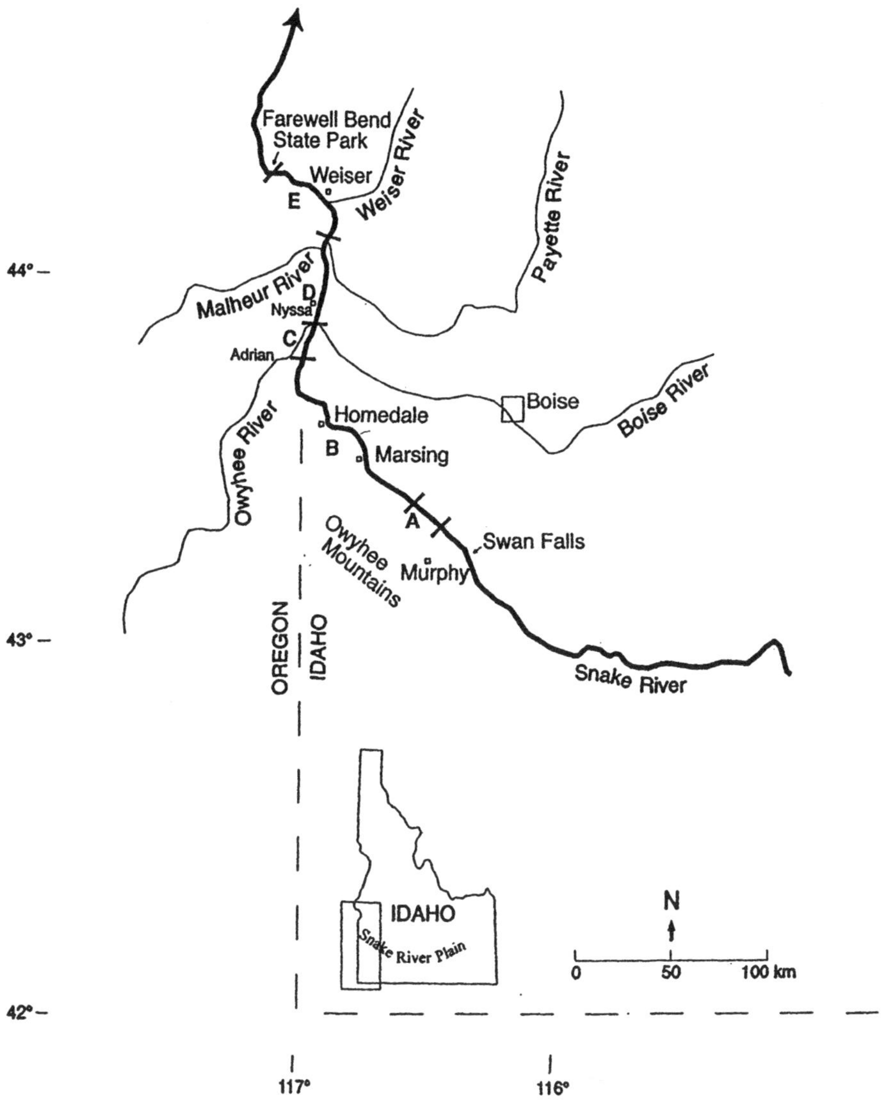

Figure 1. Map of middle Snake River showing river reaches A to E, major tributaries, and selected cultural features.

variation over periods of decades to centuries than do regime islands. Some Snake River islands are relics of the Bonneville Flood and others that post-date the flood may occupy positions of dunes or other flood bedforms. Relict islands also occur where shifts in channel position and incision of Holocene deposits isolated a surface that subsequently became higher than flood-plain level. Because no known Holocene floods of the Snake River have significantly exceeded historic floods, island formation by avulsion may be rare.

Streamflow records start as early as 1911 at Snake River gaging stations near Murphy, ID, at Nyssa, OR, and at Weiser, ID (Figure 1). Regulation began in 1927 at American Falls Dam, 420 km above Swan Falls, and records before then suggest that discharge was steady, averaging about 310 m^3s^{-1} and varying an order of magnitude in reaches A, B, and C [*U.S. Geological Survey*, various years]. This range suggests that snowmelt and high infiltration rates support base-flow and low flood peaks. Peak flows of major tributaries also are largely from snowmelt, but fewer exposures of basalt in their basins cause runoff to these streams to span 2 to 3 orders of magnitude. Reservoirs throughout the Snake River Basin also stabilize discharges.

Above reach D, tributaries contribute about 1 percent of the streamflow. In reaches D and E, however, tributaries provide up to 35 percent of the discharge. Combined with inputs from above, these flows account for about 96 percent of the discharge in reach E. The difference, minus extractions, is from small tributaries, springs, and irrigation return flow. In

Table 1.– Physical characteristics of the middle Snake River channel and islands[a]

Reach	Islands/km	Sinuosity	Gradient (m/m)	Mean Length (m)	Mean Width (m)	Mean W:L	Percent Relict Islands	Mean Length (m)	Mean Width (m)	Mean W:L
A	1.00	1.00	0.000365	570	150	0.29	80	640	170	0.29
B	0.88	1.01	0.000250	460	100	0.25	87	500	110	0.25
C	0.66	1.03	0.000222	550	90	0.21	71	690	100	0.14
D	0.72	1.22	0.000412	960	460	0.46	71	1120	570	0.49
E	0.84	1.10	0.000332	710	180	0.28	39	1070	250	0.25

[a] Sinuosity and gradient based on 1:24,000 scale topographic maps of the U. S. Geological Survey, various dates through 1992; island dimensions based on 1:9600 scale aerial photography, August 24, 1998.

reaches A, B, and C, the lack of significant tributary inflow helps effect an unusually steady discharge, whereas tributary inflows to reaches D and E increase flow variability.

Meager data indicate that suspended-sediment transport in the Snake River Basin is low compared to other parts of North America. Sediment yields, based on tabulations of the *U.S. Geological Survey* [various years], were about 9.1, 4.6, and 4.2 t km^{-2} yr^{-1} (metric tons per square kilometer per year), respectively, at Heise, ID (about 60 km downstream of Palisades Reservoir), Buhl, ID (about 100 km downstream of Lake Walcott), and King Hill, ID (about 100 km upstream from Swan Falls). Tributary inflow to the Snake River between King Hill and Swan Falls is relatively small, and much of the sediment is stored behind C.J. Strike and Swan Falls Dams. In comparison, yields average about 16 t km^{-2} yr^{-1} from forested areas of the western United States [*Osterkamp et al.*, 1998], 55 t km^{-2} yr^{-1} in United States streams entering the Pacific Ocean [*Curtis et al.*, 1973], and were about 160 t km^{-2} yr^{-1} in the Colorado River Basin prior to settlement by Europeans [*Meade et al.*, 1990].

Mean concentrations of suspended sediment in water samples collected during the 1980s from streams draining non-agricultural areas were generally less than 25 mg l^{-1}, whereas concentrations in streams of agricultural land were mostly less than 80 mg l^{-1} [*Clark*, 1994]. Concentrations in samples collected along the Snake River in the 1990s averaged about 45 mg l^{-1} at King Hill, ID, 65 mg l^{-1} at Nyssa, OR, and 85 mg l^{-1} at Weiser, ID. The increased concentration between King Hill and Weiser may result in part from irrigation return flows and tributary inputs such as the Boise River (66 mg l^{-1}) and the Weiser River (68 mg l^{-1}) (Figure 1).

Riparian vegetation changes longitudinally along the middle Snake River. The most striking changes are downstream increases in vegetation density and vertical structural complexity. Below Swan Falls Dam (just upstream from reach A), woody plants are thinly scattered along rocky banks. Most numerous is the short-statured hackberry (*Celtis reticulata* Torr.), which grows near the upslope margin of the riparian zone. Riparian woodland sequestering of fine sediment by Swan Falls Dam, and scarcity of tree plantings (and therefore propagules) of invasive species, results in limited species diversity [*Dixon and Johnson*, 1999].

Down-valley in reach A, woody riparian-zone vegetation diversifies, being dominated by coyote willow (*Salix exigua* Nutt. subsp. *exigua*), peach-leaf willow (*Salix amygdaloides* Anderss.), Russian olive (*Elaeagnus angustifolia* L.), and salt cedar (*Tamarix* L. spp.) [*Johnson et al.*, 1995]. Further downstream, in reaches D and E, the riparian zone becomes dominated by larger trees, particularly silver maple (*Acer saccharinum* L.), box elder (*Acer negundo* L.), and elm (*Ulmus* spp.). Increased woodland development is facilitated by sediment and nutrient inputs from tributaries and irrigation returns, increased frequencies of overbank flow, and seeds supplied by tree plantings in watershed cities, towns, and field borders.

3. PHYSICAL CHARACTERISTICS

Reach A, 10.0 km long, starts 15 km below Swan Falls where the Snake River emerges from canyons of bedded rocks capped by basalt [*Bond*, 1978]. The river has cut 150 m into these rocks and is lined by basalt boulders that were entrained and re-deposited by the Bonneville Flood. The flood plain in reach A consists of narrow strips bordering the river or islands, meandering is constrained by basalt boulders, and the sinuosity is 1.00 (Table 1). Owing to upstream reservoirs, meager tributary inflow and flooding, and a channel lined by coarse Bonneville flood deposits, little sediment enters this reach and little change in post-flood channel position has been possible.

Reach B is 68.5 km long and has a sinuosity of 1.01. The upper third has high basalt-capped escarpments on the east; to the west are the Owyhee Mountains. Downstream of Marsing, ID, are a series of terraces cut by the Snake and Boise Rivers that are veneered by stream gravel and Bonneville Flood deposits [*Othberg*, 1994]. The lowest third of reach B is bounded by gravel of the Deer Flat and other Boise River terraces [*Othberg*, 1994] and by the Owyhee

Mountains. The water-surface gradient, 0.000250, is probably controlled by Bonneville Flood deposits, and it is likely that the level and position of the channel bed have changed little since the flood [*Othberg et al.*, 1995]. The flood plain is nearly absent through most of the reach but in the lower part is present as narrow strips that support riparian-zone vegetation.

The Snake River in the upper half of reach C is confined by bedrock on the left and terrace gravel and Bonneville Flood deposits on the right. Flood deposits are widespread also in lower reach C [*Othberg*, 1994], where minor areas of flood plain are related to meandering of the lower Owyhee and Boise Rivers. Natural tributaries contribute little water and sediment, and sinuosity remains low, about 1.03. Reach C is 10.6 km in length and has a stream gradient of 0.000222 that is lower than that in reach B, perhaps due to ponding upstream from gravel deposition by the Owyhee and Boise Rivers after the Bonneville Flood. Bed load from tributaries in reach D appears too coarse to move readily in the Snake River and thus collects there. Reach C is transitional from steep, rocky conditions upstream to more mature, alluvial soils downstream. Owing perhaps to minimal inputs of sediment from tributaries, the occurrence of channel islands per length of river is low (Table 1).

Uplands bordering alluvium in reach D of the Snake River and its tributaries are soft beds veneered by glacial outwash, Bonneville Flood sediment, and eolian sand and silt [*Bond*, 1978]. The channel is 48.9 km long, has a sinuosity of 1.22, and meander scrolls of tributaries are common. Interaction between the Snake River and its tributaries has resulted in anabranching and incision by the tributaries into Holocene alluvial and eolian flood-plain deposits. Where some tributary meanders have adjoined the river, part of the Snake River has been intercepted by the tributary, causing divided flow and an island. The result has been the development of large, irregularly shaped islands. Whereas the Snake River gradient upstream of the Owyhee and Boise Rivers was largely determined by the Bonneville Flood, the gradient (0.000412) and channel shape in reach D appear adjusted to recent fluvial processes.

Riverine processes of reach E are less dependent on tributary inputs than in reach D. Through most of the reach, the river cuts friable rocks capped by fluvial, flood, glacial, and eolian deposits. Sideslopes of the valley bottom are interrupted on the east by the Weiser River and elsewhere by many smaller streams. The sinuosity of reach E, about 1.10, is lower than that of reach D presumably because of smaller tributary contributions. Channel incision into Bonneville Flood sediment and post-flood overbank and eolian deposits has formed a low terrace above the modern flood plain. The reach is 39.4 km long and has a water-surface gradient of 0.000332, 20 percent less than that of reach D (Table 1). The lower gradient may be the results of increased discharges, less influence of tributary sediment stored in the Snake River channel, and Bonneville Flood bed sediment still stored above Hells Canyon. The gradient of reach E appears less adjusted to recent flows than is that of reach D, possibly reflecting a lingering signature of the Bonneville Flood.

4. RESULTS

Channel islands and other fluvial landforms are products of the water and sediment fluxes and thus changes in these features typically reflect changes in hydrologic conditions. Because vegetation is dependent on physical habitat, it too varies with differences in water and sediment discharge.

4.1. Island Geomorphology

Reach A has 10 islands, two of which are partly formed of Bonneville Flood debris; others may occupy platforms of basalt rubble deposited as bedforms by the flood. Seven islands are capped by fine-grained relict deposits and an eighth is composed mostly of fine-grained deposits; only one upper-island surface is at regime, or flood-plain level. Reach B has 60 islands of similar size, shape, and frequency of occurrence as in reach A; 52 islands have relict surfaces higher than flood-plain level. Too few (seven) islands are in the short channel length of reach C to yield valid comparisons with reach B, but the lengths and widths of the islands, both relict and regime, are consistent with those of upstream islands. Regime islands of reach D are similar to the regime islands of other reaches, but mean length and width of relict islands increase markedly in response to inputs of tributary rivers. The percent of islands in reach D that are relict (25 of 35), however, is little changed from the mean for the upstream reaches. Effects of tributaries are less in reach E than in reach D, as indicated by decreased size of relict islands. Although island frequency per length of Snake River is closely similar to the mean in upstream reaches, the percent of regime islands more than doubles, reflecting post-Bonneville movement of the finer sizes of tributary bed load into the reach where some, as gravel bars, becomes vegetated regime islands.

The middle Snake River grades from a sediment-starved canyon environment stabilized by coarse Bonneville Flood debris, lacking erosive floods, sinuosity, and flood plain, to one of even lower energy, sparse mobile sediment, and negligible flood-plain area in reach B. Areas of slack water in reach C cause limited deposition of the meager sediment load and thus minor development of sinuosity and flood plain. Conditions change abruptly in reach D, marked by increased channel dynamics, island size, gradient, and flood-plain area as tributaries add water and sediment, part of which is stored

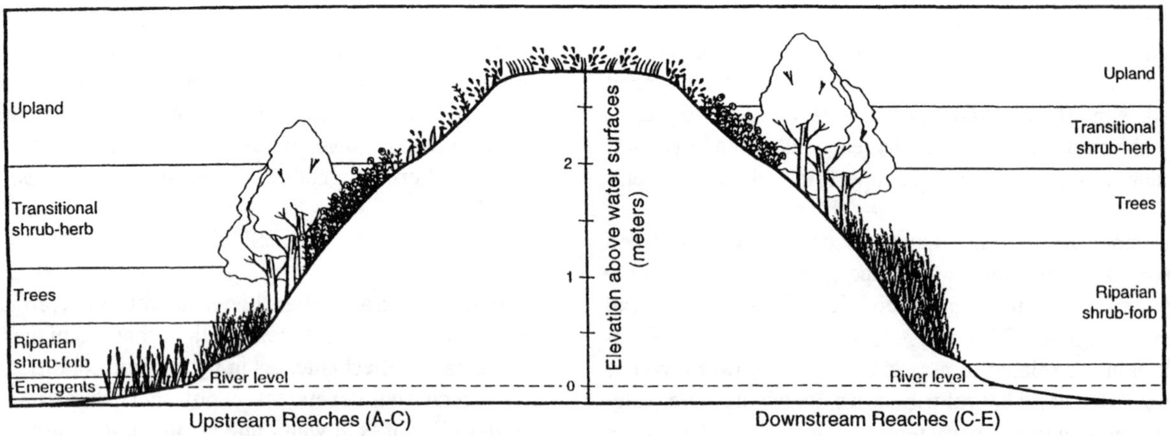

Figure 2. Diagram of Snake River banks and sections of islands in the middle Snake River comparing patterns of vegetation zonation in reaches A, B, and upstream parts of C with those of reaches D, E, and downstream parts of C.

along the channel. Fluvial sorting of the tributary sediment continues along reach E, causing an increased number of small-sized regime islands. Limited inputs of new sediment, however, result only in a partial reversal to low-energy conditions imposed by the Bonneville Flood. Relict islands, especially in the upper reaches, persist in the middle Snake River despite very low sediment loads and deposits mostly of highly erodible fine-grained sediment. Those relict islands in reach E, where the greatest discharge variability occurs due to tributary additions, have experienced the most erosion of relict sediment, in many cases exposing the underlying coarse bed material.

4.2. Island Vegetation

Low habitat and species diversity in reach A is associated with reduced flood peaks, paucity of sediment, limited seed sources, and canyon-mouth conditions of coarse channel substrate and small flood-plain area. Native riparian shrubs and trees are relatively few. Upland plants on relict islands are mostly greasewood (*Sarcobatus vermiculatus* (Hook.) Torr.), sagebrush (*Artemisia tridentata* Nutt.), rabbit-brush (*Chrysothamnus nauseosus* (Pall.) Britt.), and cheat grass (*Bromus tectorum* L.) (Figure 2).

Flow regulation and basin development are likely causes of increased numbers and diversity of both exotic and native trees in reach B, where strips of flood plain in the lower part have experienced substantial invasions of peach-leaf willow, Russian olive, and salt cedar since the 1930s [*Dixon and Johnson*, 1999]. Vegetation of reach B was sampled on four relict islands. Plants were pooled among subzones (i. e., emergent, riparian shrub-forb) and then regrouped by size structure into herb, shrub, and tree layers. Tree diversity ranged from one to seven species (Table 2, part B). Coyote willow, currant (*Ribes aureum* Pursh), and rose (*Rosa woodsii* Lindl.) dominated the shrub layer (Table 2, part A), whereas goldenrod (*Euthamia occidentalis* Nutt.), inland saltgrass (*Distichlis stricta* (L.) Greene), and peppergrass (*Lepidium latifolium* L.) were common in the herb layer (Table 2, part C). Blind Island, as an example, supports coyote willow, rose, and Russian olive, and an herb layer composed of goldenrod and summer cypress (*Kochia scoparia* (L.) Schrad.) (Table 2).

Qualitative vegetation surveys of islands in the short, near-ponded reach C found exotic purple loosestrife (*Lythrum salicaria* L.) dominant in the emergent zone. Salt cedar, Russian olive, box elder, cottonwood (*Populus deltoides* Marsh.), and coyote willow were common woody plants, whereas rose and cheat grass were representative of the herb layer.

The variety of trees present on reach-D regime islands, as shown by Boise Island (Table 2, part B), was similar to that of islands upstream except for increased cottonwood. Peach-leaf willow dominated, followed by Russian olive, cottonwood, and salt cedar. Principal plants of the shrub layer were coyote willow, currant, and rose; false indigo (*Amorpha fruticosa* L.), a native shrub, reached its maximum abundance in reach D (Table 2, part A). Plants of the herb layer on reach-D islands differed markedly from those of upstream islands. Many of the species not found upstream were obligate wetland plants. Herb-layer dominants were purple loosestrife, bulrush (*Scirpus acutus* Muhl.), and nightshade (*Solanum dulcamara* L.). Upland species were sagebrush, greasewood, rabbit-brush, and thistle (*Cirsium* spp.).

Riparian-zone stands of silver maple and box elder dominated the tree layer of all islands and gave reach E the highest tree diversity of the middle Snake River. Russian olive and salt cedar dwindled in proportion to North American species such as peach-leaf willow, silver maple, green ash (*Fraxinus pennsylvanica* Marsh.), and cottonwood. The shrub layer remained mostly coyote willow, currant, rose, and false indigo. The herb layer was diverse, with peppergrass and poison-hemlock (*Conium maculatum* L.) dominant.

Responding to down-valley physical gradients and seed supply, the aggressive, damaging exotic weed, purple loosestrife (*Lythrum salicaria* L.), is nearly absent in reach A, is prominent in reach B, and attains peak abundance in reach D. A native, false indigo, and exotics such as pepperweed and poison-hemlock sharply increase down-valley, but coyote willow is common in all reaches. The upslope zonation and vertical structure of riparian vegetation also vary down-valley due to hydrologic, sediment, and agricultural influences. The vertical range of the riparian zone is least downstream of Swan Falls owing to meager sediment supplies, narrow ranges of river stage, and a coarse substrate that limits capillary rise. This zone has a 2-m vertical range in reaches A, B, and C [*Johnson et al.*, 1995] that expands downstream to 2.5 m but is much higher on river banks supplemented by irrigation seepage (Figure 2). Seeps also affect vegetation composition; herbs such as purple loosestrife dominate seep areas whereas woody plants dominate non-seep areas.

The vertical distribution of vegetation within bottomlands of the middle Snake River, especially on relict islands of the river, also suggests biophysical gradients dependent on moisture availability and adaptation to inundation or flood damage. Vegetation on Snake River islands is grouped, in ascending and overlapping order, into emergent, riparian shrub-forb, tree, transitional shrub-herb, and upland subzones [*Johnson et al.*, 1995]. Species composition within these groups differs not only with height above (or below) mean river stage (Figure 2), but also changes downstream along physical gradients.

5. DISCUSSION

The forms and biota of most rivers are related to flow (including floods), sediment, and geology. Typically, as rivers descend from headwaters through reaches of montane, piedmont-valley, and coastal flood plain, separated by transition reaches [*Stanford et. al.*, 1996], alluvial features develop as channel bars and islands, flood-plain meander scrolls, and cut-and-fill structures. Biophysical gradients of the middle Snake River since the Bonneville Flood have been atypical owing to steady discharge and low flood peaks, meager sediment supplies, and basalt that controls infiltration, channel movement, ground-water additions, and tributary inputs of water, sediment, and seeds.

5.1. Physical Gradients, Middle Snake River

Reaches A, B, and C differ from the lower two reaches because they receive only minor inputs of ground water and tributary flow and sediment, and they rarely are subjected to highly erosive floods. Channel perimeters and profiles of the upper reaches are constrained by coarse substrate that prevents channel migration, flood-plain evolution, and sinuosity. An anomalously low channel gradient, averaging 0.000253, likely imposed by sorting of very coarse sediment during the Bonneville Flood, provides ineffective energy for bed-load transport. Bed-load supplies in the upper reaches also are limited owing to few large tributaries, storage in reservoirs, and a geologic history that includes natural damming by basalt flows [*Scott et al.*, 1982].

The Owyhee, Boise, Malheur, and Payette Rivers enter the middle Snake River along reach D, adding water and sediment that have caused (1) increased meandering, sinuosity, channel gradient, and discharge variability, and (2) the formation of flood plain and large relict channel islands below mouths of the tributaries. These changes in physical conditions from reach C to downstream reaches are documented by aerial photography, topographic maps, and hydrologic records [*U.S. Geological Survey*, various years]. Most islands in reach D are inferred to be relict, many having formed by incision of tributaries into their flood-plain deposits of the Snake River bottomland; about 30 percent of the islands are the regime type, derived largely of bed load and overbank sediment from tributaries. Owing to the geomorphic processes of formation, mean length and width of relict islands of reach D are much larger than in the upper reaches, resulting in a mean width:length ratio of 0.49, a 10-fold increase in mean area, and a corresponding increase in island surface area per unit length of river channel (Table 1). Sizes and shapes of regime islands are similar to those upstream.

In reach E, contributions of sand and gravel from reach D by tributaries increases the occurrence of regime islands (0.51 km^{-1} as opposed to 0.20 km^{-1} in reach D and a mean of 0.13 km^{-1} in reaches A, B, and C). The frequency of relict islands, however, decreases owing to a decrease in interaction of tributary channels with the Snake River. Channel gradient in reach E, in part a relic of the Bonneville Flood, declines, indicating that much of the coarse sediment load from tributaries into reach D has stalled in the lower-energy conditions of the Snake River during post-Bonneville time.

5.2. Biological Gradients, Middle Snake River

Species diversity in riverine settings is dependent on widely ranging habitats, or physical mosaics, of short channel reaches and short distances lateral to the channel [*Stanford et al.*, 1996]. Small-scale, or intra-reach habitat diversity (resulting from differences among landforms such as channel bars and islands, flood-plain features, and cut-and-fill structures), commonly is controlled by floods and fresh supplies of sediment to replenish that evacuated during floods. Ripar-

Table 2. Frequency of occurrence (percent) of plant species in size categories along transects (all subzones combined) at seven Snake River islands.

Part A. Woody plants in the shrub layer.

	ISLAND						
		Reach B		Reach D		Reach E	
TAXON	Blind	Silo	Cigar	Suzy	Boise	Ketchup	Pruitt
Coyote willow (*Salix exigua* Nutt.)	44.6	55.7	23.7	12.9	34.3	39.4	32.2
Russian olive *Elaeagnus angustifolia* L.)	17.4	1.6	8.2	---	---	---	---
Rose (*Rosa woodsii* Lindl.)	16.3	16.4	2.1	14.1	14.3	21.2	24.1
Salt cedar (*Tamarix* L. sp.)	7.6	0.8	4.1	5.3	2.9	0.4	---
Currant (*Ribes aureum* Pursh)	4.3	50.8	18.6	4.1	15.7	15.7	7.5
Poison ivy (*Toxicodendron rydbergii* (Small) Greene)	2.2	4.9	---	---	---	---	---
Rabbit-brush (*Chrysothamnus nauseosus* (Pall.) Britt.)	1.1	---	---	---	---	---	---
Skunkbush (*Rhus trilobata* Nutt.)	1.1	---	---	1.2	---	---	---
Peach-leaf willow *Salix amygdaloides* Anderss.)	---	1.6	5.2	---	2.9	---	1.1
Sagebrush (*Artemisia tridentate* Nutt.)	---	---	---	---	---	1.8	---
False indigo (*Amorpha fruticosa* L.)	---	---	---	---	9.3	1.1	2.3
Box elder (*Acer negundo* L.)	---	---	---	---	---	0.7	---
Siberian elm (*Ulmus pumila* L.)	---	---	---	---	---	0.7	1.7
Silver maple (*Acer saccharinum* L.)	---	---	---	---	---	0.4	1.1
Green ash (*Fraxinus pennsylvanica* Marsh.)	---	0.8	---	---	---	0.4	3.4
Greasewood (*Sarcobatus vermiculatus* (Hook.)Torr.)	---	---	---	5.3	---	---	---

ian-zone habitat diversity, and thus species diversity, under both pre-development and regulated flow conditions, has been minimal in much of the middle Snake River owing to exceptionally low flood peaks and a very limited sediment supply.

The upstream portions of the middle Snake River, with their scattered willows and hackberry trees, probably resemble pre-development conditions. The natural river did not support extensive riparian-zone woodlands prior to regulation because of stable streamflow and channel configuration, and coarse-textured bed and banks of the river. Extensive cottonwood and willow woodlands common to many streams of the western United States were not present along the middle Snake River.

Increasing vegetation diversity and structural complexity downstream along the modern river are largely products of

Part B. Trees (stems ≥10cm dbh).

			ISLAND				
		Reach B		Reach D		Reach E	
TAXON	Blind	Silo	Cigar	Suzy	Boise	Ketchup	Pruitt
Russian olive *Elaeagnus angustifolia* L.)	3.2	2.6	6.2	2.3	3.6	---	---
Peach-leaf willow (*Salix amygdaloides* Anderss.)	---	7.3	9.3	1.2	7.8	2.6	9.2
Salt cedar (*Tamarix* L. spp.)	---	---	1.0	2.9	0.7	---	---
Box elder (*Acer negundo* L.)	---	---	---	0.6	---	2.6	0.6
Silver maple (*Acer saccharinum* L.)	---	---	---	---	---	0.7	5.2
Green ash (*Fraxinus pennsylvanica* Marsh.)	---	0.8	---	---	---	0.7	---
Cottonwood (*Populus deltoides* Marsh.)	---	2.5	---	---	2.1	---	---
Northern catalpa (*Catalpa speciosa* Warder)	---	1.6	---	---	---	---	---
Coyote willow (*Salix exigua* Nutt.)	---	0.8	---	1.2	---	---	4.6
Honey locust (*Gleditsia triacanthos* L.)	---	---	---	1.2	---	---	---
Siberian elm (*Ulmus pumila* L.)	---	---	---	0.6	---	---	---
American elm (*Ulmus americana* L.)	---	---	---	---	---	0.7	---

human activity, although sediment and variable flows contributed by tributaries in the lower part of the middle Snake River also provided favorable conditions for the development of riparian-zone vegetation along the natural river. Beyond the natural variability, humans have influenced the riparian zone primarily by introducing many new plants pre-adapted for success along rivers in semiarid regions. Most notable in this group are Russian olive and salt cedar, which are Eurasian exotics invasive in the central and western United States. Many other North American trees not native to the Snake River Valley were introduced as well, including several species of maple, elm, willow, and cottonwood. The success of these new introductions, evident from the longitudinal patterns previously discussed, probably was enhanced by agricultural activities that contributed more sediment and nutrients to the mainstem river via tributaries. Seepage along the mainstem banks (but not on islands) from irrigated agricultural fields also fosters upland expansion of the riparian zone. About a third of the species growing in the riparian zone are exotics, a high percentage compared with other rivers [*Dixon and Johnson*, 1999].

6. CONCLUSION

The energy of unregulated streamflow creates physical change along any river continuum, and floods alter channels and alluvial surfaces along biophysical gradients. Species richness and diversity typically adjust to these physical changes in habitat, but owing to a natural paucity of sediment as well as to storage in reservoirs, the formation of riffles, gravel bars, and accretionary banks is minimal in the upper reaches of the middle Snake River. The geology, snowmelt hydrology, and numerous reservoirs in the Snake River Basin inhibit floods, further reducing processes of cut-and-fill, pool-riffle formation, and avulsion. Very coarse Bonneville sediment locally stabilizes channel banks with natural, persistent rip-rap, thereby preventing lateral channel migration and flood-plain development.

Part C. Plants in the herb layer (the five dominant species for each island).

	ISLAND						
		Reach B		Reach D		Reach E	
TAXON	Blind	Silo	Cigar	Suzy	Boise	Ketchup	Pruitt
Goldenrod (Euthamia occidentalis *Nutt.*)	38.0	44.2	30.9	5.3	4.3	0.7	1.7
Summer cypress (Kochia scoparia *(L.) Schrad.)*	35.1	13.9	23.7	---	---	---	---
Hoary cress (*Cardaria draba* (L.) Desv.)	30.4	37.7	---	34.1	---	---	---
Nightshade (*Solanum dulcamara* L.)	21.7	3.3	19.6	5.3	22.9	3.7	3.7
Inland saltgrass (*Distichlis spicata* (L.) Greene)	14.1	---	47.4	20.6	---	---	---
Dogbane (*Apocynum cannabinum* L.)	12.0	---	19.6	---	8.6	---	17.2
Goldenrod (*Solidago gigantea* Ait.)	12.0	50.0	26.8	5.3	4.3	---	---
Cheat grass (*Bromus tectorum* L.)	7.6	5.7	6.2	---	---	10.9	---
Teasel (*Dipsacus fullonum* L.)	6.5	4.1	4.1	11.2	---	0.4	---
Bulrush (*Scirpus acutus* Muhl.)	4.3	---	4.1	14.7	23.6	---	---
Peppergrass (*Lepidium latifolium* L.)	3.2	37.7	---	56.5	19.3	65.7	62.1
Cocklebur (*Xanthium strumarium* L.)	3.2	---	1.0	---	11.4	---	13.8
Nettle (*Urtica dioica* L.)	1.1	9.8	4.1	2.9	---	8.4	2.9
Purple loosestrife (*Lythrum salicaria* L.)	1.1	1.6	14.4	0.6	70.0	1.8	4.0
Poison-hemlock (*Conium maculatum* L.)	---	---	---	---	---	55.0	27.0
White sage (*Artemisia ludoviciana* Nutt.)	---	---	---	---	---	8.4	---
Umbrella Sedge (*Cyperus* L. spp.)	---	---	---	---	13.6	---	3.4
Barnyard grass (*Echinochloa muricata* (Beauv.) Fern.)	---	---	---	---	12.8	---	11.5

Islands in the middle Snake River vary in structure, shape, size, and age, yielding down-valley vertical gradients of vegetation. Many changes (Tables 1, 2) are responses to position (reach conditions) and history (processes of formation), whereas others are related to flood magnitudes. Negligible sinuosity in the upper reaches indicates that immobility of coarse flood sediment has prevented meandering and normal cut-and-fill sorting processes of most fluvial systems. Channel gradients in the two lower reaches increase from those upstream owing to inputs of water and bed sediment susceptible to sorting by the Snake River. Thus, reaches D and E exhibit adjustment to recent flow conditions whereas reaches above remain vestiges of Bonneville Flood discharges.

The streamside vegetation along the middle Snake River reflects the river's geomorphic history and its more recent modification by humans. Natural longitudinal gradients of vegetation have been intensified by exotic-plant introductions and shifts in sediment and nutrient inputs by agriculture, thereby affecting riparian-zone soils. The geomorphic development of islands influences vertical zonation as well. Relict islands with high surfaces support vegetation with both riparian and upland elements, but the former with rather simple structure and composition. In contrast, remnant islands at modern flood-plain level have little upland vegetation, but are considerably more dynamic and are dominated by more diverse and structurally complex riparian vegetation. Whereas these elevational vegetation patterns are controlled by several hydrologic and geomorphic factors, such as flow variation, sediment availability, substrate, and hydroperiod, patterns on the main banks are affected by an additional factor, ground-water seepage from irrigated agriculture that elevates riparian-zone borders. Thus, vegetation is a complex expression and integrator of the river's environmental history, in some places even reflecting extinct events (upland vegetation on relict islands) and in others modern events (plant introductions and irrigation seepage).

Acknowledgments. This study was conducted in cooperation with the Deer Flat National Wildlife Refuge, U.S. Fish and Wildlife Service, Nampa, Idaho. Todd Fenzl and Randy Aulbach of the Refuge provided superb technical and logistical assistance, and Wayne Stanley made numerous species identifications along vegetation transects. The help of each is much appreciated.

REFERENCES

Bond, J. G. (compiler), *Geologic map of Idaho,* Idaho Dep, Lands, Bur Mines and Geology, Moscow, ID, 1 sheet, 1978.

Clark, G. M., Assessment of selected constituents in surface water of the upper Snake River Basin, Idaho, and western Wyoming, water years 1975-89, *U.S. Geol. Surv., Water-Resour. Invest. Rer. 93-4229*, 49 pp., 1994.

Curtis, W. F., J. K. Culbertson, and E. B. Chase, Fluvial-sediment discharge to the oceans from the conterminous United States, *U. S. Geol. Surv., Circ. 670*, 17 pp., 1973.

Dixon, M. D., and W. C. Johnson, Riparian vegetation along the middle Snake River, Idaho: zonation, geographical trends, and historical changes, *Great Basin Naturalist, 59*(1), 18-34, 1999.

Gilbert, G. K., Lake Bonneville, *U.S. Geol. Surv. Monog. 1*, 438 pp., 1890.

Johnson, W. C., M. D. Dixon, Robert Simons, Susan Jenson, and Kevin

Larson, Mapping the response of riparian vegetation to possible flow reductions in the Snake River, Idaho, *Geomorphology, 13*(1-4), 159-173, 1995.

Leopold, L. B., and Thomas Maddock, Jr., The hydraulic geometry of stream channels and some physiographic implications, *U.S. Geol. Surv. Prof. Pap. 252*, 57 pp., 1953.

Malde, H. E., The catastrophic late Pleistocene Bonneville Flood in the Snake River Plain, Idaho, *U.S. Geol. Surv. Prof. Pap. 596*, 52 pp., 1968.

Maupin, M. A., Water-quality assessment of the upper Snake River Basin, Idaho and western Wyoming--environmental setting, 1980-92, *U.S. Geol. Surv., Water-Resour. Invest. Rep., 94-4221*, 35 pp., 1995.

Meade, R. H., R. T. Yuzyk, and T. J. Day, Movement and storage of sediment in rivers of the United States and Canada, *in The Geology of North America, Surface Water Hydrology*, edited by M. G. Wolman and H. C. Riggs, pp. 255-280, *Geol. Soc. Am. v. 0-1*, 1990.

Montgomery, D. R., Process domains and the river continuum, *J. Am Water Resour. Assoc., 35*(2), 397-410, 1999.

O'Connor, J. E., Hydrology, hydraulics, and geomorphology of the Bonneville Flood, *Geol. Soc. Am. Spec. Pap. 274*, 83 pp., 1993.

Osterkamp, W. R., Processes of fluvial-island formation, with examples from Plum Creek, Colorado, and Snake River, Idaho, *Wetlands, 18*(4), 530-545, 1998.

Osterkamp, W. R., P. Heilman, and L. J. Lane, Economic considerations of a continental sediment-monitoring program, *Internat. J. Sed. Res., 13*(4), 12-24, 1998.

Othberg, K. L., Geology and geomorphology of the Boise Valley and adjoining areas, western Snake River Plain, Idaho, *Idaho Geol. Surv. Bull. 29*, 54 pp., 1994

Othberg, K. L., J. E. O'Connor, and P. A. McDaniel, Field guide to the Quaternary geology of the Boise Valley and adjacent Snake River Valley, *Spec. Rep. Idaho Geol. Surv.*, Moscow, Idaho, 51 pp., 1995.

Schumm, S. A., *The Fluvial System*, John Wiley and Sons, New York, 338 pp., 1977.

Scott, W. E., K. L. Pierce, J. P. Bradbury, and R. M. Forester, Revised Quaternary stratigraphy and chronology in the American Falls area, southeastern Idaho, *in Cenozoic Geology of Idaho*, edited by B. Bonnichsen and R. M. Breckenridge, pp. 581-595, *Idaho Bur. Mines and Geol. Bull. 26*, 1982.

Stanford, J. A., J. V. Ward, W. J. Liss, C. A. Frissell, R. N. Williams, J. A. Lichatowich, and C. C. Coutant, A general protocol for restoration of regulated rivers, *Regulated Rivers: Res. & Manag.*, 12, 391-413, 1996.

U.S. Geol. Surv. Water Res. Data, Idaho, *U. S. Geol. Surv. Water-Data Reports*, Boise, Idaho, various years.

Vannote, R. L., G. W. Minshall, K. W. Cummins, J. R. Sedell, and C. E. Cushing, 1980, The river continuum concept, *Can. J. Fisheries and Aquatic Sci. 37*(1), 130-137, 1980.

W. R. Osterkamp, U.S. Geological Survey, 1675 West Anklam Road, Tucson, AZ 85745 (wroster@usgs.gov)

W. C. Johnson, Department of Horticulture, Forestry, Landscape, and Parks, South Dakota State University, Box 2140 A, Brookings, SD 57007-0996 (johnsonc@mg.sdstate.edu)

M. D. Dixon, Department of Zoology, University of Wisconsin, Birge Hall Madison, WI 53706-1381
(mdixon@ravel.zoology.wisc.edu)

Modeling Sediment and Wood Storage and Dynamics in Small Mountainous Watersheds

Stephen T. Lancaster and Shannon K. Hayes

Department of Geosciences, Oregon State University, Corvallis, Oregon

Gordon E. Grant

USDA Forest Service, Pacific Northwest Research Station, Corvallis, Oregon

We examine controls on supply and transport of sediment and wood in a small (approximately two square kilometers) basin in the Oregon Coast Range, typical of streams at the interface between episodic sediment and wood delivery by mass movements and frequent fluvial sediment transport. We hypothesize that wood deposited by mass movements forms dams that lead to persistent sediment storage and inhibit coherent propagation of sediment pulses. Field data show that much sediment is stored behind such dams and in terraces after the dams breach. We developed a drainage basin-scale model driven by stochastic storm and fire sequences that combines empirical, stochastic and physical models of forest growth, tree fall, wood decay, soil production and diffusion, landslide initiation, debris flow runout, and fluvial sediment transport. In a 3000-year simulation of the study area, woody debris flow deposits form dams on the main channel and lead to steps in the channel profile and terraces on the valley floor that persist in place even after nearly all deposited wood has decayed. Simulated sediment output from the network is relatively steady and shows little evidence of episodic input. Our results suggest that abundant wood plays a key role in moderating sediment flux from small basins following debris flow events. Debris flow events coincident with a lack of abundant wood, such as might occur following forest harvest, could lead to more episodic sediment flux to downstream, fish-bearing reaches.

1. INTRODUCTION

The decline of salmonid species in the Pacific Northwest has prompted increased attention to the temporal and spatial dynamics of aquatic habitat. In particular, various aspects of salmonid life history are influenced by the abundance, location, quality, and spatial distribution of gravel and wood, which create key fluvial environments used for spawning and rearing [*Reeves et al.*, 1998]. Although it is widely viewed that hillslope and channel processes influencing erosion, input, transport, or deposition of sediment and wood are inextricably linked to the quality and quantity of aquatic habitat, the degree of coupling in time and space between hillslopes and channels remains a fundamental problem. Specifically, it has proven difficult to show when accelerated mass wasting will occur, how sediment and wood introduced into steep, low-order channels will be routed downstream, and what the long-term consequences of these processes might be for channel and valley floor

Geomorphic Processes and Riverine Habitat
Water Science and Application Volume 4, pages 85-102

morphology, sediment flux, or the quality or quantity of aquatic habitat.

This study addresses sediment and wood dynamics in small (~2 km^2) drainage basins because this part of the landscape represents the interface between mass movements originating on the hillslopes and fluvial transport processes. Woody debris has a potentially strong effect on sediment dynamics and storage in these basins because of the large wood constituent in debris flows and the tendency for wood to form large wood dams [*Kochel et al.*, 1987; *Abbe and Montgomery*, 1996; *Montgomery et al.*, 1996; *Hogan et al.*, 1998; *Massong and Montgomery*, 2000]. Our goal is to better understand how both episodic and chronic inputs of sediment and wood interact with each other and with drainage network structure to produce changing volumes of stored sediment and wood in channels and on valley floors. To that end, we have developed a drainage basin-scale model that encapsulates much of our understanding of the relevant controls on sediment and wood dynamics.

We begin with a review of current approaches to modeling sediment and wood dynamics and distribution in forested mountain streams and then describe our modeling approach. We use the model to simulate sediment and wood influxes to the valley network over millennial timescales, corresponding locations and amounts of sediment and wood storage in the valleys, and sediment outflux from the basin. A comparison of model and field results reveals useful insights into locations, controls, and temporal dynamics of wood and sediment storage and dynamics in upland basins and consequences for interpreting responses of watersheds to disturbance.

1.1 *Current Perspectives and Modeling Criteria for Understanding Sediment and Wood Dynamics*

Benda and Dunne [1997a, b] used a model to explore the interaction between episodic sediment supply and the channel network in relatively large (~100 km^2) drainage basins. In their modeling they assumed that sediment moves in coherent waves, as documented by *Madej and Ozaki* [1996]. Because of the episodic delivery of sediment from debris flows, the assumption of wave-like translation resulted in the prediction that headwater basins, such as that considered in the present study, are sediment "starved" most of the time.

In contrast, other studies have identified different first-order controls on sediment storage and transport in streams. A function of drainage area and channel slope, a proxy for stream power or sediment transport capacity, discriminated many bedrock and alluvial stream reaches, but only in the absence of wood dams [*Montgomery et al.*, 1996]. Sediment supply, lithology, and woody debris have significant effects on the ability to discriminate bedrock from alluvial reaches with a stream power model [*Massong and Montgomery*, 2000]. Some of the factors affecting sediment supply are highly variable in time, such as volumes of sediment-trapping woody debris and recent landslides [*Hogan et al.*, 1998]. These studies suggest that, while local sediment volumes may vary over time, they do not necessarily migrate downstream as coherent wave forms but often disperse in place [*Lisle et al.*, 1997, 2000].

Existing models to date have only considered sediment stored or transported within the channel proper, and have generally not explicitly included dynamic interactions between wood and sediment. In our experience, in-channel sediment and wood are relatively small fractions of total valley storage, and models to explain long-term evolution of drainage basins should predict storage and dynamics in the valley as well as the channel. More generally, models predicting sediment and wood storage and dynamics in stream and valley networks should account for all processes driving input and output from those networks.

1.2 *Study Site: Oregon Coast Range*

Field and modeling work were both sited in a 2.1-km^2 tributary to Hoffman Creek in the Oregon Coast Range (Figure 1). We selected the site according to the following criteria: (a) homogeneous lithology; (b) absence of mid-slope or valley-bottom roads; (c) access to study area; (d) access to land-use records; (e) significant number of recent debris flows; and (f) basin small enough to study and model and large enough to exhibit network-scale effects. The basin is underlain by massive, gently dipping, Eocene Tyee sandstone. The topography is steep—valley side-slopes are typically 40°—and highly dissected with elevations ranging from 10 to 265 m above sea level. Soils are relatively shallow, highly porous, and have low bulk densities (about 1 kg/m^3) [*Reneau and Dietrich*, 1991] (Table 1). The climate is maritime with warm, dry summers and mild, wet winters and mean annual precipitation of approximately 1800 mm [Oregon Climate Service].

The area is forested with primary species Douglas-fir (*Pseudotsuga menziesii*) and secondary species western hemlock (*Tsuga heterophylla*) and western red cedar (*Thuja plicata*) on the hillslopes and red alder (*Alnus rubra*) in riparian areas, though these so-called riparian areas often extend upslope through the smallest hollows to the ridges, especially in immature stands. Nearly half of the basin has been harvested as recently as ca. 1965, and there is field evidence of earlier, undocumented harvest in other parts of the basin. This forest, typical of the Oregon Coast Range, results in surficial biomass that is less than but of the same order of magnitude as the mass of the soil layer, especially in mature stands, and, therefore, wood is a significant part of the mass moved by landslides and debris flows.

2. MODELING METHODS, ASSUMPTIONS, AND INITIALIZATION

Most of the sediment and much of the wood in streams and valleys originates on the hillslopes. A model encompassing sediment inputs to the valley network must account for sediment production from the parent material, i.e., conversion of bedrock to soil, and delivery of that sediment to the valley via diffusion and mass movements, i.e., landslides and debris flows. Likewise, a model encompassing wood inputs to the valley network must account for biomass growth and delivery to the valley via tree fall and mass movements. Since forest dynamics play a vital role in the timing and location of mass movements, that interaction should be included as well. Some of these processes, such as biomass growth, can be modeled relatively simply (Figure 2). Others, such as debris flows, are more complicated and require more sophisticated modeling.

Our new model is an extension of the Channel-Hillslope Integrated Landscape Development (CHILD) model [*Lancaster*, 1998; *Tucker et al.*, 2001a, b]. As such, the present model operates on a triangulated irregular network (TIN) and shares the CHILD model's drainage area calculation algorithm and stochastic precipitation and runoff generation models.

Figure 1. Location map of the Oregon Coast Range showing the Hoffman Creek watershed (outlined with dotted line) and study basin (outlined with solid line).

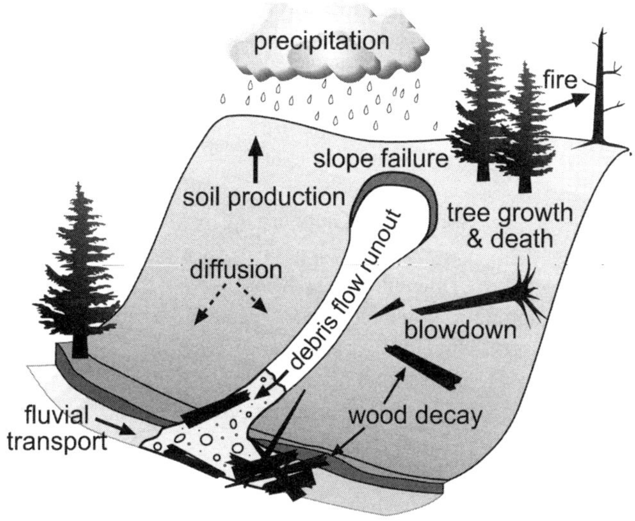

Figure 2. Conceptual figure illustrating the climatic, geomorphic, and vegetative processes simulated in the model.

2.1 *Landscape and Storm Characteristics*

The model uses gridded digital elevation model (DEM) data with 10-meter discretization to interpolate the elevations of the nodes in the TIN. Node locations are random to eliminate grid bias and form a TIN with the same average discretization as the original DEM. Additional points are added at large drainage areas to eliminate "jaggy" channels typical of interpolated TIN's. Finally, channel-adjacent nodes that would fall within channels are removed [*Lancaster*, 1998]. Nodes in the landscape are classified according to three types: (1) hillslope nodes containing vegetation and soil; (2) channel nodes with wood and sediment deposits and vegetation; and (3) valley nodes with wood and sediment deposits and vegetation (Figure 3). Elevations of hillslope nodes are static because relative elevations on hillslopes do not change much over millennial time scales and any model-driven changes would only decrease the accuracy of topographically driven transport processes [*Dietrich et al.*, 1995]. Channel and valley node elevations evolve over time in response to aggradation and evacuation of sediment and wood because fluvial processes are sensitive to these fluctuations, but bedrock elevations are held static for channel and valley nodes.

We attempt a compromise between the model's descriptiveness and the area modeled by lumping hillslope areas into aggregates. Calculations on each hillslope node (e.g., ~100 m^2) result in a fine-scale map of landscape character-

TABLE 1. Sources and values for landscape, forest, and debris flow parameters

Parameter	Value	Source
Soil storage porosity	0.624	*Reneau and Dietrich*, 1991
Soil flow porosity	0.05	hypothesized
Alluvial porosity	0.40	*Hough*, 1957
Soil cohesion, C_s	500 Pa	*Schroeder and Alto*, 1983
Soil saturated bulk density, ρ_s	1620 kg/m^3	*Reneau and Dietrich*, 1991
Sediment and soil grain density	2660 kg/m^3	*Reneau and Dietrich*, 1991
Soil and alluvial diffusivity	3.2 x 10^{-3} m^2/yr	*Reneau et al.*, 1989; *Roering et al.*, 1999[a]
Soil production rate for zero soil depth, decay scale	2.8 x 10^{-4} m/yr, 0.3 m	*Heimsath et al.*, 2001
Soil saturated hydraulic conductivity, K_s	1.0 x 10^{-3} m/s	calibrated
Alluvial saturated hydraulic conductivity	1.0 x 10^{-4} m/s	*Montgomery et al.*, 1997
Mean rainfall intensity and duration	1.7 mm/hr, 20 hrs	*Benda and Dunne*, 1997a
Mean interstorm duration	5 days	*Duan*, 1996
Downstream hydraulic width exponent, coefficient	0.5, 7.0 m/(m^3/s)$^{1/2}$	Oregon Coast Range flow vs. width data
At-a-station hydraulic width exponent	0.25	*Leopold and Maddock*, 1953
Downstream hydraulic roughness exponent, coefficient	-0.01, 0.03	*Leopold and Maddock*, 1953
At-a-station hydraulic roughness exponent	-0.21	*Leopold and Maddock*, 1953
Channel drainage area threshold	1.0 x 10^4 m^2	field verified
Fluvial transport exponents, m_f, n_f, p_f	0.6, 0.7, 3.0	*Tucker et al.*, 2001b
Fluvial transport coefficient, K_f	1.0 x 10^{-3} m^{5}s^5/kg^3	hypothesized
Critical shear stress for fluvial transport, τ_c	0	hypothesized
Internal and bed slip friction angles, ϕ_i, ϕ_b	42°, 28°	*Iverson*, 1997
Time step for debris flow motion	0.1 s	calibrated
Deposit erodibility, K_e	1.0 x 10^{-4} m/s-Pa	hypothesized
Friction factor, C_f	0.02	hypothesized
Critical shear stress for debris flow scour, τ_{cr}	2000 Pa	hypothesized
Macimum root strength	14 kPa	*Burroughs and Thomas*, 1977
Ratio of lateral and vertical root strength, m	2.33	*Burroughs*, 1984; *Hammond et al.*, 1992
Root and biomass growth constants	0.95, 19.05, -0.05	*Sidle*, 1992
Root growth time constant	0.25 yr^{-1}	*Sidle*, 1992
Root decay time constant and exponent	0.5 yr^{-1}, 0.73	*Burroughs and Thomas*, 1977; *Benda and Dunne*, 1997a
Root strength depth constant	2.0 m^{-1}	*Benda and Dunne*, 1997a
Tree height index	40 m	*Means and Sabin*, 1989
Maximum biomass weight	3.00 kPa	*Sidle*, 1992
Biomass time constant	0.12 yr^{-1}	*Sidle*, 1991
Tree diameter at breast height constants, b_0, b_1, and b_2	74.0 m, -0.0105 m^{-1}, 0.911	*Garman et al.*, 1995
Wood decay constant	0.031 yr^{-1}	*Harmon et al.*, 1986
Wood density	450 kg/m^3	
Blowdown parameter	1.0 x 10^{14} kg/m^3	calibrated
Mean time between fires	200 yr.	*Long et al.*, 1998

[a] For reference, *Roering et al.* [1999] obtained the best fit of the linear diffusion model to data with a constant of 5.0 x 10^{-3} m^2/yr.

istics, e.g., soil depth, vegetation age, and landslide susceptibility. Channel source and channel-adjacent areas form the lumped aggregates (Figure 3b). Nodes are classified and grouped according to the topographic component of landslide susceptibility [*Montgomery and Dietrich*, 1994] such that each aggregate contains several "bins" of nodes with similar susceptibility. During storms each bin, rather than each hillslope node, is considered against a failure criterion using bin-average values of soil depth and root strength. This method saves processing time and mimics

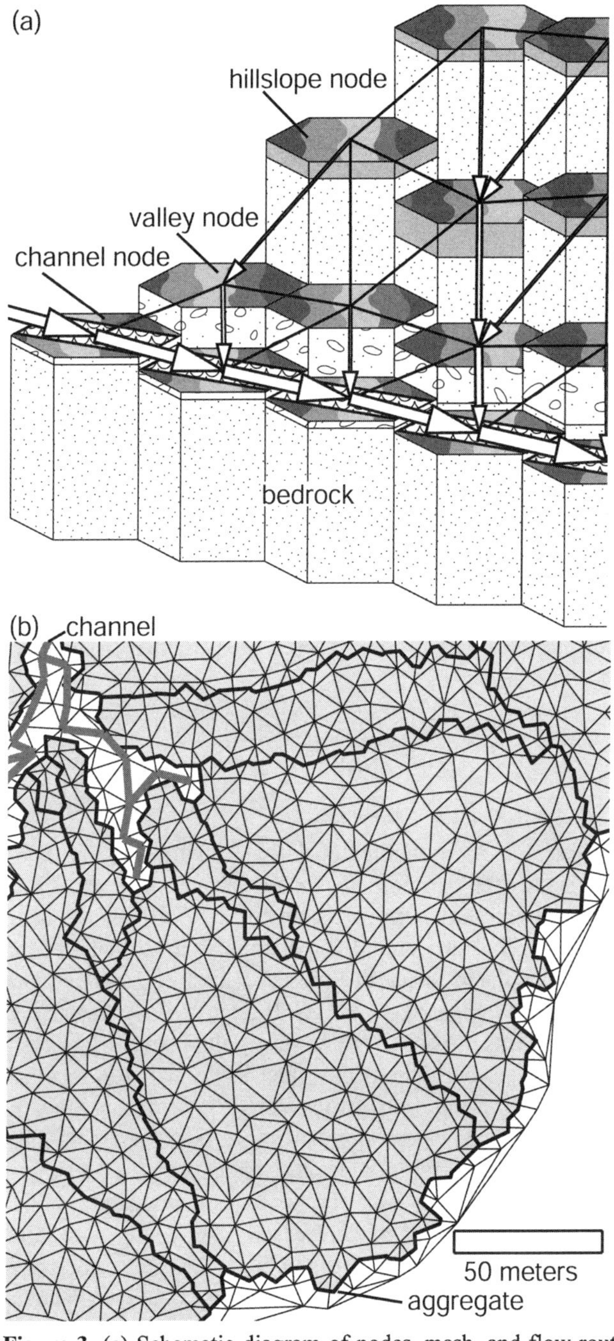

Figure 3. (a) Schematic diagram of nodes, mesh, and flow routing. Nodes are connected by edges of Delaunay triangular mesh and have associated Voronoi areas, i.e., area closer than to any other node. Flow follows steepest edge to neighboring node (arrows). Hillslope nodes have vegetation and soil overlying bedrock; channel and valley nodes have vegetation and alluvium overlying bedrock; and channel nodes contain a channel segment. (b) Part of irregular mesh showing aggregates of hillslope nodes (shaded gray) and channels (thick gray line). Nodes neither within an aggregate nor connected to a channel segment are valley nodes.

the way that whole hollows evacuate during a failure but tends to average out local heterogeneities in soil depth and root strength. The hillslope nodes making up the aggregates are updated upon evacuation of soil and wood during landslides.

The model is fed a stochastic time series of storms based on the work of *Eagleson* [1978], as in *Benda and Dunne* [1997a], *Duan et al.* [1998], *Tucker and Bras* [2000], and *Tucker et al.* [2001b]. The parameters of the stochastic model were derived from storm data for the Oregon Coast Range [*Benda and Dunne*, 1997a] or elsewhere in western Oregon [*Duan et al.*, 1998] (Table 1). In response to each storm event, landslides may initiate and resulting debris flows propagate downstream until they deposit on channel and valley nodes. The storms also drive tree fall. The model records both the runout path and the deposited depths of wood and sediment at each point along the stream. The history of previous events bears directly on later ones: sliding hollows are evacuated of soil, runout paths are scoured of wood and sediment or have wood and sediment deposited, and subsequent debris flows and fluvial transport encounter previous deposits, which may change channel and valley gradients and act as barriers.

The sediment deposited in the valley network by debris flows must originate as hillslope soil, defined here as material lacking the structure of underlying bedrock. Soil depths on hillslopes are governed by soil production and diffusion, where the soil production rate at a point decreases exponentially with the soil depth [*Heimsath et al.*, 1997, 2001]. Diffusion [*Reneau et al.*, 1989; *Roering et al.*, 1999] and soil production [*Heimsath et al.*, 2001] parameter values have been measured in the Oregon Coast Range (Table 1). Though hillslope elevations do not change, soil depths evolve over time. Soil production is only active on the hillslopes, but diffusion acts on all landscape nodes and, thus, may transport material among hillslope, valley and channel nodes. Diffusion of material from a node is contingent on supply—bedrock does not diffuse. The maximum time steps for soil production and diffusion are the storm and inter-storm durations but may be dynamically shortened if necessary for numerical stability [*Tucker et al.*, 2001b].

Channel nodes are those with a drainage area exceeding a threshold of 10^4 m^2, determined from analysis of slope-area plots [*Tarboton et al.*, 1991; *Ijjasz-Vasquez and Bras*, 1995] and trial and error. Channel source area values from field measurements [*Montgomery and Dietrich*, 1988, 1992] produce a channel network with "feathered" extremities on relatively coarse DEMs such as ours. Through field reconnaissance we found that this threshold may exclude some small channels but effectively marks the transition from bowl-shaped hollows to V-shaped valleys. Drainage area is determined by routing each node's area downstream in the direction of steepest descent. As debris flows deposit sediment and wood and those deposits are re-worked by subsequent debris flows and fluvial processes, the location of the channel may change. When that happens, hillslope and valley nodes may become channel nodes, and abandoned channel nodes become valley nodes.

Sediment deposited in the channel network is transported according to a power law of excess shear stress, where shear stress is represented as a power law of unit discharge and slope derived from continuity and the Manning equation:

$$Q_s \leq K_f b\left[\rho g\left(\frac{Qn}{b}\right)^{m_f} S^{n_f} - \tau_c\right]^{p_f} \tag{1}$$

where Q_s is potential sediment discharge, i.e., contingent on supply; K_f, m_f, n_f, and p_f are constants; b is hydraulic width; Q is water discharge; n is Manning's hydraulic roughness; S is hydraulic slope; ρ is water density; g is gravitational acceleration; and τ_c is critical shear stress [*Tucker et al.*, 2001b] (Table 1). Discharge is generated by saturation overland flow, such that alluvial depth in the channel affects discharge, and hydraulic width and roughness are calculated from empirical power laws of discharge, both downstream and at-a-station [Tucker *et al.*, 2001b] (Table 1). Equation (1) represents total load, i.e., both suspended and bed load, for values of p_f in the range of 2.5-3 [*Engelund and Hansen*, 1972; *Vanoni*, 1975; *Govers*, 1992]. This transport law has not been calibrated for streams in the Oregon Coast Range. Rather, it is simple and generic. The most salient feature of equation (1) for the present study is the dependence on local hydraulic slope, which changes during the simulation as a result of sediment and wood aggradation and scour. Our observations indicate that the streams in the study area are not competent to remove wood from debris flow deposits, and these observations are consistent with findings of *Lienkaemper and Swanson* [1987]. Therefore, we assume that wood cannot be transported by fluvial processes, so until they decay, wood deposits act as barriers to sediment transport by decreasing upstream hydraulic slope.

2.2 *Landslide Susceptibility*

Landslide initiation is stochastic, where the probability of a landslide, i.e., sites that fail and become debris flows, exists when rainfall exceeds a duration-dependent intensity. Our formulation of this critical precipitation is based on *Montgomery and Dietrich* [1994] and *Dietrich et al.* [1995] and lies somewhere between the two in complexity, similar to *Montgomery et al.* [2000]. We assume uniform saturated hydraulic conductivity within the soil layer and zero conductivity beneath that layer. The critical precipitation, P_{cr}, is given by

$$P_{cr} = \frac{K_s hb\cos\theta\sin\theta\rho_s}{A_{eff}\rho_w}\left[1 - \frac{1}{\tan\phi_i}\left(\tan\theta - \frac{C_r + C_s}{h\rho_s g\cos^2\theta}\right)\right] \tag{2}$$

where K_s is saturated soil hydraulic conductivity; h is vertical soil thickness; b is flow width; θ is slope angle; ρ_s is soil saturated bulk density; ρ_w is water density; A_{eff} is effective area contributing to flow and is dependent on storm duration; ϕ_i is internal friction angle; C_r is cohesive root strength; and C_s is soil cohesion (Table 1). We derived a simple expression for the effective area contributing to flow, A_{eff}, by solving the Darcy equation for an effective upslope length contributing to subsurface flow during a storm of known duration, t_d, and squaring that length to get A_{eff}:

$$A_{eff} = min[\left(\frac{t_d K_s \sin\theta}{n_{eff}}\right)^2, A] \tag{3}$$

where A is topographically defined contributing area; and n_{eff} is the effective porosity for subsurface flow, which field experiments have shown is much smaller than the actual porosity (i.e., the actual travel time is much smaller than that calculated from the actual porosity [*Iverson et al.*, 1997; *Montgomery et al.*, 1997]), most likely because the experiments are actually measuring the arrival of a peak in discharge rather than of the water itself. Using A_{eff} accounts for greater saturation during longer storms but does not account for transient pore pressure increases from short, intense rainfall periods during longer storms. These transient increases may be responsible for many natural failures [*Montgomery et al.*, 1997; *Iverson*, 2000]. We made failure stochastic to account for some of the great uncertainty around predicting actual failure at a site. Greater landslide susceptibility corresponds to lower critical precipitation, P_{cr}, and the probability for stochastic failure during a given storm of intensity, P, is given by

$$Prob(failure) = 1 - \frac{P_{cr}}{P},\ P \geq P_{cr} \tag{4}$$

$$= 0,\ P < P_{cr}$$

Note that saturated hydraulic conductivity, K_s, may vary over orders of magnitude between sites and even within relatively small regions in the field, and critical precipitation is directly proportional to this highly variable and uncertain parameter [*Duan*, 1996]. If this parameter is too small (or too large), failures will occur too often (or too seldom). Because the amount of sediment delivered to the channel network by debris flows is ultimately limited by the soil production rate, K_s will mainly affect soil depths by allowing, or not, realistic amounts of soil to accumulate before failing. We calibrated, albeit roughly, saturated hydraulic conductivity by running several simulations on a small part of the study area while varying K_s's order of magnitude within the range of reported values [*Montgomery et al.*, 1997] and selecting the value that produced reasonable average soil depths on the hillslopes (~0.5 m) [*Montgomery et al.*, 1997; *Heimsath et al.*, 2001] (Table 1).

2.3 *Debris Flow Runout*

When landslides occur the sediment and wood from hillslope nodes in the failing bin(s) are delivered to the nearest channel where they continue to move as debris flows. To describe debris flow runout, we use a physically-based debris flow model, summarized here, that is simplified to run within the network scale model. *Iverson* [1997] and *Iverson et al.* [2000] developed a model describing debris flow runout with mixture theory and depth-averaged conservation of mass and momentum in two dimensions. We neglect smaller terms and convective accelerations in the momentum balance and treat debris flow motion as a one-dimensional point process, where that point moves with the front of the flow [R. Iverson, U.S. Geological Survey, personal communication, 1999]. Conservation of momentum in the flow direction is then given by:

$$h\frac{\partial v}{\partial t} + v\frac{\partial h}{\partial t} = -\operatorname{sgn} v\left(hg\cos\theta - \frac{p_b}{\rho_m} + hv^2\frac{\partial \theta}{\partial s}\right)\tan\phi_b + hg\sin\theta \qquad (5)$$

where h is slope-normal debris flow depth; v is slope-parallel debris flow velocity; p_b is pore pressure at the bed; ρ_m is debris flow mixture density; s is the slope-parallel direction; ϕ_b is bed friction angle (Table 1); and the factor, -sgn v, indicates the direction opposite that of the debris flow velocity. The terms on the left-hand side represent changes in flow velocity and depth, respectively. The first group of terms on the right-hand side resists motion and is therefore multiplied by friction slope. The terms within parentheses represent normal forces on the flow: normal gravitational force, basal pore pressure, and centripetal acceleration due to changes in slope angle, respectively. To get the basal pore pressure in equation (5), hydrostatic pressure is multiplied by 1.8, consistent with the experimental observations of *Iverson et al.* [2000]. The last term on the right-hand side represents the slope-parallel component of the gravitational force.

Debris flow velocity conforms to changes in flow direction, i.e., $v_{new} = v_{old}\cos\alpha$, where α is the angle between the new and old downstream directions. This angle, α, is calculated over a spacing of several nodes (>30 m) both up- and downstream. Debris flow length is held constant. Depth must conform to changes in flow width, i.e., $h_{new} = h_{old}(b_{old}/b_{new})$. Width can cover as many as three nodes and is determined by the flow depth and local channel or topographic geometry. When width and, therefore, depth change, velocity changes as if no forces were acting on the flow, i.e., equation (5) is solved for the change in velocity with the right-hand side set to zero.

Iverson [1997] and *Iverson et al.* [2000] assumed constant debris volume with time. But, we need to model the effects on runout of increases in debris flow volume during runout—*May* [1998] found that on the order of half of debris flow deposit volumes that she measured in the Oregon Coast Range were from entrainment during runout. To model depth changes through scour, we employ a shear-excess law similar to equation (1) with p_f equal to 1, as scour has often been considered to be proportional to shear stress (e.g., *Howard and Kerby* [1983]):

$$\left.\frac{\partial h}{\partial t}\right|_e = K_e(\rho_m C_f v^2 - \tau_{cr}) \qquad (6)$$

where K_e is deposit erodibility; C_f is a friction factor; τ_{cr} is critical shear stress (Table 1); t is time; and the rate of scour is constrained to be positive or zero. We assume that bedrock is not erodible. Although the physics of scour by debris flows are poorly understood and equation (6) and its parameters are essentially a hypothesis for that physics, momentum conservation and bedrock's non-erodibility enforce rigorous physical bounds on scour.

Debris flows are processed sequentially with a separate time step (Table 1) and travel from node to node in the direction of steepest descent. Initial width depends on mesh discretization and is approximately 9 m in the simulation presented here, and length is defined by the number of nodes contributing to the initiating landslide. Debris flows are divided into two parts, head and tail, and equations (5) and (6) and the other rules are applied at the head. Head length is total length divided by the number of nodes contributing to the landslide. Scoured and incorporated material is added to the head, and when it stops the front of the tail becomes the new head with the old head's prior velocity. This scheme accounts for two observations. First, scoured material and accumulated debris usually remain at the front. Second, tails often bypass the more debris-laden heads when the latter stop. Our scheme retains a feasible simplicity while allowing process dynamics to determine the final deposit geometry.

Unlike any other debris flow runout model that we are aware of, ours incorporates all three major constituents observed in the field: sediment, water, and wood. Debris flows must incorporate all surface wood in their paths or stop. As with sediment, we use equation (6) to model scour of wood from deposits. We neglect any other effects of wood, such as the resistance by standing trees to uprooting or breakage and any extra resistance due to the shape and strength of the wood pieces, because a previous series of model experiments determined that the simulated runout length distribution was closest to an observed distribution without these other effects.

2.4 *Forest Growth, Root Strength, Blowdown, and Wood Decay*

As indicated above, the presence of wood in channels affects fluvial sediment transport; the strength of tree roots affects landslide susceptibility; and the presence of wood on hillslopes and in valleys affects debris flow momentum. Therefore, we must account for these effects by modeling: (a) growth and decay of tree roots; (b) growth and decay of wood biomass; (c) stochastic movement of wood, e.g., from riparian areas to channels, by treefall; and (d) stochastic events resulting in forest death, i.e., fires.

The evolution of several variables describing the forest is governed by a set of empirical equations, the parameters of which vary according to species. We have chosen parameter values that are representative of Douglas-fir (*Pseudotsuga menziesii*) because it is the dominant species in the field area. Root strength, C_r, evolves according to exponential decay of root strength after stand death and sigmoid-increasing strength, as in *Sidle* [1992] and *Duan* [1996], and partitioning of root strength between vertical and lateral components, with the vertical component decreasing exponentially with soil depth. Some parameter values used in root strength calculation were derived specifically for the Oregon Coast Range, while others are generic (Table 1). The lateral and vertical components of root strength are summed to get the total root cohesion, C_r, which is added to soil cohesion in equation (2). In our model, root strength can decay from an arbitrary value rather than being constrained to decay from the maximum value. Also, we use a differential form so that root strength at the next time step evolves from the present value. Upon stand death, the constants representing "initial" lateral and vertical root strength, C_{V_0} and C_{L_0}, respectively, are reset from the total root strength at the time of death, C_{r0}, according to a partitioning coefficient, m:

$$C_{V_0} = \frac{C_{r_0}}{1+m}, \quad C_{L_0} = mC_{V_0} \tag{7}$$

This root strength model neglects scale effects. In reality, larger failure perimeters should have larger lateral root strength [*Montgomery et al.*, 2000], but, in practice, the model does not calculate failure perimeter.

Wood volume grows as the stand ages according to the sigmoid function of *Sidle* [1992]. Again, our model employs a differential form during evolution so that biomass at the next time step evolves from the present value. Parameter values for this relationship are generic (Table 1). Maximum tree height is determined by *Richards*'s [1959] equation on a 5-parameter base as used by *Duan* [1996] and evolves with time according to a differential form of that equation. The tree height index used in the maximum tree height relationship was derived for Douglas-fir in the Oregon Coast Range [*Means and Sabin*, 1989] (Table 1).

Tree diameter at breast height (D_{bh}, height = 1.37m) is calculated by inverting an empirical relationship describing height as a function of D_{bh} [*Garman et al.*, 1995] to solve for it as a function of maximum tree height, H_w:

$$D_{bh} = \frac{1}{b_1}\ln\left[1-\left(\frac{H_w-H_b}{b_0}\right)^{\frac{1}{b_2}}\right], \; H_b < H_w \le b_0 \tag{8}$$

$$= 0, \; H_w \le H_b$$

where b_0, b_1, and b_2 are empirical coefficients determined for Douglas-fir in the Oregon Coast Range [*Garman et al.*, 1995], and H_b is breast height, 1.37 m. In order that the argument of the logarithm does not become negative, tree height may not exceed b_0.

Trees fall via a stochastic blowdown model. The number of trees falling at a given landscape node during each storm is exponentially distributed, and the mean, or expected, number of blowdowns, μ_N, is given by the ratio of the drag force from wind to the resisting strength of roots:

$$\mu_N = \frac{P^2}{C_r}\left(\frac{\rho_a C_d V_R^2}{2B_T}\right) \tag{9}$$

where P is the storm precipitation rate; C_r is the root strength; ρ_a is the density of air; C_d is the drag coefficient; V_R is the ratio of storm wind velocity to precipitation rate; B_T is the ratio of tree crown width (i.e., the cross-sectional area presented to the wind divided by tree height) to height. Shelter or exposure effects are neglected. The term in parentheses is lumped into a single "blowdown" parameter for model input (Table 1). The order of magnitude of this parameter is calibrated to provide slightly decreasing live biomass over time for old-growth stands, as has been observed in the Oregon Coast Range [T. Spies, U.S. Forest Service, pers. comm., 2000]. As in *Van Sickle and Gregory* [1990] and *Robison and Beschta* [1990], fall direction for each blowdown is chosen at random. Wood is distributed over the nodes on which the tree falls as if it were a perfect cone with the maximum tree height and D_{bh} calculated from equation (8), and biomass is conserved, i.e., a tree cannot fall from a node unless it has enough live biomass. In this way, wood is contributed to the channel from riparian zones and, depending on the tree height, may come from several nodes' distance. Fallen and deposited wood decay over time according to a single exponential [*Harmon et al.*, 1986] with a rate derived for Douglas-fir in western Oregon (Table 1).

Fires occur at exponentially distributed intervals and kill the entire forest, whereupon all trees fall. In nature, fires have variable size and intensity, and many trees are left standing, but, for simplicity, we assume we may neglect these variations. Neglecting size variation is justified by the finding that nearly all fires are larger than the basins we model (i.e., < 5 km^2) [*Wimberly et al.*, 2000; M. Wimberly, U.S. Forest Service, pers. comm., 2000]. As stand-killing fires typically burn only a small fraction of existing biomass [*Huff*, 1984; *Harmon et al.*, 1986; *Spies et al.*, 1988], we assume that fires consume no wood.

2.5 Model initial conditions from digital elevation models and surveyed channel profiles

The initial topography for the model simulations was generated from a 10-meter digital elevation model (DEM) and characteristics of the longitudinal channel profile. The DEM was generated from 7.5 minute topographic maps. For the purposes of modeling locations of sediment storage, the DEM-based valley topography is inappropriate because it results in a longitudinal channel profile with large steps and intervening "flats" as long as several hundred meters such that sediment accumulates on the flats. The problem is exacerbated by the fact that the model considers this initial profile to be bedrock and, therefore, not erodible. To remedy this problem we used characteristics of the longitudinal channel profile surveyed in the field to make a smooth initial bedrock profile.

It is often observed that stream gradient, or slope, and contributing area are related as,

$$S = KA^{-\theta} \tag{10}$$

where θ is the concavity index; and K is the steepness index [*Flint*, 1974]. This relationship has been used in many studies to characterize streams [e.g., *Hack*, 1957; *Tarboton et al.*, 1991; *Willgoose*, 1994; *Moglen and Bras*, 1995; *Tucker and Bras*, 1998]. By finding contributing areas with the DEM and matching the longitudinal profiles from the DEM and field survey, we found the contributing area at every point along the surveyed profile. We then used the surveyed profile and the DEM contributing areas to derive K and θ (Table 2). We used the method of *Snyder et al.* [2000], in which the slopes are calculated between 10-meter elevation intervals from the surveyed profile. To extrapolate a bedrock surface from the outlet up every branch of the network with equation (10), we "tuned" the steepness and concavity indexes to transition smoothly with the DEM elevations along the main channel (Table 2). This method resulted in steps along some tributary channels, but these steps are unlikely to affect the results for the main channel.

In order to avoid an entrenched bedrock profile only one node wide, we repeatedly determined drainage directions according to a probabilistic criterion such that the probability of flowing to any downslope neighbor is proportional to the relative magnitude of the slope in that neighbor's direction (whereas at all other times in the simulation flow direction is deterministic and follows steepest descent). The bedrock elevation was calculated for every channel node each time flow directions were re-determined, but node elevations were not changed until the end, when elevations at all nodes that had been channels, i.e., channel and valley nodes, were changed. This method resulted in some elevated bedrock "terraces" with thick soil adjacent to the channel, especially in the lower reaches of the main channel (Plate 1a). The profile-smoothing procedure successfully eliminated the main channel steps and flats that were artifacts of the DEM.

Finally, we developed a procedure to provide both heterogeneous soil depths in landslide-prone hollows and realistic soil depths on ridges and side-slopes. An initial soil layer evolved by diffusion and soil production over 6000 years (as in, e.g., *Dietrich et al.*, [1995]). The storm model ran in isolation for 100 years to find the maximum intensity and duration during that time. Assuming a 6-year-old forest, when root strength is at a minimum, we determined failure areas given a storm with that intensity and duration, and the soil was removed from these areas. At this point, most of the landslide-prone hollows were emptied of soil, but soil remained on the ridges and planar slopes. In order to refill the hollows to different depths to mimic different

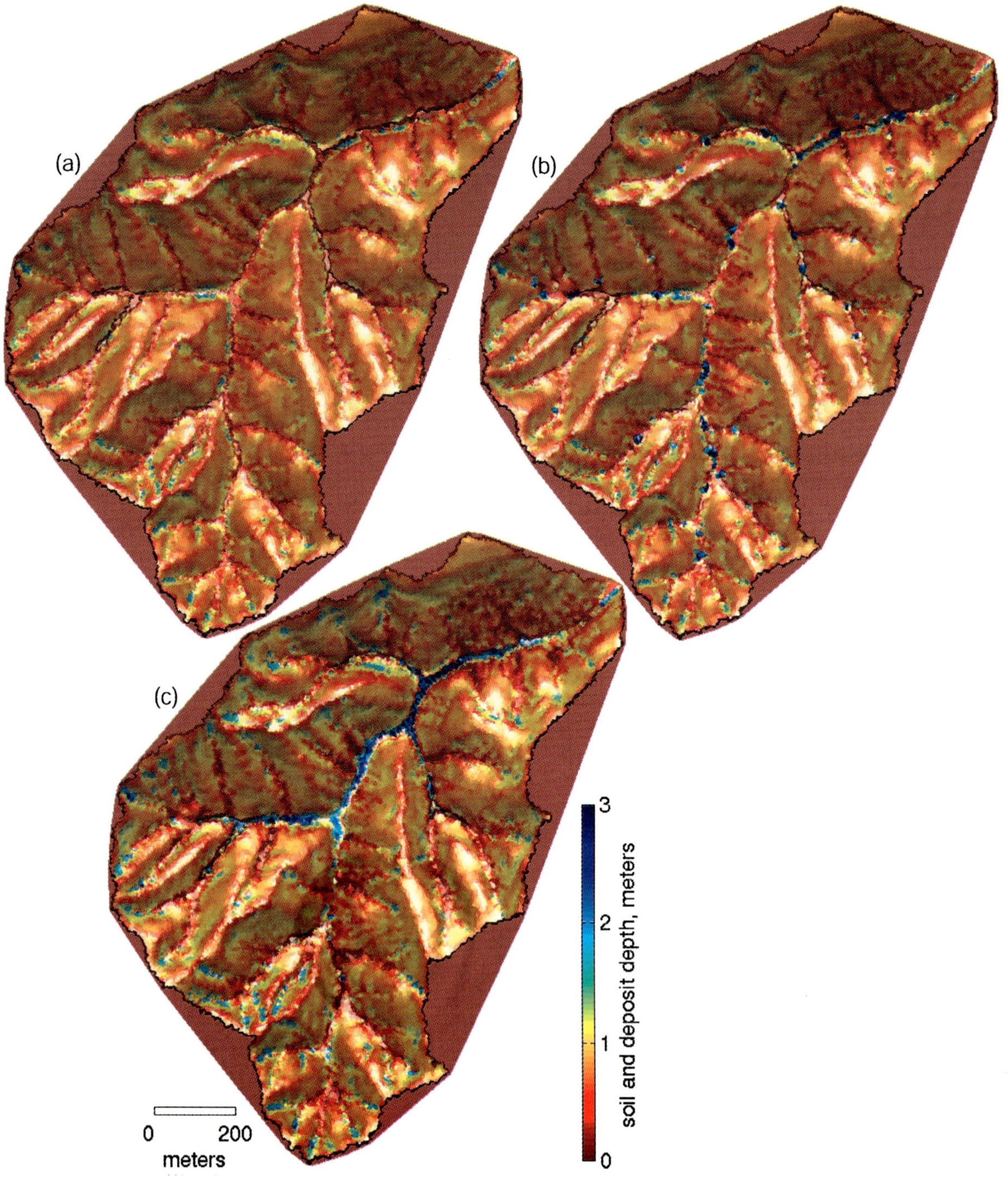

Plate 1. Shaded relief maps colored according to soil or sediment depth of (a) initial condition for model simulation, (b) after 200 years (18 years after the first fire), and (c) after 3000 years (the end of the simulation). The color scale is compressed to show variations in soil depth and to highlight all deposits greater than 3 meters in depth.

TABLE 2. Parameters of slope-area relationship, both derived from data and "tuned"

Range of contributing area, A (m^2), used for derivation	Range of contributing area, A (m^2), to which tuned profile was applied	Derived concavity index, θ	Tuned concavity index, θ	Derived steepness index, K	Tuned steepness index, K
$10^5 \leq A < 10^6$	$8.5\text{x}10^4 \leq A < 10^6$	0.407	0.41	14.8	13.5
$A \geq 10^6$	$A \geq 10^6$	1.41	1.41	$1.63\text{x}10^7$	$1.6\text{x}10^7$

times since failure, evolution of the soil layer then proceeded for different random times between 0 and 2000 years in each aggregate. The forest at each failure site was regrown for the lesser of 300 years or the randomly chosen time of soil evolution to provide an old forest on all nodes except those that had recently failed. Finally, the landscape evolved for 10 more years while it was submitted to stochastic storm input, and during this time failed soil was removed from the system. This procedure produced a heterogeneous, realistic initial soil layer (Plate 1a) and prevented a massive initial influx of debris flows to the valley network.

3. PREDICTING SEDIMENT AND WOOD FLUXES

3.1 *Simulation and Field Methods*

Beginning with the initial condition described in the previous section, we simulated a period of 3000 yrs. in the study basin (Plate 1). In order to adequately represent the spatial distribution of sediment and wood inputs to the valley network, the model must adequately represent observed runout lengths.

In order to locate all recent debris flows in the study basin, we attempted to walk the entire channel network defined by a 10^4 m^2-contributing area threshold. We identified 35 debris flow deposits, and were able to measure the runout length from source area to deposit for 28 of them. Horizontal runout length was then measured between failure source and deposit terminus on a DEM using a GIS. We grouped all measured debris flows into approximate age classes based on aerial photographs (1945-1997) and the age of trees growing on the deposits. We also measured total deposit and wood constituent volumes and down wood volumes in several of the smaller channels [*Harmon et al.*, 1986]. Although we tried to capture the full failure history of the study basin, evidence of many smaller events is erased over time as new failures occur, and therefore the older part of our record is skewed towards large events.

3.2 *Simulation Results and Comparison to Field Data*

We test how reasonable the modeled sediment and wood fluxes are by comparing simulation results to field data. We check modeled debris flow runout length, size, sediment and wood input rates, and longer term denudation rates. Our main concern is correctly simulating the spatial distribution of sediment input to the channel network from debris flows.

The cumulative distribution function (CDF) of simulated debris flows mimics the CDF of the 28 debris flow runout lengths measured in the study basin (Figure 4). The spatial distribution of debris flow inputs to the valley network is therefore reasonably accurate.

Simulated debris flow sizes are also similar to measured failure volumes. The average debris flow volume for the simulation is 165 m^3, well within the range of reported values (Table 3). Our own field measurements of wood in debris flow deposits also indicate that wood volumes predicted by the model are reasonable.

The simulated failure rate per unit area is similar to the measured rate in the study area. From 35 debris flows in the study area within the last 50 years, we calculate a failure rate of 0.33 km^{-2}yr^{-1}. Since the mean fire recurrence interval for this area is approximately 200 yrs [*Long et al.*, 1998], or one basin-wide disturbance in 200 years, and nearly half the basin was clearcut in 50 years, approximately equivalent to one basin-wide disturbance in 100 years, the measured debris flow rate may exceed the long-term "natural" rate by a factor of about 2 such that the "correct" rate is closer to 0.17 km^{-2}yr^{-1}. During the 3000-year simulation, 1182 debris flows occurred, for a failure rate of 0.191 km^{-2}yr^{-1}, which is similar to the estimated natural rate for the study area. This value is low compared to short-term landslide frequencies reported in the literature (Table 3), but higher than the long-term failure rate, based on measured lowering rates, of 0.01- 0.03 km^{-2}yr^{-1} calculated by Montgomery *et al.* [2000].

Other comparisons suggest that the rate and timing of landslides may not be realistic. Extended periods of several hundred years pass in the simulation without a landslide event, and the time series of simulated debris flow events shows that, except before the first fire, all debris flows occur shortly after fires (Figure 5a). *Schmidt et al.* [2001] found that landslides in older forests occurred in significant gaps between trees. The model's binning procedure averages out such heterogeneities in root strength and makes failures in older forests unlikely. Soil storage increases

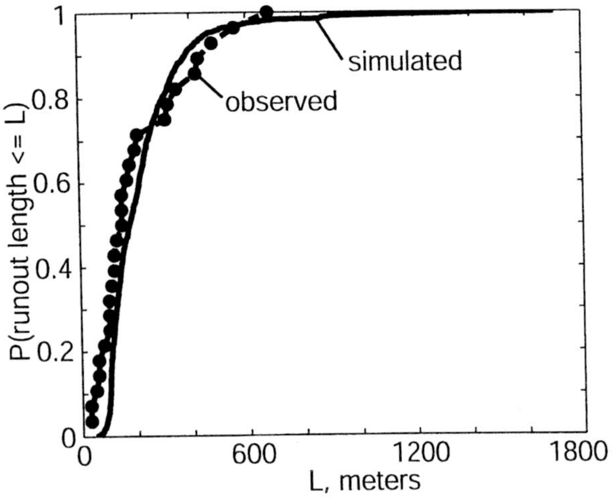

Figure 4. Simulated and observed cumulative distribution functions of debris flow runout length, i.e., sample probability of runout length less than or equal to some length, *L*.

during the simulation (Figure 5b), and the simulated denudation rate from landslides, 0.0158 mm yr^{-1}, is low relative to the bedrock lowering rate determined from colluvial transport of 0.061 ± 0.025 mm yr^{-1} [*Reneau and Dietrich*, 1991] and from soil production of 0.1 mm yr^{-1} [*Heimsath et al.*, 2001]. The coarseness of the DEM-derived topography should lower effective diffusion rates below those observed in the field, and errors in hillslope gradient will affect landslide susceptibility. Both of these factors could lower the denudation rate from landslides. Neither of these possible shortcomings affect the spatial distribution of sediment input by debris flows.

In contrast to the event-based sediment input dominated by debris flow deposition (Figure 5b), simulated sediment output from the basin is relatively smooth because it is controlled primarily by fluvial transport (Figure 5c). Occasional debris flows reaching the outlet lead to small steps in the cumulative output coincident with debris flow occur-

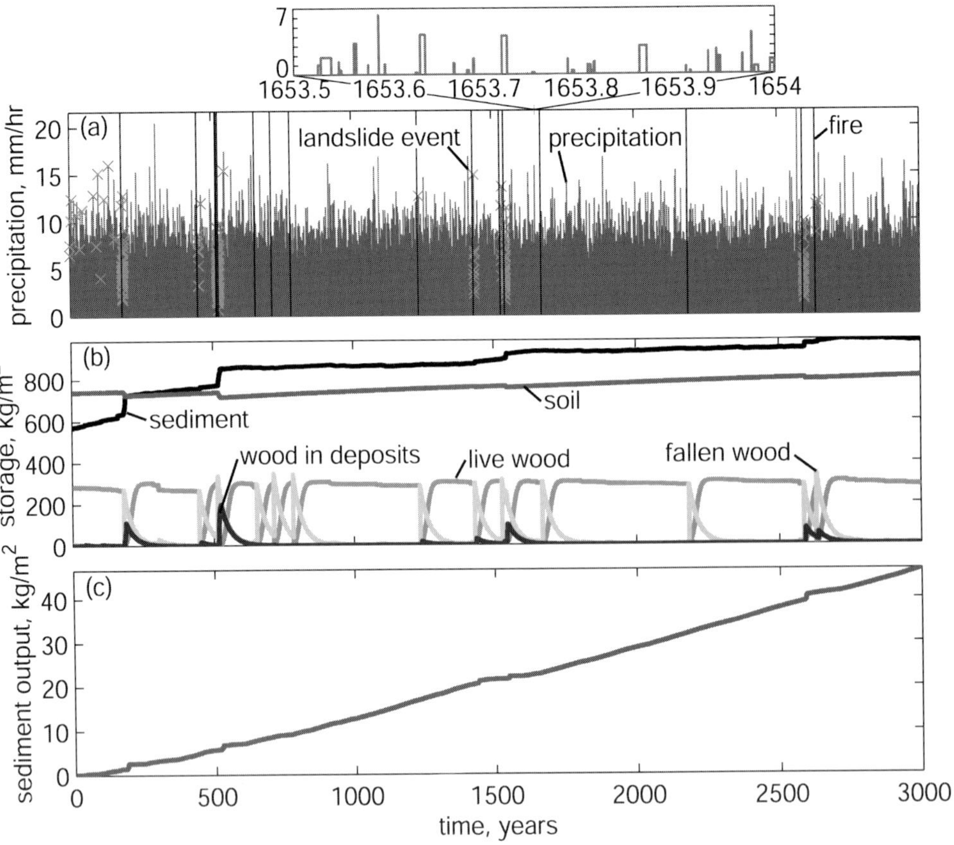

Figure 5. (a) Precipitation intensity *vs.* simulated time, indicating fires and landslide events. Part of the precipitation series (0.5 years) is shown with a dilated time axis to illustrate variations in storm duration and intensity and interstorm periods. (b) Normalized storage masses *vs.* simulated time: sediment and wood in deposits normalized by valley area (sum of channel and valley node areas); soil normalized by hillslope area (sum of hillslope node areas); live wood and fallen (dead) wood (excluding wood in deposits) normalized by basin area (sum of all hillslope, valley and channel node areas). (c) Mass of cumulative sediment output normalized by basin area *vs.* simulated time.

TABLE 3. Summary of the present study and landslide and debris flow studies in areas geologically similar to the Hoffman Creek site in the Oregon Coast Range

Average landslide volume (m^3)	Landslide frequency (km^{-2} yr^{-1})	Number	Period of record (yr)	Reference
610	N/A	73	single storm	*May*, 1998
450	N/A	36	N/A	*Benda and Cundy*, 1990
54	0.533	39	15	*Swanson et al.*, 1977[a]
110	1.03	317	10	*Swanson et al.*, 1977[b]
250	8.0	35	10	*Montgomery et al.*, 2000
	5.8	25	10	*Montgomery et al.*, 2000[c]
20	N/A	92	single storm	*Robison et al.*, 1999[d]
115	N/A	76	single storm	*Robison et al.*, 1999[e]
	0.33	35	50	field data, present study
165	0.191	1182	3000	simulation, present study

[a] field-based survey in mature forest
[b] air photo-based survey in recent clear-cut
[c] non-road-related slides only
[d] landslide initiation site only, Mapleton, Oregon, site only
[e] landslide and non-channelized debris flow, Mapleton, Oregon, site only

rences, and impoundment of sediment by woody debris causes a relative flattening of sediment output in periods following these occurrences and lasts for approximately 100 years. This period of flattened sediment output is coincident with the period over which wood deposits decay (Figure 5b). The sediment output during the simulation is equivalent to a denudation rate of 7.6 x 10^{-3} mm yr^{-1}, indicating that roughly half of the denudation by debris flows is stored in the valley network at the end of the simulation.

4. PREDICTING LOCATIONS AND AMOUNTS OF SEDIMENT AND WOOD STORAGE

4.1 *Simulation and Field Methods*

To quantify simulated sediment and wood storage, we calculated the cross-sectional area of simulated deposits at each point along the main channel. The down-valley direction was determined by fitting a line to the channel node and the next three downstream nodes. From the channel node, sediment and wood depths were read, or "measured", at 0.1-meter intervals along line segments perpendicular to the down-valley direction to the right and left as long as the measurement points were still within the valley, i.e., belonged to a channel or valley node, and were within a maximum of 20 meters from the channel node. This latter criterion was based on the maximum valley widths measured in the study area and was necessary to keep the simulated valley transects from extending up tributary valleys. These "surveys" provide snapshots of simulated valley storage at each model output time, every 20 years. We also calculated average storage along the main channel by putting each of the instantaneous measurements into 50-meter bins and calculating the average for each bin.

We determined sediment and wood storage in the field by a similar method. We surveyed the longitudinal profile of the main channel of the study basin with a hand level and stadia rod. This survey was relatively detailed, with usually <5 channel widths between survey points. At longer intervals, ~10 channel widths, we surveyed valley transects. Using the valley transects and the assumption of 40° valley side slopes, we calculated the cross-sectional area of sediment stored above the elevation of the channel. This information combined with the channel survey allowed us to calculate the sediment stored in the valley above the elevation of the channel (Figure 6). Where the channel bed is bedrock, sediment stored above the channel is equal to the sediment stored in the valley. Where valley floor deposits formed distinct wedges, i.e., relatively flat surfaces followed by downward steps, we estimated the cross-sectional area of valley floor storage with a line connecting the beginning of the flat surface to the bottom of

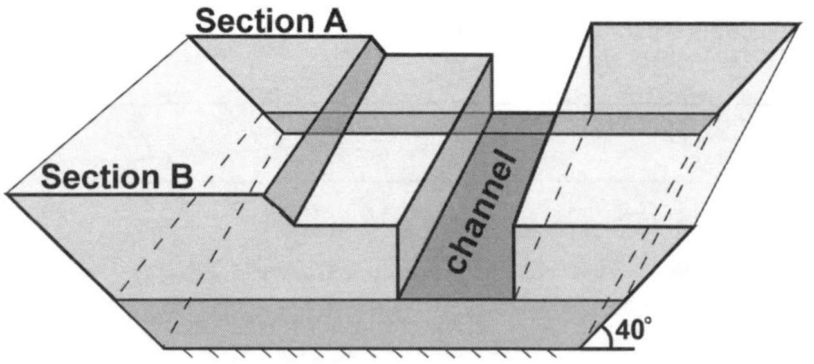

Figure 6. Schematic diagram of the sediment volume estimation method. Storage is calculated for sediment and wood both above the channel bed and including some sediment and wood below the channel bed in distinct wedges detectable from the longitudinal channel profile.

the step. Including these wedges yielded a second, larger estimate of valley storage (Figure 6).

4.2 *Simulation Results and Comparison with Field Data*

Comparing simulation and field results following a similar disturbance, fire in the case of the simulation and partial clearcutting in the case of the field area, indicates that the model accurately predicts many of the observed sediment and wood accumulation features both in terms of spatial distribution and magnitude (Figures 7 and 8a).

Cross-sectional areas of valley floor sediment and wood storage show that local storage maxima are often associated with steps in the longitudinal channel profile (Figure 7). Some storage maxima, however, are not associated with channel steps, e.g., the storage maximum near the bedrock points in the middle of the profile. This fact indicates, and our observations confirm, that much of the storage on the valley floor is in fans and terraces that are incised by the channel.

As sediment and wood influx in our model are dominated by debris flows, the pattern of debris flow deposition dominates the pattern of storage early in the simulation (Plate 1b) and many of the storage peaks are dominated by wood (Figure 8a). These storage peaks are associated with large steps in the longitudinal channel profile, many of which actually form dams (Figure 8a, Figure 9a).

Over 100 years of simulated time, wood deposits decay, and sediment deposits are reworked by fluvial transport (Figure 5b, Figure 8b, Figure 9b). The same peaks are evident at the same locations, and the dams have become steps (Figure 8b, Figure 9b). For example, the location of maximum storage is the same, approximately 600 m downstream, at 200 and 300 years, and at 300 years the channel step associated with this maximum is still pronounced. Dams at 200 years lead to fluvial deposition upstream such that stored sediment volume at some places actually increases between 200 and 300 years. For example, the magnitude of the storage peak at 1200 m downstream decreases slightly between 200 and 300 years, but the wood component decays while the sediment component increases (Figure 8). This increase must be due to fluvial deposition because no debris flows occurred in the interval (Figure 5a).

None of the storage peaks move between 200 and 300 years, and persistent peaks in storage are also evident in the average storage profile (Figure 10). The valley cross-sections in the neighborhood of the location of maximum storage show that at 300 years the channel has incised the deposit and formed a high terrace that accounts for most of the storage in the cross-section (Figure 9). If this cross-section is typical of others, then most of the storage in the valley at 300 years is in terraces, resulting in persistent stationary storage peaks. The actual amount of storage in terraces is influenced by bank erosion or channel avulsion due to local wood input, which will tend to increase terrace erosion. Conversely, lower decay rates for buried wood and decay-resistant species would tend to increase the longevity of dammed deposits. These effects are not represented in the current model.

Toward the end of the simulation, storage is dominated by an area of fluvial deposition in the middle of the lower half of the profile (Figure 10) because the fluvial model created a graded profile to smooth out the difference in concavity between the upper and lower parts of the initial bedrock profile (Table 2). Note that, because the elevation at the basin outlet is fixed, the reach near the outlet cannot aggrade.

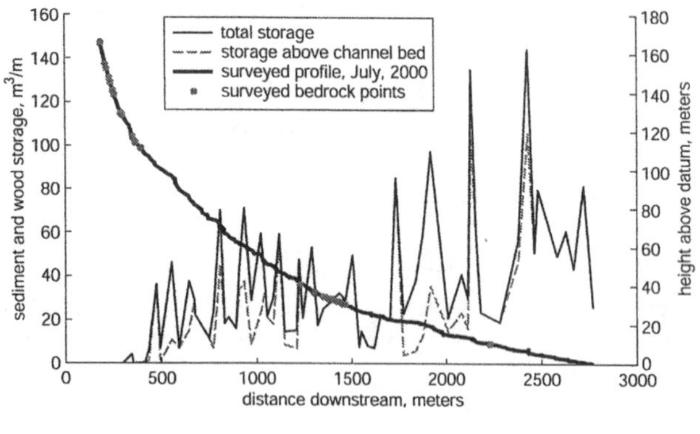

Figure 7. Surveyed Hoffman Creek channel profile with bedrock points highlighted and cross-sectional areas of sediment stored in surveyed valley transects *vs.* distance from divide. Two storage estimates are shown: sediment stored above the channel bed, representing a minimum estimate; and a more realistic estimate including sediment "wedges".

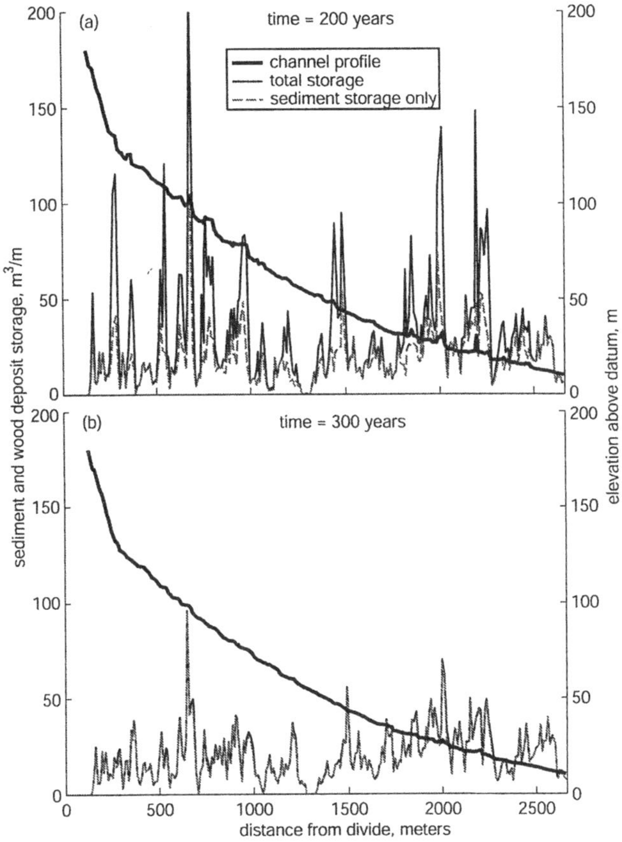

Figure 8. Simulated main channel profile and cross-sectional areas of sediment and wood deposit storage at every point along the main channel *vs.* distance downstream at two times, (a) 200 years (18 years after the first fire) and (b) 300 years (118 years after the first fire and before the second fire).

5. DISCUSSION

In small mountain drainage basins where debris flows are common, such as the study area, large deposits composed of both debris flow and fluvial deposits are the rule rather than the exception, as overlapping parts of the valley network receive debris flows and fluvial deposition from diverse sources. The simulation reproduces this phenomenon.

The model predicts that relatively large quantities of sediment and wood are stored high in the network, in small, headwater streams and their valleys. The amounts of storage predicted by the model are comparable to the volumes observed in the field. This prediction was contingent on reproducing the correct spatial distribution of debris flow inputs to the network, i.e., mimicking the observed distribution of debris flow runout lengths. These large storage volumes predicted by the model are composed of both debris flow and fluvial deposits, similar to what we observe in the field.

The model correctly predicts the existence of nodal points of sediment and wood accumulation, and, in the short term, the relative magnitudes of these storage peaks are close to those of the data (figures 7 and 8a). In the simulation these nodal points of larger storage fluctuate volumetrically but remain stationary and persist over time such that even the average storage has distinct maxima and minima along the profile (Figure 10). Storage nodal points are stationary because deposition leads to more deposition. Even as wood dams decay, impounded sediment is not released all at once. These wood dams lead to depositional zones because deposition behind dams reduces stream gradient and, thus, transport capacity in these zones.

These depositional zones negate the possibility of coherent, wave-like sediment movement. Rather, sediment tends to accumulate at and move between these "sticky spots" in the network that correspond to locations of frequent debris flow deposition, e.g., at changes in valley width and flow direction, both of which often occur at tributary junctions. These deposits can grow over time as they trap sediment

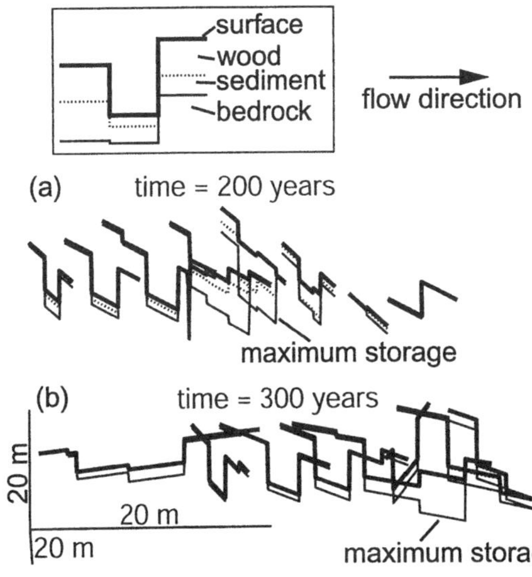

Figure 9. Simulated valley cross sections of the maximum sediment and wood deposit storage at (a) 200 years and (b) 300 years. At the earlier time (a), wood and sediment fill the valley floor and form a dam at the location of maximum storage. Later (b), almost all wood has decayed, but a sediment wedge remains. The channel is incised into the sediment such that most of the sediment at the location of maximum storage is in a terrace next to the channel. Note that the channel takes a different route through this part of the valley between the two times shown.

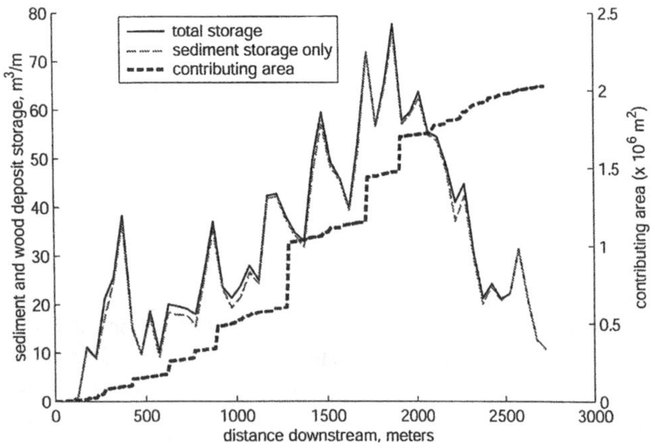

Figure 10. Cross-sectional areas of sediment and wood deposit storage averaged over 50-meter channel segments and samples every 20 years for the 3000 year simulation, and contributing area at 3000 years *vs.* distance downstream. Step-increases in contributing area occur at tributary junctions.

transported from upstream, and shrink as the valley-spanning wood jams that help dam sediment decay.

Small mountain drainage basins, then, add capacitance to the system. The pulsed sediment input to the system does not cause pulsed output, as we might expect if sediment deposits translated as waves, as posited by *Benda and Dunne* [1997b]. Rather, these small basins absorb the sporadic and abrupt inputs of sediment from debris flows and output a relatively smooth fluvial sediment transport signal, as posited by *Massong and Montgomery* [2000]. Management practices that remove wood from the system would likely remove much of this capacitance and dramatically alter the sediment output of small basins to larger, fish-bearing streams.

Complex interactions among sediment, wood, climate, and disturbance regime necessitate use of a complex model, albeit with simple parts, to explore sediment and wood delivery to the channel network, storage patterns, and subsequent transport. We do not presume to test the model as a predictive tool. Rather, the model represents a kind of hypothesis of what processes and phenomena control sediment and wood dynamics and storage in small mountainous drainage basins. Simulation results presented here illuminate some aspects of interactions among these controls while raising new questions, some of which may be addressed by future field studies. For example, our results show a strong interaction between wood and sediment, and we have pointed out that burial of wood may slow its decay. By dating buried wood deposits we hope to quantify this effect.

6. CONCLUSIONS

The simulation indicates, and our field observations confirm, that wood and sediment dynamics are strongly linked. The coupling between sediment and wood implies a strong coupling between forests and channels. Through this coupling, forest conditions "drive" channel conditions, and forests, sediment dynamics, and channels are inextricably linked. Forest management practices that change forest conditions will inevitably change channel conditions and must therefore be carefully tailored to mitigate adverse impacts on riverine habitat.

Acknowledgements. This research was funded by the Coastal Landscape Analysis and Modeling Study (CLAMS), USDA Forest Service, Pacific Northwest Research Station. John Green, Simon Mudd, and Christine May assisted us in the field. Richard Iverson offered helpful advice concerning the debris flow runout model. Arjun Heimsath graciously shared his findings on soil production rates in the Oregon Coast Range. Michael Wimberly offered advice for modeling forest fires. Discussions with Lee Benda, Mauro Casadei, William Dietrich, Daniel Miller, David Montgomery, and Jonathan Stock yielded useful insight. Gregory Tucker, Nicole Gasparini, and Rafael Bras are continuing collaborators with the CHILD model and its offspring. Thorough reviews by William Dietrich, Faith Fitzpatrick, David Montgomery, and Gregory Tucker greatly improved the paper.

REFERENCES

Abbe, T.B., and D.R. Montgomery, Large woody debris jams, channel hydraulics and habitat formation in large rivers, *Reg. Rivers: Res. Mgmt.*, *12*, 201-221, 1996.

Benda, L.E., and T.W. Cundy, Predicting deposition of debris flows in mountain channels, *Can. Geotech. J.*, *27*(4), 409-417, 1990.

Benda, L., and T. Dunne, Stochastic forcing of sediment supply to channel networks from landsliding and debris flow, *Water Resour. Res.*, *33*(12), 2849-2863, 1997a.

Benda, L., and T. Dunne, Stochastic forcing of sediment routing and storage in channel networks, *Water Resour. Res.*, *33*(12), 2865-2880, 1997b.

Burroughs, E.R., Landslide hazard rating for portions of the Oregon Coast Range, in *Proceeding of Symposium on Effects of Forest Land Use on Erosion and Slope Stability*, pp. 265-274, East-West Cent., Univ. of Hawaii, Honolulu, 1984.

Burroughs, E.R., and B.R. Thomas, Declining root strength in Douglas fir after felling as a factor in slope stability, *Res. Pap. INT Intermt. For. Range Exp. Stn. INT-190*, 27 pp., 1977.

Dietrich, W.E., R. Reiss, M.-L. Hsu, and D.R. Montgomery, A process-based model for colluvial soil depth and shallow landsliding using digital elevation data, *Hydrol. Proc.*, *9*, 383-400, 1995.

Duan, J., *A Coupled Hydrologic-Geomorphic Model for Evaluating Effects of Vegetation Change on Watersheds*, Ph.D. Thesis, 133 pp., Oregon State Univ., Corvallis, 1996.

Duan, J., J. Selker, and G.E. Grant, Evaluation of probability density functions in precipitation models for the pacific northwest, *J. Am. Wat. Resour. Assoc.*, *34*(3), 617-627, 1998.

Eagleson, P.S., Climate, soil, and vegetation, 2, the distribution of annual precipitation derived from observed storm sequences, *Water Resour. Res.*, *14*(5), 713-721, 1978.

Engelund, F., and E. Hansen, *A Monograph on Sediment Transport in Alluvial Streams*, Teknisk Forlag, Copenhagen, 1972.

Flint, J.J., Stream gradient as a function of order, magnitude, and discharge, *Water Resour. Res.*, *10*, 969-973, 1974.

Garman, S.L., S.A. Acker, J.L. Ohmann, and T.A. Spies, Asymptotic height-diameter equations for twenty-four tree species in western Oregon, Forest Research Laboratory, Oregon State University, Corvallis, *Research Contribution 10*, 22 pp., 1995.

Govers, G., Evaluation of transporting capacity formulae for overland flow, in *Overland Flow*, ed. By A.J. Parsons and A.D. Abrahams, 243-273, Chapman & Hall, New York, 1992.

Hack, J.T., Studies of longitudinal stream profiles in Virginia and Maryland, *U.S. Geol. Surv. Prof. Pap.*, *294-B*, 1957.

Hammond, C., D. Hall, S. Miller, and P. Swetik, Level 1 stability analysis (LISA), documentation of version 2.0, *U.S. For. Ser. Gen. Tech. Rep. INT-285*, 190 pp., 1992.

Harmon, M.E., J.F. Franklin, F.J. Swanson, P. Sollins, S.V. Gregory, J.D. Lattin, N.H. Anderson, S.P. Cline, N.G. Aumen, J. Sedell, G.W. Lienkaemper, K. Cromack, Jr., and K.W. Cummins, Ecology of coarse woody debris in temperate ecosystems, *Adv. Ecol. Res.*, *15*, 133- 302, 1986.

Heimsath, A.M., W.E. Dietrich, K. Nishiizumi, and R.C. Finkel, The soil production function and landscape equilibrium, *Nature*, *388*, 358-361, 1997.

Heimsath, A.M., W.E. Dietrich, K. Nishiizumi, and R.C. Finkel, Stochastic processes of soil production and transport: erosion rates, topographic variation, and cosmogenic nuclides in the Oregon Coast Range, *Earth Surf. Processes Landforms*, in press, 2001.

Hogan, D.L., S.A. Bird, and M.A. Hassan, Spatial and temporal evolution of small coastal gravel- bed streams: Influence of forest management on channel morphology and fish habitats, in *Gravel-Bed Rivers in the Environment*, ed. by P.C., Klingeman, R.L. Beschta, P.D. Komar, and J.B. Bradley, pp. 365-392, Water Resources Publications, Highlands Ranch, CO, 1998.

Hough, B.K., *Basic Soils Engineering*, 513 pp., Ronald Press, New York, 1957.

Howard, A.D., and Kerby, G., Channel changes in badlands, *Geol. Soc. Amer. Bull.*, *94*, 739-752, 1983.

Huff, M.H., *Post-Fire Succession in the Olympic Mountains, Washington: Forest Vegetation, Fuels, and Avifauna*, Ph.D. dissertation, Univ. Washington, Seattle, 1984.

Ijjasz-Vasquez, E.J., and R.L. Bras, Scaling regimes of local slope versus contributing area in digital elevation models, *Geomorphology*, *12*(4), 299-311, 1995.

Iverson, R.M., The physics of debris flows, *Rev. Geophys.*, *35*(3), 245-296, 1997.

Iverson, R.M., Landslide triggering by rain infiltration, *Water Resour. Res.*, *36*(7), 1897-1910, 2000.

Iverson, R.M., M.E. Reid, and R.G. LaHusen, Debris-flow mobilization from landslides, *Annu. Rev. Earth Planet. Sci.*, *25*, 85-138, 1997.

Iverson, R.M., R.P. Denlinger, R.G. LaHusen, and M. Logan, Two-phase debris flow across 3-D terrain: model predictions and experimental tests, *Proceedings, Second International Conference on Debris-flow Hazards Mitigation*, Taiwan, 2000.

Kochel, R.C., D.F. Ritter, and J. Miller, Role of tree dams in the construction of pseudo-terraces and variable geomorphic responses to floods in Little River valley, Virginia, *Geology*, *15*, 718-721, 1987.

Lancaster, S.T., *A Nonlinear River Meandering Model and its Incorporation in a Landscape Evolution Model*, Ph.D. Thesis, 277 pp., Massachusetts Institute of Technology, Cambridge, 1998.

Leopold, L.B., and T. Maddock, Jr., The hydraulic geometry of stream channels and some physiographic implications, *U.S. Geol. Surv. Prof. Pap. 252*, 57 pp., 1953.

Lienkaemper, G.W., and F.J. Swanson, Dynamics of large woody debris in streams in old-growth Douglas-fir forests, *Can. J. For. Res.*, *17*, 150-156, 1987.

Lisle, T.E., J.M. Nelson, J. Pitlick, M.A. Madej, and B.L. Barkett, Variability of bed mobility in natural, gravel-bed channels and adjustments to sediment load at local and reach scales, *Water Resour. Res.*, *36*(12), 3743-3755, 2000.

Lisle, T.E., J.E. Pizzuto, H. Ikeda, F. Iseya, and Y. Kodama, Evolution of a sediment wave in an experimental channel, *Water Resour. Res.*, *33*(8), 1971-1981, 1997.

Long, C.J., C. Whitlock, P.J. Bartlein, and S.H. Millspaugh, A 9000-year fire history from the Oregon Coast Range, based on a high-resolution charcoal study, *Can. J. For. Res.*, *28*, 774-787, 1998.

Madej, M.A, and V. Ozaki, Channel response to sediment wave propagation and movement, Redwood Creek, California, USA, *Earth Surf. Processes Landform, 21*(10), 911-927, 1996.

Massong, T.M., and D.R. Montgomery, Influence of sediment supply, lithology, and wood debris on the distribution of bedrock and alluvial channels, *Geol. Soc. Amer. Bull.*, *112*(5), 591-599, 2000.

May, C.L., *Debris Flow Characteristics Associated with Forest Practices in the Central Oregon Coast Range*, M.S. Thesis, 121 pp., Oregon State Univ., Corvallis, 1998.

Means J.E., and T.E. Sabin, Height growth and site index curves for Douglas-fir in the Siuslaw National Forest, Oregon, *Western Journal of Applied Forestry*, *4*(4), 136-142, 1989.

Moglen, G., and R.L. Bras, The effect of spatial heterogeneities on geomorphic expression in a model of basin evolution, *Water Resour. Res.*, 31(10), 2613-2623, 1995.

Montgomery, D.R., and W.E. Dietrich, Where do channels begin?, *Nature*, *336*, 232-234, 1988.

Montgomery, D.R., and W.E. Dietrich, Channel initiation and the problem of landscape scale, *Science*, *255*, 826-830, 1992.

Montgomery, D.R., and W.E. Dietrich, A physically based model for the topographic control on shallow landsliding, *Water Resour. Res.*, *30*(4), 1153-1171, 1994.

Montgomery, D.R., T.B. Abbe, J.M. Buffington, N.P. Peterson, K.M. Schmidt, and J.D. Stock, Distribution of bedrock and alluvial channel in forested mountain drainage basins, *Nature*, *318*, 587-589, 1996.

Montgomery, D.R., W.E. Dietrich, R. Torres, S.P. Anderson, J.T. Heffner, and K. Loague, Hydrologic response of a steep unchanneled valley to natural and applied rainfall, *Water Resour. Res.*, *33*, 91-109, 1997.

Montgomery, D.R., K.M. Schmidt, H.M. Greenberg, and W.E. Dietrich, Forest clearing and regional landsliding, *Geology*, *28*(4), 311-314, 2000.

Reeves, G.H., P.A. Bisson, and J.M. Dambacher, Fish communities, in *River Ecology and Management: Lessons from the Pacific Coastal Ecoregion*, ed. by R.J. Naiman and R.E. Bilby, pp. 200-234, Springer-Verlag, New York, 1998.

Reneau, S.L., W.E. Dietrich, M. Rubin, D.J. Donahue, and A.J.T. Jull, Analysis of hillslope erosion rates using dated colluvial deposits, *J. Geol.*, *97*, 47-63, 1989.

Reneau, S.L., and W.E. Dietrich, Erosion rates in the southern Oregon Coast Range: Evidence for an equilibrium between hillslope erosion and sediment yield, *Earth Surf. Processes Landforms*, *16*, 307-322, 1991.

Richards, F.J., A flexible growth function for empirical use, *J. Exp. Biol.*, *10*, 290-300, 1959.

Robison, E.G., and R.L. Beschta, Identifying trees in riparian areas that can provide coarse woody debris to streams, *For. Sci.*, *36*(3), 790-801, 1990.

Robison, E.G., K. Mills, J. Paul, L. Dent, and A. Skaugset, Storm impacts and landslides of 1996: final report, *For. Pract. Tech. Rep. No. 4*, 145 pp., Oregon Department of Forestry, 1999.

Roering, J.J., J.W. Kirchner, and W.E. Dietrich, Evidence for nonlinear, diffusive sediment transport on hillslopes and implications for landscape morphology, *Water Resour. Res.*, *35*(3), 853-870, 1999.

Schmidt, K.M., J.J. Roering, J.D. Stock, W.E. Dietrich, and D.R. Montgomery, The variability of root cohesion as an influence on shallow landslide susceptibility in the Oregon Coast Range, *Can. Geotech. J.*, in press, 2001.

Schroeder, W.L., and J.V. Alto, Soil properties for slope stability analysis: Oregon and Washington Coastal Mountains, *For. Sci.*, *29*, 823-833, 1983.

Sidle, R.C., A conceptual model of changes in root cohesion in response to vegetation management, *J. Environ. Qual.*, *20*, 43-52, 1991.

Sidle, R.C., A theoretical model of the effects of timber harvesting on slope stability, *Water Resour. Res.*, *28*(7), 1897-1910, 1992.

Snyder, N.P., K.X Whipple, G.E. Tucker, and D.J. Merritts, Landscape response to tectonic forcing: Digital elevation model analysis of stream profiles in the Mendocino triple junction region, northern California. *Geol. Soc. Amer. Bull.*, *112*(8), 1250-1263, 2000.

Spies, T.A., J.F. Franklin, and T.B. Thomas, Coarse woody debris in Douglas-fir forests of western Oregon and Washington, *Ecology*, *69*(6), 1689-1702, 1988.

Swanson, F.J., M.M. Swanson, and C. Woods, *Inventory of mass erosion in the Mapleton Ranger District Siuslaw National Forest*, a cooperative study of the Siuslaw National Forest and the PNW Forest and Range Experiment Station, 1977.

Tarboton, D.G., R.L. Bras, and I. Rodriguez-Iturbe, On the extraction of channel networks from digital elevation data, *Hydrol. Processes*, *5*, 81-100, 1991.

Tucker, G.E., and R.L. Bras, Hillslope processes, drainage density, and landscape morphology, *Water Resour. Res.*, *34*, 2751-2764, 1998.

Tucker, G.E., and R.L. Bras, A Stochastic Approach to Modeling the Role of Rainfall Variability in Drainage Basin Evolution, *Water Resour. Res.*, *36*(7), 1953-1964, 2000.

Tucker, G.E., S.T. Lancaster, N.M. Gasparini, R.L. Bras, and S.M. Rybarczyk, An object-oriented framework for hydrologic and geomorphic modeling using triangulated irregular networks, *Comp. Geosci.*, in press, 2001a.

Tucker, G.E., S.T. Lancaster, N.M. Gasparini, and R.L. Bras, The channel-hillslope integrated landscape development (CHILD) model, *Landscape Erosion and Landscape Evolution Modeling*, GSA special volume, ed. by R.S. Harmon and W.W. Doe, III, in press, 2001b.

Van Sickle, J., and S.V. Gregory, Modeling inputs of large woody debris to streams from falling trees, *Can. J. For. Res.*, *20*(10), 1593-1601, 1990.

Vanoni, V.A., *Sedimentation Engineering*, A.S.C.E., New York, 1975.

Willgoose, G., A statistic for testing the elevation characteristics of landscape simulation models, *J. Geophys. Res.*, *99*(B7), 13,987-13,996, 1994.

Wimberly, M.C., T.A. Spies, C.J. Long, and C. Whitlock, Simulating historical variability in the amount of old forests in the Oregon Coast Range, *Conservation Biology*, *14*(1), 167-180, 2000.

Stephen T. Lancaster, Shannon K. Hayes and Gordon E. Grant, Forestry Sciences Laboratory, 3200 SW Jefferson Way, Corvallis, OR 97331

Section 3

Anthropogenically Altered Systems

Stream Geomorphology and Fish Community Structure in Channelized and Meandering Reaches of an Agricultural Stream

Kelly M. Frothingham

Department of Geography and Planning, Buffalo State College, Buffalo, New York

Bruce L. Rhoads

Department of Geography, University of Illinois, Urbana, Illinois

Edwin E. Herricks

Department of Civil and Environmental Engineering, University of Illinois, Urbana, Illinois

In environments dominated by human activity, such as the agricultural Midwest, channel morphology is strongly influenced by anthropogenic factors. Past research has shown that human-induced channel modifications, including stream channelization, affects both the abiotic and biotic components of a stream. However, connections between changes in geomorphological form and function and ecological conditions are still poorly understood. Knowledge of these connections is critical for successful approaches to stream naturalization. The objective of this research is to explore the linkage between geomorphological variability and ecological conditions in adjacent channelized and unchannelized reaches of an agricultural stream in Illinois. Geomorphological and fisheries data were collected in the headwaters of the Embarras River where channel maintenance for agricultural drainage produced a straight channelized reach adjacent to a meandering unchannelized reach. Site characterization involved topographic mapping of each reach and establishing permanent cross sections to document the full range of morphologic variability at the sites. Fish sampling was conducted approximately monthly to provide information on community structure in the meandering and channelized reaches. The meandering reach has greater morphological variability, both over time and space, than the channelized reach. The meandering reach also contains more fish and larger individual fish than the channelized reach, suggesting that increased geomorphological complexity results in increased fish abundance and total biomass. Fisheries analysis also indicates that fish-community structure varies temporally in both reaches with the greatest contrast occurring seasonally.

INTRODUCTION

Human modification of watersheds has produced profound changes in streams throughout the agricultural Midwest. A growing emphasis on environmental quality in this

Geomorphic Processes and Riverine Habitat
Water Science and Application Volume 4, pages 105-117

region of the United States has generated concern about the effects of humans on stream ecosystems. The concept of stream naturalization has been proposed as a management concept to address human and biophysical dimensions of environmental-quality issues in this human-dominated environment [*Rhoads* and *Herricks*, 1996; *Rhoads* and *Monahan*, 1997; *Rhoads et al.*, 1999]. Naturalization aims to establish morphologically and hydraulically varied, yet dynamically stable [*Rhoads*, 1995] stream systems that are capable of supporting biologically diverse aquatic ecosystems. Successful implementation of naturalization strategies requires an improved understanding of the geomorphological and ecological dynamics of human-modified streams [*Frothingham et al.*, in press], as well as the effects of social interactions between scientists and non-scientists in community-based environmental decision-making [*Rhoads et al.*, 1999].

Fluvial geomorphologists and aquatic ecologists both are interested in stream dynamics; however, the focus of these two disciplines differs in important ways. Geomorphologists primarily are interested in the interaction among flow properties, sediment transport, and channel morphology [*Dietrich*, 1987; *Rhoads* and *Welford*, 1991], whereas ecologists typically are concerned about the structure and function of aquatic communities [*Osborne* and *Wiley*, 1992; *Aadland*, 1993]. The potential linkage between geomorphology and ecology centers on a common interest in the physical properties of streams where geomorphological conditions define the amount, diversity and structure of physical habitat. Ecologists recognize that spatial variability of physical properties, such as mean velocity, flow depth and substrate characteristics, has a strong influence on relationships between habitat and community structure and function [*Aadland*, 1993; *Beebe*, 1996; *Gore* and *Hamilton*, 1996; *Fisher*, 1997]. Channel morphology also has been acknowledged as an important factor in habitat diversity and species richness in streams [*Southwood*, 1977; *Frissell et al.*, 1986; *Gelwick*, 1990; *Osborne* and *Wiley*, 1992; *Aadland*, 1993; *Fisher*, 1997; *Kemp et al.*, 1999; *Montgomery et al.*, 1999]. Fluvial geomorphologists, on the other hand, have developed theories and techniques related to the dynamic interaction between fluvial forms and processes [*Hooke* and *Harvey*, 1983; *Dietrich*, 1987; *Rhoads* and *Welford*, 1991]. These theories provide a foundation for an improved understanding of physical habitat and time-related changes in habitat conditions, but integration of geomorphological and ecological analysis has yet to be achieved for low-gradient streams in the agricultural Midwest.

Human-induced channel modification, referred to as stream channelization, affects both the abiotic and biotic components of a stream. Stream channelization results in decreased spatial variation of channel morphology and 3-D flow structure [*Dietrich*, 1987; *Rhoads* and *Welford*, 1991; *Rhoads* and *Urban*, 1997]. Physical processes impacted by channelization include pool-riffle development and alternate bar formation [*Rhoads* and *Welford*, 1991; *Gregory et al.*, 1994]. Channelization also changes aquatic communities, reducing species richness and altering local community structure and function [*Brookes*, 1988; *Swales*, 1988]. The reduced species richness in channelized streams has been attributed mainly to decreased physical habitat diversity associated with uniform channel morphology [*Frissell et al.*, 1986; *Gelwick*, 1990; *Osborne* and *Wiley*, 1992; *Aadland*, 1993].

In watersheds of East Central Illinois, stream channelization typically occurs in a piecemeal fashion, whereby only a few kilometers of a stream are modified in any particular project. As a result, streams are impacted most severely at the planform scale, i.e. the scale at which the planform characteristics of streams can be clearly discerned [*Rhoads* and *Herricks*, 1996; *Frothingham et al.*, in press]. Differences in the spatial extent and timing of channelization has produced a diverse array of channel morphologies [*Rhoads* and *Herricks*, 1996], but the exact relation between geomorphological form and function and ecological conditions is still poorly understood. Such knowledge is critical for naturalization of these stream systems, which depends on an integrated eco-geomorphological approach to watershed management [*Frothingham et al.*, in press].

The objective of this research is to explore how channel morphology is related to stream ecology using fish-community characteristics as an indicator of stream ecosystem condition or state. To achieve this objective, geomorphological and fish-community data are compared for an adjacent straight channelized reach and an unmodified meandering reach of an agricultural stream in East Central Illinois. The central hypothesis of this study is that increased spatial and temporal variability in stream geomorphology results in increased fish species abundance, biomass and richness.

STUDY AREA

Geomorphological and fisheries data were collected from adjacent meandering and channelized reaches in the headwaters of the Embarras River in Champaign County, IL (Figure 1). The Embarras River originates on the Urbana moraine at the southern edge of the Champaign-Urbana metropolitan area and flows south between the West Ridge and Philo moraines. The two study sites are located approximately 10 km south of Champaign-Urbana.

Channelization of the Embarras River began in the late 19th century and over the last 100 years many sections have been subject to repeated maintenance. The meandering reach is situated within an 800-m section of the Embarras

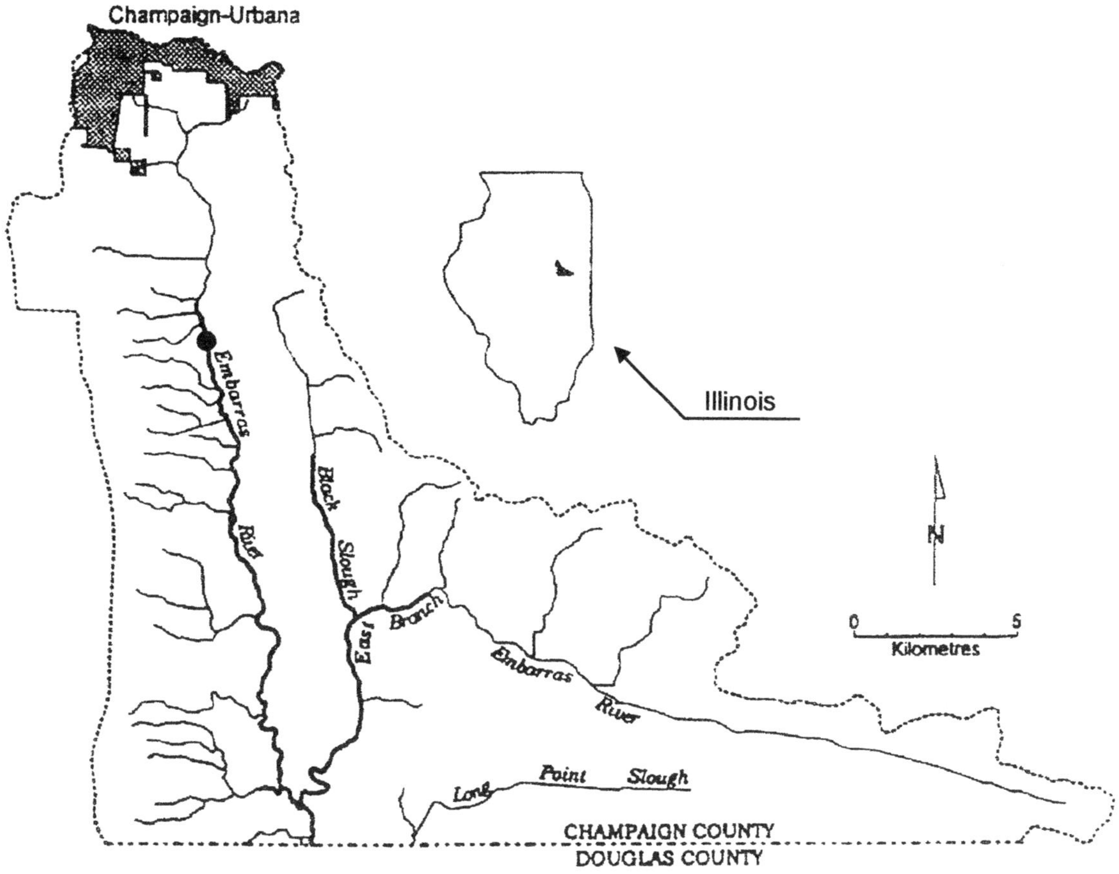

Figure 1. The Embarras River showing the location of the meandering and channelized reaches (•).

River that has not been channelized since at least the 1930s. This 800-m section represents unmodified channel conditions in the sense that channel dimensions and planform characteristics have fully adjusted to any past channelization and the river channel has been dynamically stable over the past several decades. The study reach within this unmodified section consists of a high-amplitude meander bend approximately 70 m in length (Figure 2). Bankfull channel width ranges from 13.0 to 18.0 m and bankfull depth is between 1.0 and 2.0 m. The channelized reach is located immediately downstream of the 800-m long unchannelized section. It is located at the upstream end of a 275-m section of the river that was straightened between 1973 and 1982 [*Rhoads* and *Herricks*, 1996; *Rhoads* and *Urban*, 1997]. This 275-m channelized section subsequently remained straight, but in 1996 channel maintenance was conducted to deepen the channel by 0.65 m and to regrade the channel banks to 3:1 side slopes. The channelized reach is 50 m long and has a trapezoidal channel cross section. Bankfull width varies between 13.0 and 16.0 m and bankfull depth is approximately 2.0 m (Figure 3).

METHODOLOGY

The field measurement program for each reach included: (1) mapping in detail the channel morphology, (2) documenting changes in channel form, and (3) electrofishing to assess fish community structure. Channel morphology was mapped in June 1997 using a total station. Two local benchmarks were established at each site for horizontal and vertical control; survey data were reduced to Cartesian (X,Y,Z) coordinates. Contour maps for each reach were generated from the coordinate information using the computer software package Surfer, which defines contours for a uniform grid of elevation data interpolated via kriging.

Changes in channel morphology were determined by conducting repeat surveys at monumented cross sections oriented orthogonally to the planform alignment of the bankfull channel (Figures 2 and 3). The cross sections encompass the full range of morphologic variability in the two reaches. Repeat surveys of the cross sections were conducted three times between October 1997 and June 1998.

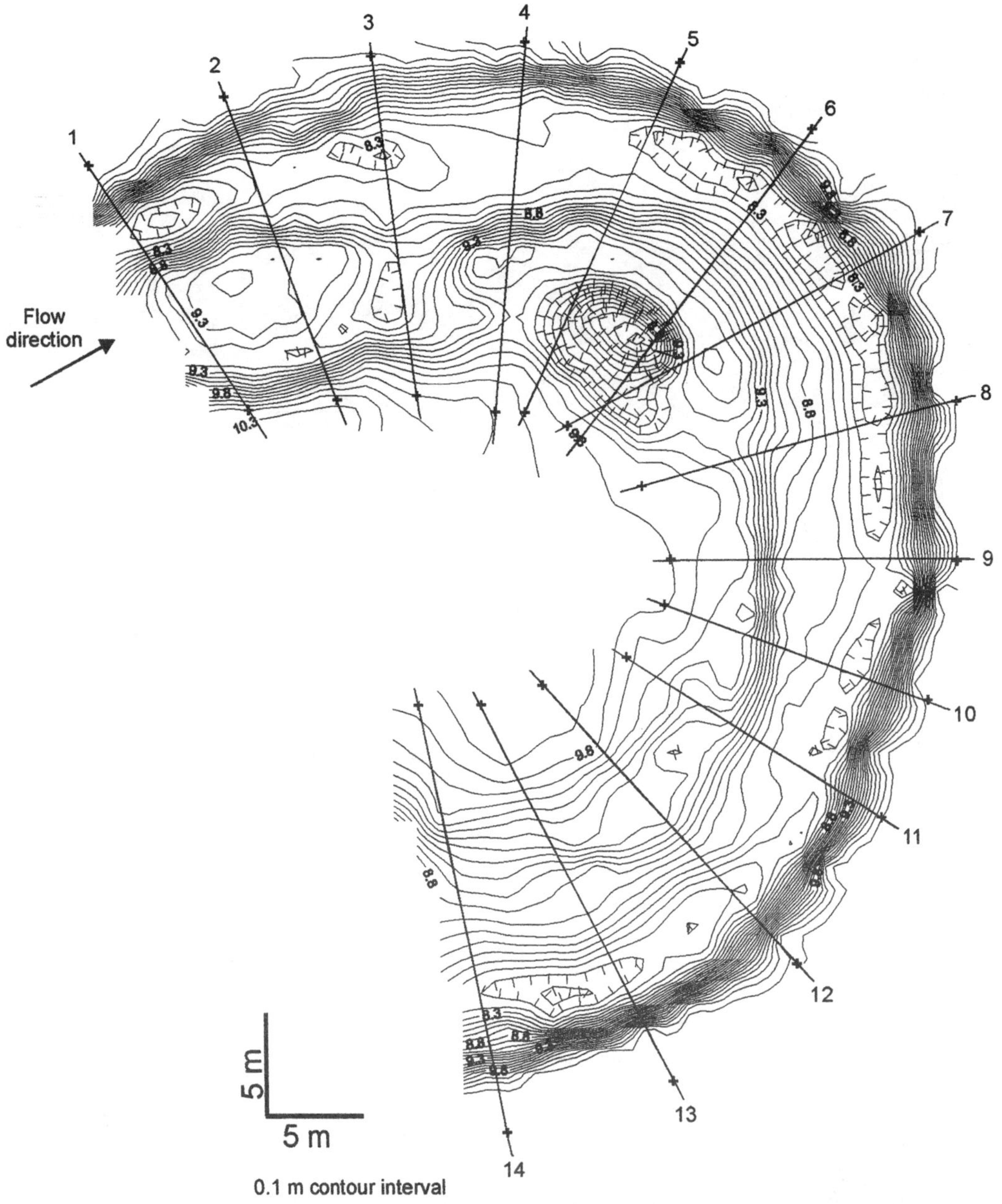

Figure 2. Bed topography of the meandering reach and locations of permanent cross-sections. Cross sections are oriented orthogonally to the local channel direction as defined by an arc of curvature along the centerline of the bankfull channel.

Spatial variability of channel morphology within each reach was evaluated qualitatively by visually inspecting variations in bed elevation on contour maps. Temporal variability in channel morphology was defined by changes in the profiles of surveyed cross sections. Sediment dynamics were quantified in terms of net sediment accumulation or removal by volume within each reach between channel surveys. Calculations of net volumetric sediment flux were performed in Surfer by interpolating a uniform grid of elevation data from the survey information using kriging, and then differencing the interpolated elevation values for each grid node for successive surveys. Negative flux volumes indicate net deposition and positive flux volumes indicate net erosion.

Fish community sampling was conducted on a quasi-monthly basis from July 1997 through December 1998 at the straight and meandering reaches. Ten collections were obtained in the meandering and straight reaches [*Tompkins*,

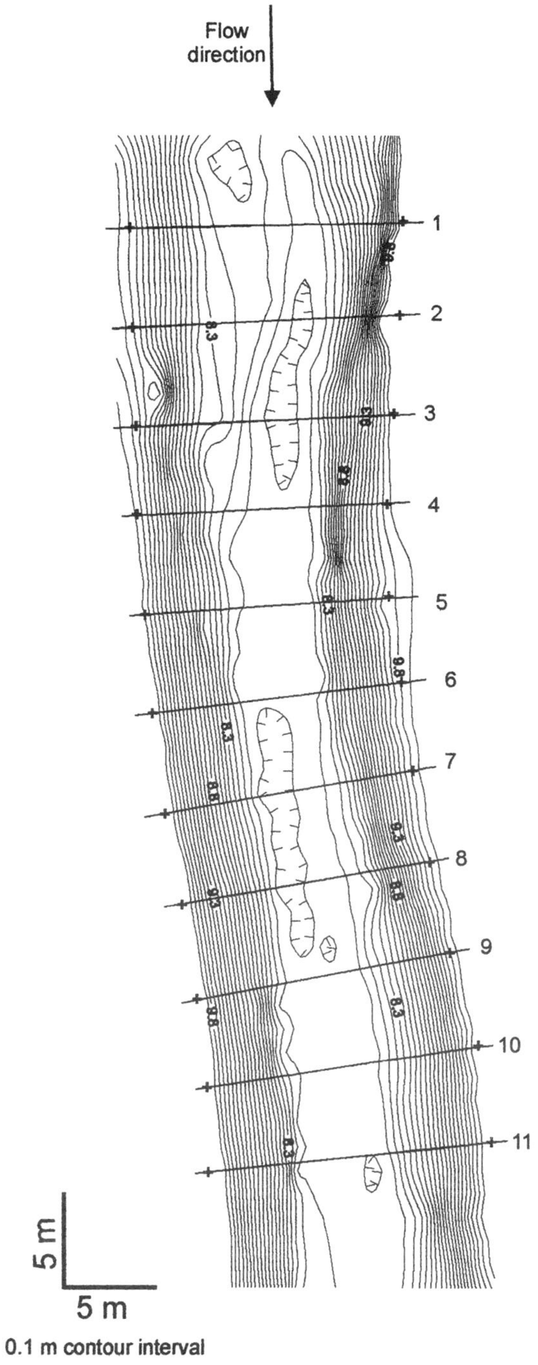

Figure 3. Bed topography of the channelized reach and locations of permanent cross-sections. Cross sections are oriented orthogonally to the local channel direction as defined by the linear path of the channel centerline.

1998; *Hauser*, 1999]. Samples were collected using a Coffelt Mark 10 backpack electro-shocker and block nets were installed at the upstream and downstream ends of a reach to ensure that fish could not escape. Sampling campaigns were conducted only during sub-bankfull conditions and all fish collections were obtained between 8:00 a.m. and 5:00 p.m. Fish were identified, weighed and measured. Fisheries data from both reaches were analyzed through general assessment of species number and length.

RESULTS

Spatial Variability in Channel Morphology

Topographic mapping indicates that the meandering reach has greater spatial variability in channel morphology, and thus physical habitat, than the channelized reach (Figures 2 and 3). Topographic highs and lows occur throughout the meandering reach (Figure 2). Point bars are present along the inner bank of the upstream (cross-sections 5-8) and downstream (cross-sections 12-14) parts of the reach. Pools exist at the bend entrance (cross-section 1-3) and opposite the point bars at bend apices (cross-sections 6-8 and 12-14). The point bar-pool morphology produces pronounced asymmetry of channel cross sections, which are characterized by a nearly vertical cut bank and a gently sloping point-bar face (Figure 2). Cross-stream relief of the channel bed is as much as 0.75 m at the bend apices. Riffles between the pools have fairly symmetrical channel geometry (cross-sections 3-5 and 9-11). Longitudinal variation in the thalweg elevation between the riffles and pools is about 0.5 m.

The geometry of the channel in the channelized reach is essentially trapezoidal, reflecting recent channelization (Figure 3). Channel banks have uniform side slopes that extend to the edge of a relatively flat channel bottom. The elevation of the channel bed varies less than 0.1 m, except at the upstream end of the reach where a bar along the west bank produces local cross-stream relief of about 0.45 m. Longitudinal variation in bed elevation along the thalweg is less than 0.25 m throughout the reach.

Spatial Variability in the Fish Community

The fish collections were analyzed to assess differences in community structure between the meandering and the channelized reaches (Table 1). Species richness is similar in the channelized and meandering reaches, but fish abundance and biomass is greater in the meandering reach than in the channelized reach (Figures 4 and 5, Table 1). The fish community in these two reaches is composed mainly of mixed cyprinids, including various minnows and carp, and some additional benthic species like yellow bullhead and golden redhorse (Figures 4 and 5). Average fish abundance for the meandering reach exceeds that of the channelized reach by about 25%. Fish in the meandering reach also are, on average, twice as large as those in the channelized reach (Table 1).

Table 1. Fish collection summary results

Site	Meandering reach	Channelized reach
Total number of species	28	27
Range of number of fish collected per sampling campaign	192-850	151-495
Average number of fish collected per sampling campaign	434	346
Average weight of fish weighing greater than one gram	15 grams	6 grams
Percent of fish weighing less than one gram	4%	11%

Further analysis of fish collection data provides additional insight into the effects of spatial variability in geomorphological conditions on fish community structure. This detailed analysis examines the average number of fish, average length and difference ratios for number and length for the six most common species in the two reaches (Table 2). A positive value for the difference ratio indicates a greater value in the meandering reach; correspondingly, a negative value indicates a greater value in the channelized reach. For all species evaluated, with the exception of the Creek chub, more specimens were collected in the meandering reach and the average length is larger in the meandering reach than in the channelized reach. Although there is little difference in the annual community composition in morphologically varied and morphologically uniform reaches, there are clear differences in fish size/biomass characteristics. Fish in the meandering reach are larger, and generally occur in greater numbers than fish in the channelized reach.

Temporal Variability in Channel Morphology

The repeat surveys of monumented cross sections show that the meandering reach exhibits greater temporal change in channel morphology than the channelized reach (Figures 6 and 7). At the entrance to the meandering reach (cross-section 1), erosion of a bar platform along the inner bank and deposition on the channel bed has transformed the geometry of the pool at this location (Figure 6). The most prominent channel change in the meandering reach involves systematic erosion of the outer cut bank, especially at bend apices. Bank erosion at the two apices (cross-sections 6 and 12) exceeds one meter per year (Figure 6). Maximum erosion has occurred at the downstream end of the reach (cross-section 14) where the outer bank has moved by about 2 m between 1997 and 1998. The cut bank erodes mainly through slab failures that introduce large blocks of bank material into the stream. These blocks alter local flow conditions near the bank and enhance physical-habitat diversity for fish. Net deposition is evident on point bars along the inner banks of the upstream (cross-section 6) and downstream (cross-sections 12) bends.

Changes in channel cross sections in the channelized reach are small and localized compared to those in the meandering reach (Figure 7). Minor systematic erosion is evident at the toe of the right (south) bank where flow is deflected laterally toward this side of the channel by the bar at the upstream end of the reach (cross-section 5). Overall, however, channel form in this reach is highly stable.

Temporal Variability in Sediment Dynamics

Calculations of net sediment flux within each reach confirm that the meandering reach is more dynamic over time than the channelized reach (Table 3). Two time intervals were used to illustrate temporal variability in sediment flux: October 1997 to early June 1998 (June 1 and June 5 for the straight and highly sinuous reach, respectively) and early June to late June 1998 (June 25). The first eight-month period provided information for the entire winter/spring period, which included a variety of hydrological events ranging from baseflow to overbank conditions. The second, twenty-day interval was characterized by sustained near-bankfull to bankfull flow in the two reaches.

Erosion was the dominant process in both reaches; however, the volume of net erosion is greater in the meandering reach than in the channelized reach (Table 3). During the first interval net erosion in the meandering reach was an order of magnitude greater than in the channelized reach, whereas during the second interval it was about two times greater than in the channelized reach. The positive sediment flux for the meandering reach suggests that, over the period of record, erosion of cut banks exceeded deposition on point bars. Net sediment flux in the meandering reach for the twenty-day period in June 1998 is only slightly less than the net flux for the eight-month period from October 1997 until June 1998. This similarity in net flux illustrates the important influence of sustained bank-

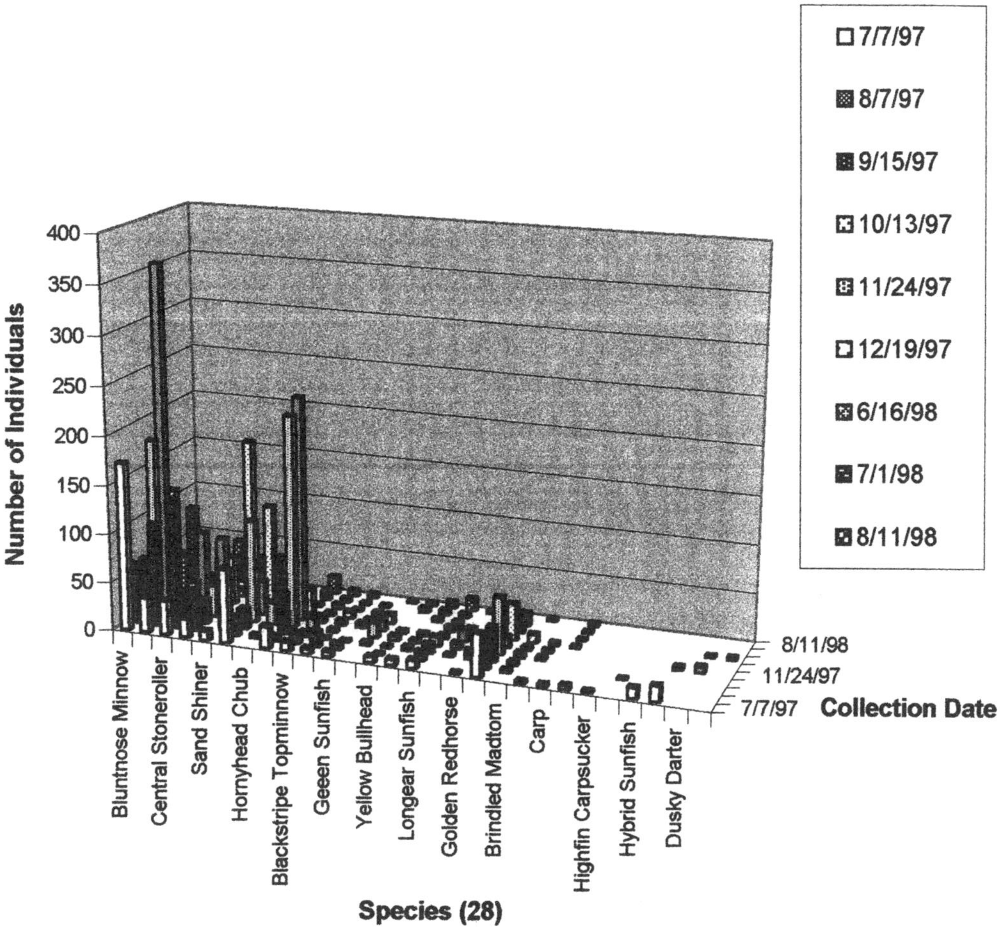

Figure 4. Fish community composition of the meandering reach.

full flow on the rate of channel-shaping processes. The net flux is nearly zero for the eight-month period at the channelized reach, suggesting that erosion and deposition approximately were balanced during this period of variable flow conditions. The channelized reach apparently was able to transfer downstream sediment evacuated from the meandering reach upstream. The comparatively large positive net flux during the twenty-day period confirms that bankfull events most strongly influence the rate of sediment dynamics in the channelized reach. The difference in net sediment flux between the channelized and meandering reaches indicates that the greater spatial variability of channel morphology in the meandering reach is associated with higher temporal rates of erosional and depositional processes.

Temporal Variability in Fish Community Structure

Given the similarity in community composition between the meandering and channelized reaches (Table 1), analysis of temporal variability of fish-community composition was performed for the meandering reach only. In this analysis, the fisheries collections from the meandering reach were used to cluster different collection dates based on the similarity in community composition (Jaccard coefficient of similarity for presence/absence data) and fish abundance (Euclidean distance coefficient for presence/abundance data). Both dendrograms indicate that the fish community changes through time (Figures 8 and 9). Adjacent months often are the most similar, whereas the largest dissimilarities generally are associated with different seasons. Therefore, the changes observed in community structure in the meandering reach are likely related to short term phenomenon such as changing flow stage and water temperature and seasonal migration or movement of fish [*Chambers*, 1994].

DISCUSSION

The findings of this study generally are consistent with those from other investigations comparing stream mor-

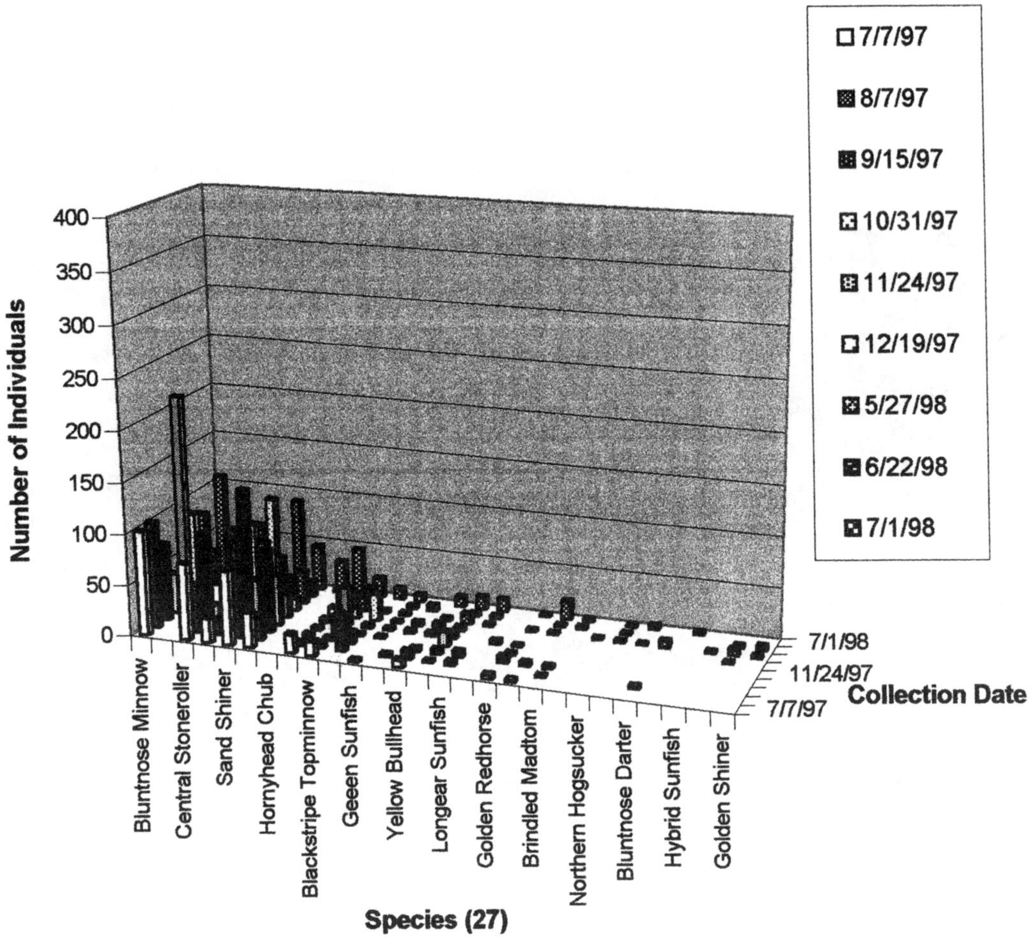

Figure 5. Fish community composition of the channelized reach.

Table 2. Fish collection and difference analysis comparison for adults

Species	Average # collected unmodified meandering	Average # collected straight trapezoidal	Average # collected difference ratio (S:N)	Average length un-modified meandering (mm)	Average length straight trapezoidal (mm)	Average length ratio (S:N)
Bluntnose Minnow	158	103	**0.53**	57	45	0.79
Central Stoneroller	56	45	**0.20**	66	57	0.86
Creek Chub	39	44	**-0.13**	98	63	0.64
Greenside Darter	4	1	**0.75**	71	60	0.85
Longear Sunfish	9	1	**0.89**	89	71	0.80
Striped Shiner	30	3	**0.90**	100	70	0.70

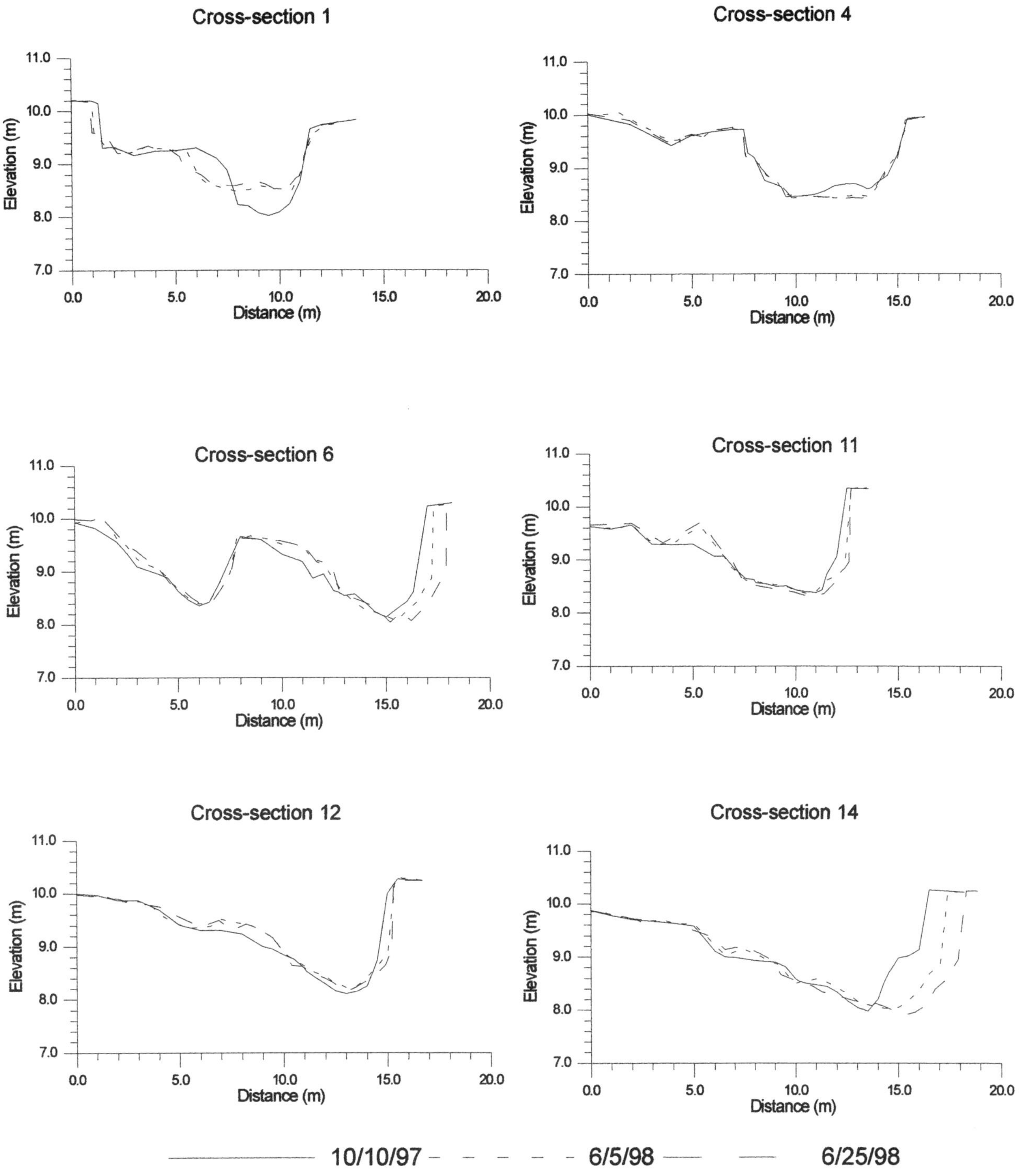

Figure 6. Changes in channel cross sections at the meandering reach for the period October 1997 through June 1998.

phologic variability and fish-community characteristics [*Gorman* and *Karr*, 1978; *Schlosser*, 1982; *Swales*, 1988; *Gelwick*, 1990; *Lobb* and *Orth*, 1991]. Results support the hypothesis that increased spatial variability in channel morphology results in improved habitat for fish. The meandering reach of the Embarras River, which has greater spatial variability in channel morphology than the channelized reach, contained more and larger fish than the channelized reach. This association suggests that availability of physical habitat, defined on the basis of stream morphol-

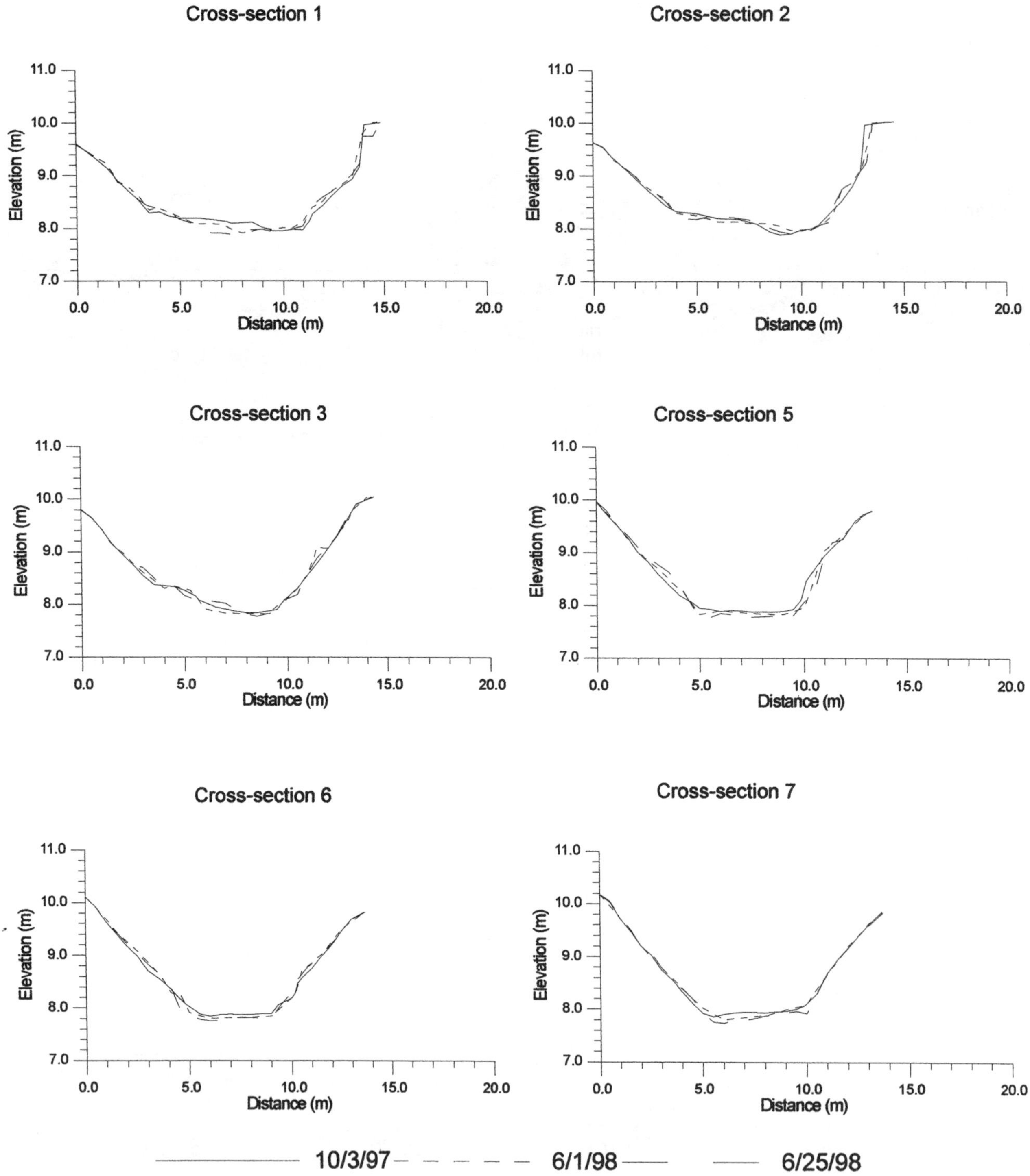

Figure 7. Changes in channel cross sections at the channelized reach for the period October 1997 through June 1998.

ogy, has a detectable influence on ecological conditions in human-modified agricultural streams. No substantial difference in species richness was observed between the channelized and meandering sections of the Embarras River. Other work has shown that species richness in the Embarras River has remained relatively stable over the last 100 years [*Hauser*, 1999]. The similarity in community composition in this study reflects the mobility of fish and, considering that the sampled reaches are adjacent to one another, is not surprising. While the fish community in the

Table 3. Sediment flux results

Site	Date	Volume of erosion (m^3)	Volume of deposition (m^3)	Net volume change (m^3m^{-1})
Meandering reach	10/10/97-6/5/98	83.7	36.7	0.67
	6/5/98-6/25/98	55.8	17.3	0.55
Channelized reach	10/3/97-6/15/98	16.8	17.5	0.01
	6/1/98-6/25/98	26.1	11.0	0.29

meandering reach changed over time, the relationship of this temporal variability to channel morphology is uncertain. Fish community conditions were affected by short-term factors (i.e., weather-related hydrologic variability) and longer-term phenomenon (i.e., seasonal hydrological variability and temperature regimes). Although channel morphology/habitat changed during the period of fish sampling, the effects of these changes on fish-community structure could not be discerned from the type of information collected in this study.

The linkage between physical-habitat and fish community structure documented in this investigation represents a first step in the development of a knowledge base to guide integrated eco-geomorphological approaches to stream naturalization in East Central Illinois [*Frothingham et al.*, in press]. Because naturalization may involve myriad management strategies, ranging from passive recovery to ecologically sensitive channel maintenance, a critical need exists for adequate knowledge to define the ecological benefits that will accrue from a particular style of naturalization. Without such knowledge, both the formulation of ecological goals and assessments of the capacity to attain these goals will be uninformed, increasing the risk of project failure. Concern about failure is especially acute in East Central Illinois where most naturalization projects are likely to be undertaken by drainage-district commissioners, who are concerned first and foremost about drainage, but who also have an interest in being environmental stewards [*Rhoads et al.*, 1999]. The money to support drainage-district projects is raised through taxes imposed on landowners within a district. The commissioners feel obligated to use their neighbors tax dollars wisely and to demonstrate that expenditures produce tangible benefits.

The findings of this study also support the contention of drainage-district commissioners that a management trade-off exists between the static stability of a channelized reach and the dynamic stability of unmodified stream channels. Streams in East Central Illinois have low values of stream power [*Rhoads* and *Herricks*, 1996] and generally do not exhibit dramatic post-channelization adjustment, such as downcutting and channel widening [e.g., *Harvey* and *Watson*, 1986; *Simon* and *Hupp*, 1990] or rapid re-establishment of meanders [e.g. *Brookes*, 1988]. Many channelized streams in this region are statically stable [*Rhoads* and Urban, 1997; *Urban*, 2000]--a consequence that is viewed as desirable from a land-drainage perspective, but as undesirable ecologically. The lack of geomorphological recovery from human-induced disturbance prevents the development of physical habitat within morphologically homogenous channelized streams.

The findings here show that the dynamic stability of the unmodified meandering reach provides diverse physical habitat that locally enhances the abundance and biomass of fish communities. One approach to naturalization in East Central Illinois is to preserve meandering segments as "habitat islands" within extensively modified agricultural stream systems [*Frothingham et al.*, in press]. This approach was used to protect the 800-m meandering segment from channelization in 1996 [*Rhoads* and *Herricks*, 1996]. However, drainage–district commissioners still perceive erosion rates in this section of the river as excessive [*Rhoads et al.*, 1999]. This study confirms that the mean-

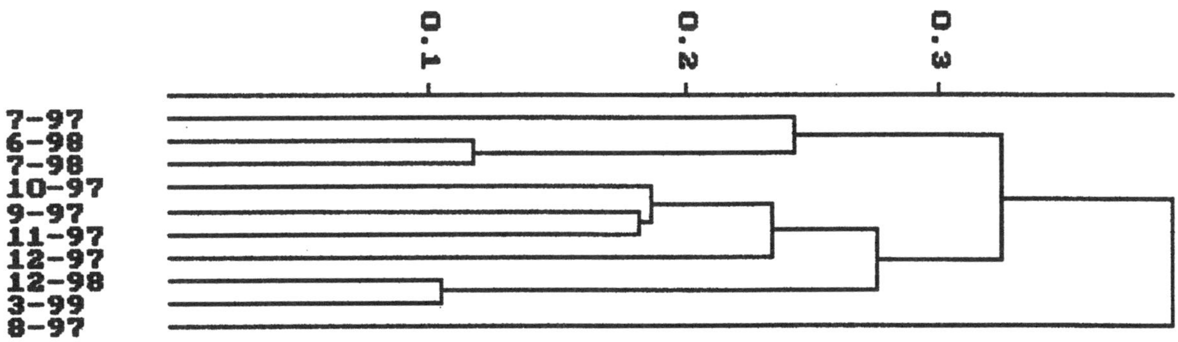

Figure 8. Meandering reach fisheries presence/absence dendrogram (Jaccard Dissimilarity).

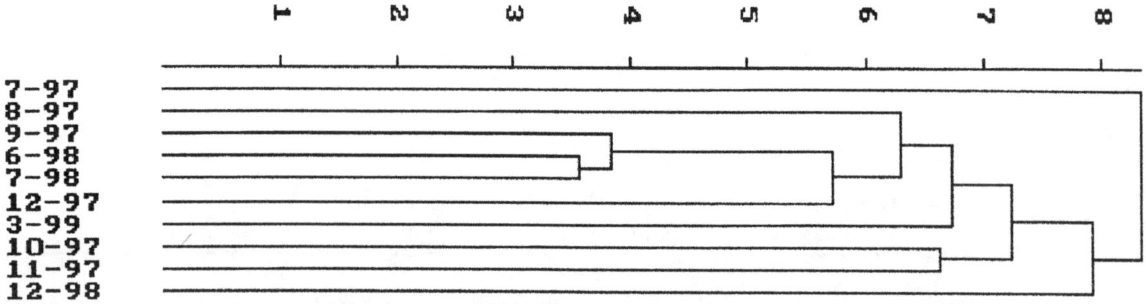

Figure 9. Meandering reach fisheries presence/abundance dendrogram (Euclidean Distance).

dering reach is migrating laterally; however, the rates of migration documented in this short-term field investigation are consistent with erosion rates in unmodified meandering segments of the Embarras River over the last 60 years [*Rhoads* and *Urban*, 1997; *Urban*, 2000] and with rates of erosion for other channels of similar size and drainage area [*Hooke*, 1980; *Hooke* and *Redmond*, 1992]. Those responsible for stream management in East Central Illinois, if they endeavor to improve environmental quality, should understand the dynamic aspects of river channels and the linkage of these dynamics, via processes that maintain spatial heterogeneity of physical habitat, to ecological conditions. Such understanding depends not only on the availability of accurate scientific information on eco-geomorphological linkages, an issue this study has aimed to address, but also on meaningful social interaction between scientists and local people [*Rhoads et al.*, 1999].

CONCLUSION

This study has provided detailed data on the linkages between ecological conditions and morphological variability in a straight channelized reach and an unmodified meandering reach of an agricultural stream in East Central Illinois. The meandering reach has greater morphological variability over time and space, resulting in greater habitat diversity, fish abundance and fish biomass than the channelized reach. This finding is consistent with previous research that has emphasized availability of physical habitat as a key factor governing fish community characteristics in human-modified streams in the Midwest [*Schlosser*, 1982; *Terhaar* and *Herricks*, 1989]. Ongoing work is examining how the difference in morphological complexity between the two reaches influences three-dimensional flow characteristics, thereby creating a difference in the diversity of hydraulic habitat [*Frothingham*, 2001].

This study is part of a comprehensive research program that seeks to develop a knowledge base to guide stream-naturalization activities in East Central Illinois. The comprehensive program is based on a multiscale eco-geomorphological perspective [*Frothingham et al.*, in press]. It combines historical analysis of fisheries [*Chambers*, 1994; *Hauser*, 1999] and stream-dynamics [*Urban*, 2000] with field-based analysis of current geomorphological and ecological conditions [*Frothingham*, 2001] to better understand eco-geomorphological linkages in agricultural stream systems. The knowledge emerging from this research will underpin efforts to develop stream-management alternatives that promote dynamically stable stream systems capable of supporting biologically diverse aquatic ecosystems.

Acknowledgments. Michael Urban, Melinda Newell, Marta Graves, Matthew Ladewig, Perry Cabot, Alexander Sukhodolov, and Kathleen Gregory assisted in the collection of geomorphological data. Timothy Hauser, Mark Tompkins, John Schwartz, Todd Looby, Monica Perez, and Beth Wentzel assisted in the collection of fisheries data. David R. Montgomery, E. Eric Knudsen, and an anonymous reviewer helped the manuscript considerably by providing many constructive comments. This work was supported by grants from the National Science Foundation (SBR-9811322), the U.S. Environmental Protection Agency (R82-5306-010), the University of Illinois Research Board, and the Department of Geography at the University of Illinois.

REFERENCES

Aadland, L. P., Stream habitat types: their fish assemblages and relationship to flow, *North American Journal of Fisheries Management*, *13*, pp. 790-806, 1993.

Beebe, J. T., Fluid speed variability and the importance to managing fish habitat in rivers, *Regulated Rivers: Research and Management*, *12*, pp. 63-79, 1996.

Brookes, A., Physical effects, in *Channelized Rivers*, pp. 81-110, John Wiley and Sons, Chichester, 1988.

Chambers, M. D., *Spatial and temporal variation of fish communities in the headwaters of the Kaskaskia River*, Master of Science thesis, University of Illinois, Urbana, 1994.

Dietrich, W. E., Mechanics of flow and sediment transport in river bends, in *River Channels & Environment and Process*, edited by K. Richards, pp. 179-224, Basil Blackwell, Oxford, 1987.

Fisher, S. G., Creativity, idea generation, and the functional morphology of streams, *Journal of the North American Benthological Society*, *16*, pp. 305-318, 1997.

Frissell, C. R., W. J. Liss, C. E. Warren, and M. D. Hurley, A hierarchical framework for stream habitat classification: viewing streams in a watershed context, *Environmental Management, 10*, pp. 199-214, 1986.

Frothingham, K. M., *Geomorphological Processes in Meandering and Straight Reaches of an Agricultural Stream in East Central Illinois: Relations to Aquatic Habitat*, Ph.D. thesis, University of Illinois, Urbana, 2001.

Frothingham, K. M., B. L. Rhoads, and E. E. Herricks, A multiscale conceptual framework for integrated eco-geomorphic research to support stream naturalization in the agricultural Midwest, *Environmental Management*, in press.

Gelwick, F. P., Longitudinal and temporal comparisons of riffle and pool fish assemblages in a northeastern Oklahoma Ozark stream, *Copeia, 4*, pp. 1072-1082, 1990.

Gore, J. A., and S. W. Hamilton, Comparison of flow-related habitat evaluations downstream of low-head weirs on small and large fluvial ecosystems, *Regulated Rivers: Research and Management, 12*, pp. 459-469, 1996.

Gorman, O. T., and J. R. Karr, Habitat structure and stream fish communities, *Ecology, 59*, pp. 507-515, 1978.

Gregory, K. J., A. M. Gurnell, C. T. Hill, and S. Tooth, Stability of the pool-riffle sequence in changing river channels, *Regulated Rivers: Research and Management, 9*, pp. 35-43, 1994.

Harvey, M. D., and C. C. Watson, Fluvial processes and morphological thresholds in incised channel restoration, *Water Resources Bulletin, 22*, pp. 359-368, 1986.

Hauser, T. A., *Descriptive comparison of resident and transient fish communities in headwater streams in East Central Illinois*, Master of Science thesis, University of Illinois, Urbana, 1999.

Hooke, J. M., Magnitude and distribution of rates of river bank erosion, *Earth Surface Processes, 5*, pp. 143-157, 1980.

Hooke, J. M., and A. M. Harvey, Meander changes in relation to bend morphology and secondary flows, *Special Publications of the International Association of Sedimentologists, 6*, pp. 121-132, 1983.

Hooke, J. M., and C. E. Redmond, Causes and nature of river planform change, in *Dynamics of Gravel-bed Rivers*, edited by P. Billi, R. D. Hey, C. R. Thorne, and P. Tacconi, pp. 557-571, John Wiley & Sons, Chichester, 1992.

Kemp, J., D. M. Harper, and G. A. Crosa, Use of "functional habitats" to link ecology with morphology and hydrology in river rehabilitation, *Aquatic Conservation Marine Freshwater Ecosystems, 9*, pp. 159-178, 1999.

Lobb III, M. D., and D. J. Orth, Habitat use by an assemblage of fish in a large warmwater stream, *Transactions of the American Fisheries Society, 120*, pp. 65-78, 1991.

Montgomery, D. R., E. M. Beamer, G. R. Pess, and T. P. Quinn, Channel type and salmonid spawning distribution and abundance, *Canadian Journal of Fisheries and Aquatic Science, 56*, pp. 377-387, 1999.

Osborne, L. L., and M. J. Wiley, Influence of tributary spatial position on the structure of warmwater fish communities, *Canadian Journal of Fisheries and Aquatic Science, 49*, pp. 671-681, 1992.

Rhoads, B. L., and M. R. Welford, Initiation of river meandering, *Progress in Physical Geography, 15*, pp. 127-156, 1991.

Rhoads, B. L., Stream power: a unifying theme for urban fluvial geomorphology, in *Stormwater Runoff and Receiving Systems: Impact, Monitoring, and Assessment*, edited by E. E. Herricks, pp. 65-75, Lewis, Boca Raton, 1995.

Rhoads, B. L., and E. E. Herricks, Naturalization of headwater streams in Illinois, in *River Channel Restoration*, edited by A. Brookes, and D. F. Shields Jr, pp. 331-367, John Wiley and Sons, Chichester, 1996.

Rhoads, B. L., and K. M. Monahan, Geomorphological principles for 'naturalizing' streams and rivers in Illinois, in *Governor's Conference on the Management of the Illinois River System*, edited by A. M. Strawn, pp. 79-86, Water Resources Center, University of Illinois, Urbana, 1997.

Rhoads, B. L., and M. A. Urban, Human-induced geomorphic changes in low-energy agricultural streams: an example from east central Illinois, in *Management of Landscapes Disturbed by Channel Incision*, edited by S. S. Y. Wang, E. J. Langendoen, and F. D. Shields, Jr., pp. 968-973, The Center for Computational Hydroscience and Engineering, The University of Mississippi, Oxford, 1997.

Rhoads, B. L., D. Wilson, M. A. Urban, and E. E. Herricks, Interaction between scientists and nonscientist in community-based watershed management: emergence of the concept of stream naturalization, *Environmental Management, 24*, pp. 297-308, 1999.

Schlosser, I. J., Fish community structure and function along two habitat gradients in a headwater stream, *Ecological Monographs, 52*, pp. 395-414, 1982.

Simon, A., and C. R. Hupp, The recovery of alluvial systems to imposed channel modifications, West Tennessee, USA, in *Vegetation and Erosion*, edited by J. B. Thornes, pp. 145-160, John Wiley and Sons, Chichester, 1990.

Southwood, T. R. E., Habitat, the templet for ecological strategies?, *Journal of Animal Ecology, 46*, pp. 337-365, 1977.

Swales, S., Fish populations of a small lowland channelized river in England subject to long-term river maintenance and management works, *Regulated Rivers: Research and Management, 2*, pp. 493-506, 1988.

TerHaar, M. J., and E. E. Herricks, *Management and Development of Aquatic Habitat in Agricultural Drainage Systems*. UILU-WRC-89-212, 145 pp., Water Resources Center, University of Illinois, Urbana, 1989.

Tompkins, M. R., *Analysis of the fisheries in maintained streams using the PHABSIM Model and a habitat driven population model*, Masters of Science thesis, University of Illinois, Urbana, 1998.

Urban, M. A., *Conceptualizing anthropogenic change in fluvial systems: drainage development on the upper Embarras River, Illinois*, Ph.D. thesis, University of Illinois, Urbana, 2000.

Kelly M. Frothingham, Buffalo State College, Department of Geography and Planning, 1300 Elmwood Avenue, Buffalo, NY 14222

Bruce L. Rhoads, University of Illinois, Department of Geography, 607 South Mathews Avenue, 220 Davenport Hall, Urbana, IL 61801

Edwin E. Herricks, University of Illinois, Department of Civil and Environmental Engineering, 1105 Newmark Civil Engineering Laboratory, 205 North Mathews Avenue, Urbana, IL 61801

Mass Wasting and Sediment Storage in a Small Montane Watershed: an Extreme Case of Anthropogenic Disturbance in the Humid Tropics

Matthew C. Larsen

U.S. Geological Survey, Guaynabo, Puerto Rico, USA

Abigail Santiago Román

Geology Department, Binghampton University, Binghampton, New York, USA

By the peak of land-use conversion for subsistence cropping and plantation agriculture in Puerto Rico in the 1940's, 94 percent of the original forest cover had been eliminated. In a small (26.4 km^2) upland watershed that typifies this land-use history, field surveys and examination of aerial photographs indicate that more than 2,000 landslides have occurred since about 1820 when forest clearing began. The landslides are attributable to a combination of three factors: a highly weathered bedrock (Cretaceous granodiorite), episodic heavy rainfall, and almost two centuries of intense land-use practices. On average, landslide scars number 140/km^2 in the Cayaguás watershed, equal to 80 landslide scars/km^2/100 y. The volume of hillslope material eroded by landsliding is estimated at 660,000 m^3/km^2 (870,000 Mg/km^2). If all colluvium was transported from the catchment, then the volume is equivalent to a mean surface lowering of the entire watershed by 660 mm, or 3.8 mm/y. Soil augering, field observations at construction sites, road cuts and stream banks, mapping from aerial photographs, and GIS-based estimates of the surface area of footslopes, indicate that colluvium may total 149,000 Mg/km^2. If mobilized, this would be sufficient stored material to supply the annual average fluvial sediment yield for as long as 129 years. The great availability of colluvial and alluvial sediment on footslopes, floodplains, and in channels will maintain high sediment yield well into the 21st century in spite of government efforts to reforest hillslopes and institute other hillslope soil conservation measures.

INTRODUCTION

Understanding the processes and the rates by which land surfaces are altered by natural and human geomorphic agents is a central focus of geomorphological research and a fundamental requirement for effective management of natural resources. In environments that have undergone or are undergoing rapid land-surface alteration for agriculture, construction, or logging, case studies and conceptual geomorphic models enable land managers to attempt to mitigate landscape degradation. This information base can assist planners in other similar settings to predict outcomes of development strategies.

Researchers working in many settings have quantified the sources, as well as the routes and rates of movement of detrital material through a drainage basin to evaluate watershed status and sustainability and to interpret landform

Geomorphic Processes and Riverine Habitat
Water Science and Application Volume 4, pages 119-138

change [*Caine and Swanson*, 1989; *Dietrich and Dunne*, 1978; *Graf*, 1987; *Kelsey*, 1980; *Kelsey et al.*, 1987; *Madej and Ozaki*, 1996; *Montgomery et al.*, 2000; *Phillips*, 1986; *Sutherland and Bryan*, 1991]. The common approach is to determine annual fluxes of hillslope and fluvial sediment and make estimates of residence time of sediment that is in storage between these two [*Costa*, 1975; *Jacobson and Coleman*, 1986; *Meade*, 1982; *Trimble*, 1981]. Although these types of studies must use estimates of the quantity and spatial distribution of stored sediment, they provide valuable insight into the geomorphic behavior of a catchment.

The important role of stored colluvial and alluvial sediment with respect to fluvial sediment yield has been documented in temperate and subtropical environments. The stored material, derived from anthropogenic or natural sources, *e.g.* glaciation, can maintain higher than predisturbance sediment yields well after the initial disturbance is complete [*Haggett*, 1961; *Costa*, 1975; *Trimble*, 1977; *Meade*, 1982; *Church and Slaymaker*, 1989]. Sustained high sediment yields may degrade water quality, reservoir capacity, and estuarine and near-shore marine environments until stored sediment is immobilized or exported, and an equilibrium land-use practice is achieved.

Landsliding triggered by earthquakes or rainfall is the dominant process of hillslope erosion in many montane humid tropical environments [*Garwood et al.*, 1979; *Jones*, 1973; *Larsen et al.*, 1999; *Maharaj*, 1993; *Reading et al.*, 1995; *Scatena*, 1995; *Simonett*, 1967]. Rainfall during intense and (or) prolonged storms has been the principal cause of landslides in Puerto Rico during the 20th century [*Larsen and Simon*, 1993; *Larsen and Torres-Sánchez*, 1998]. Because of relatively short distances to channels in upland watersheds, hillslope material eroded by landslides is commonly delivered directly to stream channels resulting in high sediment delivery ratios [*Ahmad et al.*, 1993; *Gupta*, 1988; 1995; *Larsen and Parks*, 1997]. If hillslope angles are low or footslopes and floodplains are wide enough, some of the colluvium may reside in temporary storage before being entrained by fluvial processes. It is this buffer area of sediment storage that, because of its great spatial and temporal variability, confounds much of the research on hillslope erosion and fluvial transport of sediment.

Long-term data sets of hillslope erosion and fluvial sediment transport are difficult and expensive to develop. As a consequence, relatively few definitive studies exist, particularly in tropical environments. Because of extensive Federal and Commonwealth government-sponsored data collection programs, Puerto Rico serves as a data-rich natural laboratory for such studies. A large network of surface-water gaging stations, automatic water samplers, and rainfall gages is operated by the U.S. Geological Survey [*USGS*, 1999]. In addition, regular aerial photographic surveys have been conducted and land cover has been quantified for a number of years during the 20th century.

These data sets provide an opportunity to address several research questions in a small, montane catchment in Puerto Rico: How much hillslope soil has been eroded by mass wasting, principally by landsliding, at what rate has erosion occurred, and what has been the effect of anthropogenic disturbance in this process? How much sediment is sequestered on footslopes and in and near river channels, and finally, what is the rate of fluvial export of sediment from this watershed?

CAYAGUAS WATERSHED DESCRIPTION

The Cayaguás watershed (26.4 km^2) is in southeastern Puerto Rico, about 35 km south of San Juan (Fig. 1). The watershed drains to the north and is bordered on the south by the Cayey Mountains. The climate is humid tropical with a mean annual rainfall of 1,930 mm and a mean annual temperature of 25°C [*U.S. Department of Commerce*, 1999]. Hillslopes are steep in the watershed; the mean slope is 19% (10.8°) and total relief is 293 m (Table 1). The average (map) channel gradient of the Cayaguás river is 12% (6.8°). The Cayaguás watershed is a tributary to the Río Grande de Loíza, which has the largest drainage basin in Puerto Rico, and flows into the Loíza Reservoir (Fig. 1). Constructed in 1953, this 267-ha impoundment supplies about 50% of public water to the 1.6 million inhabitants of metropolitan San Juan. Bathymetric surveys completed in 1994 indicate that sediment infilling of the reservoir, discussed below, had eliminated 47% of the original capacity [*Webb and Soler-López*, 1997].

The Upper Cretaceous Granodiorite of San Lorenzo, a batholith which occupies a 500-km^2 area of southeastern Puerto Rico, underlies 88% of the Cayaguás watershed [*Broedel*, 1961; *Rogers et al.*, 1979]. The granodiorite is medium to dark gray, medium-grained, and contains 51% plagioclase, 21% quartz, 11% each of potassium feldspar and hornblende, and 5% biotite. About 11% of the southeastern part of the Cayaguás watershed was mapped as quartz monzonite. This rock type ranges locally in composition to a granodiorite, but the median composition is 35% plagioclase, 33% quartz, 28% potassium feldspar and 2% each hornblende and biotite. The remaining 1% of the watershed is mapped as Quaternary alluvium, described as clay to boulder-sized material, including remnants of low terraces, and having a thickness of more than 10 m locally.

The high rate of chemical weathering of the crystalline granodiorite bedrock produces abundant material for transport due to the highly erodible nature of intrusive rock in this humid tropical watershed. As feldspars are weathered during the year-round wet and warm environment, the rock

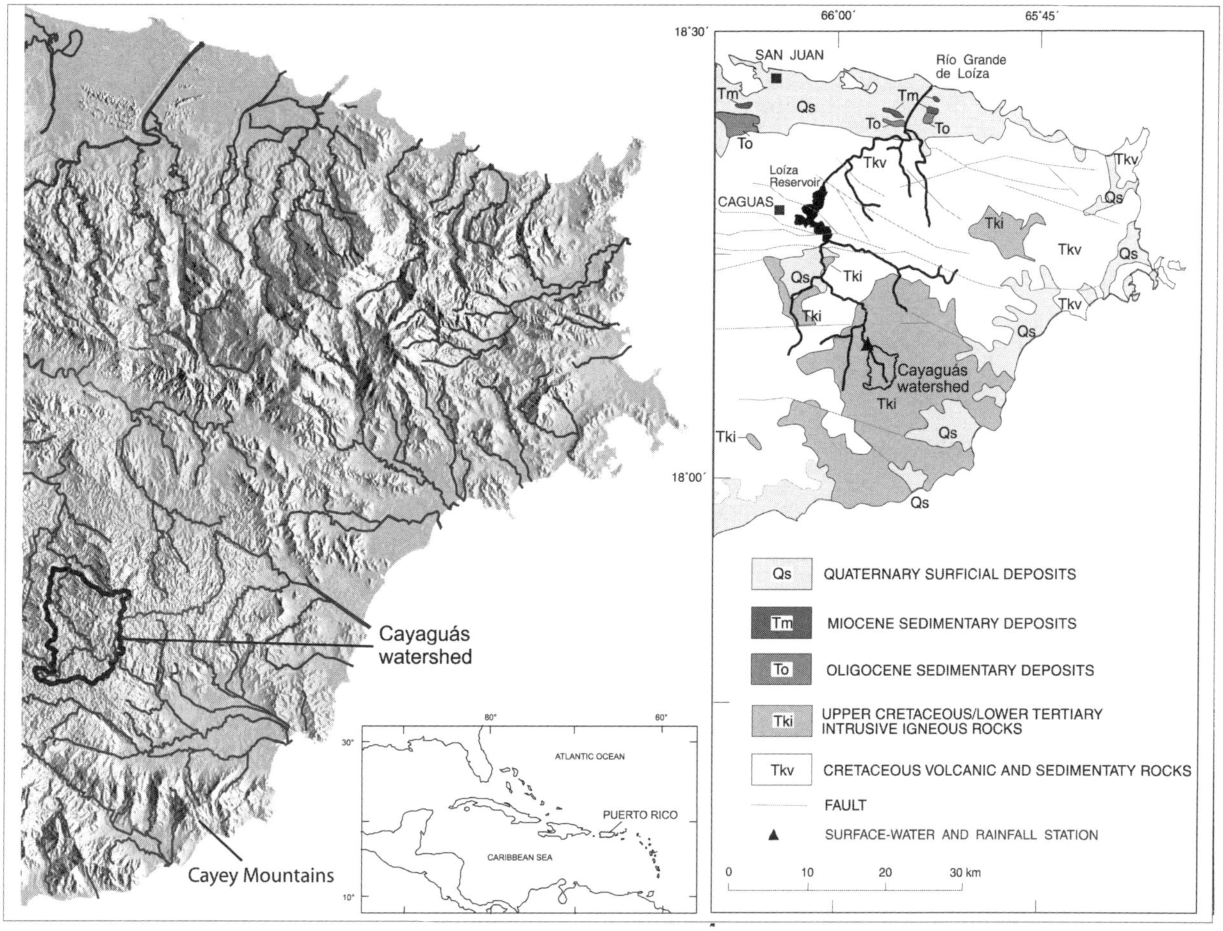

Figure 1. Shaded relief and simplified geology of eastern Puerto Rico showing location of the Cayaguás watershed (geology modified from Monroe, 1980).

matrix disaggregates and weathers to saprolite at depth and to an easily transported grus (medium- to coarse-grained sand) visible at the soil surface [*Gerrard*, 1994]. Judging from saprolite thickness in nearby undisturbed forest in the same bedrock type, saprolite may have had a pre-forest clearing thickness of several meters or more on ridgetops and hillslopes [*Simon et al.*, 1990; *Larsen*, 1997]. The thick layer of saprolite and regolith have provided abundant material for landsliding.

Landslide basal surfaces vary with rainfall intensity: short-duration, high-intensity storms frequently mobilize only a few 10's of cm of regolith above the saprolite surface; long-duration storms with deeper infiltration of rainwater may induce landslides that fail at the weathered rock-saprolite boundary [*Larsen and Simon*, 1993; *Larsen*, 1997].

Although large corestones are abundant on many hillslopes and in some channel reaches, most material delivered to drainage channels is sand sized (D_{50} = 0.6 mm)[*Larsen*, 1997]. Consequently, fluvial erosion and transport has been efficient at the millennial to geologic time scale, enhancing channel extension, as manifested by high drainage density seen on topographic maps of the area (Table 1). Modern evidence can be seen in active slumping at the heads of tributaries of some zero-order valleys. In addition to extension, widening of main channel valleys results from mass wasting on valley sides. Headward extension of these has enabled the stream system to approach within 12 km of the south (Caribbean) coast of the island, in spite of being part of a north coast draining river that crosses a straight-line distance of more than 50 km to reach the Atlantic Ocean (Fig. 1).

Ultisols (Tropohumults, Tropodults) and Inceptisols (Eutropepts) are the dominant soil orders in the Cayaguás watershed, typical of the upland areas of Puerto Rico [*U.S. Department of Agriculture*, 1977]. Alfisols (Tropaqualfs) are mapped along some valley floors. Hillslope soils are predominately sandy clay loam. Soil erosion by slopewash has been estimated to be 10 Mg/km^2/y at sites in secondary

Table 1. Geomorphic, geographic, and streamflow characteristics for the Cayaguás watershed [*U.S. Geological Survey*, 1999].

Mean annual runoff, 1975 to 1995, mm	1480
Mean annual rainfall, mm	1930
Watershed drainage area, km^2	26.4
Stream order at surface water gaging station, Strahler	5
Total watershed relief, m	293
Mean slope of watershed, dimensionless	0.19
Mean channel slope, dimensionless	0.12
Main channel (5th order) length, km	10.1
Total channel length, km	172
Drainage density, km/km^2	6.5

forest (15° mean slope angle) and in pasture (24° mean slope angle) in the Río Grande de Loíza watershed, which includes the Cayaguás watershed [*Gellis et al.*, 1999]. In addition, *Gellis et al.* [1999] reported that slopewash ranges from 31 to 86 Mg/km^2/y on hillslopes with active agriculture (16° mean slope angle). In a slopewash erosion study in undisturbed forest in the Luquillo Experimental Forest, a preserve administered by the U.S. Forest Service, in eastern Puerto Rico, *Larsen et al.* [1999] estimated erosion rates of 46 Mg/km^2/y on steep hillslopes (41° mean slope angle) underlain by intrusive bedrock, comparable to the bedrock in the Cayaguás watershed. Soil creep is estimated to contribute 11 Mg/km^2/y to stream channels [*Larsen*, 1997; *Lewis*, 1974]. Gullying was observed on steep sections of unpaved roads, on areas of cropland, and on landslides scars in the Cayaguás watershed. However, the frequency and the magnitude of observed gullies does not indicate that gullying is an important agent of hillslope erosion in this watershed [*Gellis et al.*, 1999; *Larsen*, 1997].

Specific 19th-century land-use data for the Cayaguás watershed are limited. It can be assumed that land-use history until the 20th century was generally comparable to island-wide averages (Fig. 2). According to *Morales-Muñoz* [1943], settlers migrated up the Río Grande de Loíza valley from San Juan as early as the 16th century. Cattle were introduced to roam the unfenced forest, and were periodically rounded up for consumption or sale in San Juan. A large tract of land (about 100 km^2) that extended from the present-day city of Caguas southeast to the Cayey Mountains, and included all of the Cayaguás watershed, was ceded by the Spanish crown to Miguel Muñoz de Oneca *ca.*1700. The land remained in the Muñoz de Oneca family for most of the following century and agricultural development was limited until the early 19th century. During the 19th and early 20th centuries, common agricultural practices included the use of alternating cash (tobacco and sugar cane) and subsistence (beans, corn, roots) crops [*Beishlag*, 1955]. Individual farm size generally varied with hillslope; areas with gentle slopes had farms averaging 10 ha whereas areas with the steepest slopes had farms averaging only 2 ha. Many farms, however, were very small; for example, a government resettlement program in the 1940's installed families on plots of 0.1 to 1 ha [*Beishlag*, 1955]. The small average plot size on steep slopes meant that areas most vulnerable to soil erosion were likely to be farmed intensively by families whose survival depended on the limited land area. This resulted in a negative feedback mechanism whereby the productivity of the eroded hillslopes declined forcing families to keep the land in constant production (Fig. 3). On numerous small farms in the mountainous interior of the island, decreasing soil fertility resulted in using the land to graze cattle. For example, by mid 20th century, out of 6,059 farms in southern Puerto Rico, 76% had up to 5 head of cattle each. After the 1940's, the amount of land used for agriculture began a marked decline as rural population sought work in Puerto Rico and U.S. industrial centers (Fig. 2). Off-island migration rates were high enough to cause a dip in population growth during the 1950's. A rapid rise in forest cover also occurred, and by the 1970's, the percentage of land in forest was the highest it had been in 100 years.

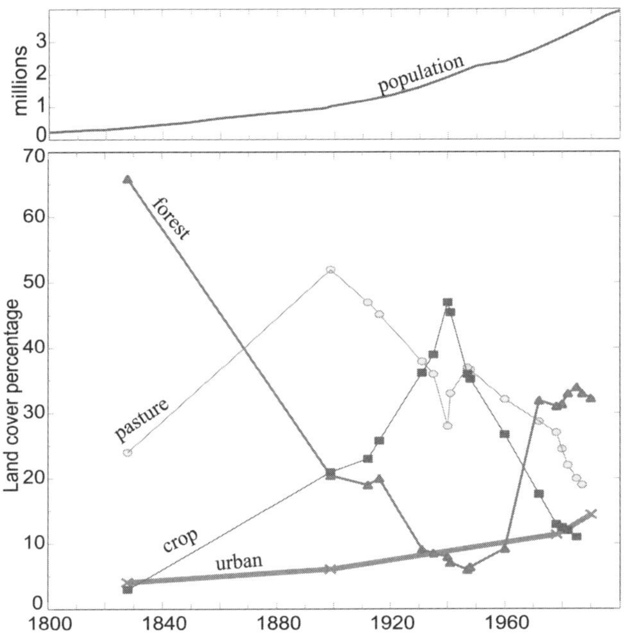

Figure 2. Generalized 19th-20th century land-use classification and population for Puerto Rico (land-use data from Alberts, 1940; Wadsworth, 1950; Birdsey and Weaver, 1987; Franco et al., 1997).

At present, vegetation in the Cayaguás watershed is dominated by actively-grazed pasture and secondary forest. A wide variety of broad-leaf tropical evergreen trees is present. The principal crop species are *Colocasia esculenta*, (known locally as malanga) also known as taro, grown on wet bottomland soils, *Dioscorea rotundata* (locally known as ñame), which is planted in raised mounds on hillslopes and ridgetops where drainage is good, and *Musa sp.* (plantains, bananas), also planted where drainage is good.

Population statistics for the Cayaguás watershed can be approximated by using U.S. Census data for the municipality of San Lorenzo, within which the watershed is contained. In 1990, population density was 97 inhabitants/km^2, the annual per capita income was $2,888, and 72% of the families were below the U.S. government poverty threshold [*U.S. Department of Commerce*, 1991].

METHODS

Landslides and Land use

Stereo pairs of black and white aerial photographs dating from 1937, 1951, and 1995 (scales ranging from 1:17,000 to 1:20,000) were examined with 2- to 4-power magnifying stereo viewers to determine the generalized land-use categories (for 1937 and 1995), and the location and extent of landslide scars. Land-use classification was also derived for 1950 from published maps [*Brockman*, 1952]. Field personnel from the Puerto Rico Agrigultural Extension Service and the U.S. Natural Resources Conservation Service were interviewed to determine agriculture practices and history. Although mapping of landslides in the entire watershed was planned at the outset of this study, the extremely high number of landslide scars present resulted in a scaled down approach in which some areas of the watershed were not mapped.

Landslides were classified as recent or historic. Recent scars were defined as those that occurred during the previous 10 years and were observed on aerial photographs with a sharp break or disruption in vegetation type, little to no vegetation regrowth, bare soil or soil with little vegetation regrowth, and other geomorphic characteristics such as a clearly definable steep head and side scarps, hummocky soil and downslope debris deposits [*Larsen and Torres Sánchez*, 1998; *Wieczorek*, 1984]. Because of extensive area in actively-grazed pasture in the watershed, old landslide scars are easily recognized [*Gellis et al.*, 1999]. These 'historic' landslide scars were defined as having occurred approximately within the past 100 years (but possibly dating to the early 1800's) and having some vegetation regrowth, subdued head and side scarp boundaries, and eroded or absent debris deposits. Each landslide was assigned an identification number, and the length (both initiation area and runout) and width were estimated by using a calibrated ruler. Length measurements based on aerial photographs are necessarily shortened in proportion to slope angle. This was corrected by using the cosine function and slope angle determined with geographic information system (GIS) software [*ESRI*, 1993]. An approximate scar depth was estimated by using the typical rural Puerto Rican house height as a gage; 3 m in 1937 and in 1951, and 3 to 6 m in 1995. Using the product of the average soil density and individual landslide volume, an eroded mass was estimated for each scar. This value is affected in many cases by continued erosion of head and side scarps after the initial failure occurrence. These values were summed and divided by the watershed area to normalize the erosion rate. The landslide type was classified according to scar morphology into debris flow, slump, soil slip, debris avalanche, earthflow, and complex, using the terminology of *Cruden and Varnes* [1996] and *Campbell* [1975]. Analysis of aerial photography allows the cataloging of landslide scars in an extensive area; however, the technique is limited by the poor quality of some black and white photographs taken in 1937 and in 1951. During field reconnaissance, landslide scars not visible on aerial photographs were commonly observed where forest canopy and shadows on steep hillslopes mask landslide features. The number of landslides identified in the watershed is therefore presumed to be underestimated.

A total of 203 of the landslides mapped in this study were field checked to evaluate the accuracy of dimensions, classification, and location as determined in the photographs. In addition, soil samples were collected to determine bulk density. Field measurements of landslide length and width were accomplished by using a metric tape or a rangefinder and a stadia rod to measure scar depth. The landslide depths recorded were the average along the scar. Landslide lengths were defined as the distance from head scarp to the slope base. Hillslopes in the Cayaguás watershed are relatively short (10's to ~150 m), and the erosive portion of landslide scars was usually continuous across this distance. Because only accessible, visible scars were chosen for field checking, it is possible that the correspondence between field- and aerial photograph-determined measurements may be poorer than that discussed below. However, landslide scars that are difficult to measure because of forest cover or remoteness would reduce the quality and quantity of field data. This is because scar morphology is difficult to recognize and measure under broadleaf tropical evergreen forest canopy, particularly with the dense understory typical of second-growth forest. Simple linear regression models relate the dimensions of landslides mapped in the field to the dimensions estimated from aerial photographs. The models were used to adjust all the dimensions of landslide scars cataloged from aerial photographs in this study.

Sediment Storage

Sediment resides in various storage compartments in a watershed before it is either entrained by fluvial and mass-wasting processes or sequestered for long time scales. These compartments can be described as active, semiactive, or inactive depending on how potentially mobile the deposits may be [*Madej*, 1995]. In the Cayaguás watershed, channel bars and the channel bed were defined as active, floodplains and colluvium were defined as semiactive, and Quaternary alluvium was defined as inactive.

Perennial stream-channel length is substantially under represented on the 1:20,000-scale USGS topographic maps that cover the study area (maps date from 1960 and 1967). This is the result of several factors. The maps were made from aerial photographs with limited resolution. First-order streams are very narrow and in many cases, are located where forested and shrub-covered hillslopes and shadows obscure them from the view of the map maker. Black and white aerial photographs have an approximate resolution of 5 to 10 m, depending on local variation in photograph quality, magnification, illumination, air quality, and humidity when the photo was taken. As a result, the exact location of first-order stream channels is beneath the resolution of the technology used to make the topographic maps in the 1960's, particularly with respect to the determination of where water is flowing. An expanded hydrographic network was approximated following standard techniques using a digital elevation model (DEM) with a 5 m cell width that was derived from digital 1:20,000-scale topography (Table 1) [*Tarboton et al.*, 1991]. Using an approximate minimum contributing area for first-order perennial streams of 7,500 m^2, a stream channel network was created. The minimum contributing area was determined by field mapping the location of perennial flow in first-order channels using a Global Positioning Satellite unit. The contributing area upstream was then measured using GIS software. Although local heterogeneity in such factors as groundwater flow paths, bedrock weathering, and slope results in variable contributing area across a watershed [*Montgomery and Dietrich*, 1989], this approximation is an improvement over the under-represented perennial stream channel lengths shown in the topographic maps for the Cayaguás watershed. The stream order, total channel length, and drainage density listed in table 1 were derived from the expanded hydrographic network developed with the methodology described above.

Watershed sediment storage was estimated by using field surveys of channel bars and channel width. The length, width, and height of unvegetated channel bars that are subaerial at base flow were measured in three reaches of the Cayaguás River. The bar surface area per length of measured

Figure 3. Rural family and residence, 1940's, eastern Puerto Rico, with deforested hillslopes in background. (Photograph taken by F.H. Wadsworth, International Institute of Tropical Forestry, U.S. Forest Service, Río Piedras, Puerto Rico).

channel was proportioned to Strahler stream order and multiplied by the overall perennial stream-channel length to estimate the total bar surface area [*Strahler*, 1956]. The bulk density of sediment composing channel bars was measured by collecting volumetric samples, and drying the material at 105°C for 24 hours. The mean bulk density of bar sediment was multiplied by the bar volume to calculate bar mass. GIS software was used to estimate the surface area of floodplains, all river channels, Quaternary alluvium, and footslopes with colluvial storage.

Channel bed sediment volume was quantified by calculating channel area and multiplying by a mean thickness that ranged from 0.02 to 1 meter, ranked by stream order. The bed-thickness value was based on observation of bedrock, coarse gravel, and boulders underlying the channel sand exposed by episodic degradation during the period from 1984 to 1999.

Channel surface area was determined by using the expanded perennial stream-channel length and a sample of measured channel widths at various locations between the basin outlet and headwaters. The areal extent of floodplain alluvium was calculated to a first approximation along all stream channels by using the 5-m DEM with GIS software and the following calculations. First, based on proportions observed in the field during channel bar surveys, a corridor of land having a width double that of the average channel width was created along all stream channels. The average channel width used was that determined from stream order. Next, the surface area with a slope of 4° or less was determined inside of the corridor adjacent to the channels [*Larsen*, 1997]. Finally, the area of the actual channel was subtracted from the corridor to arrive at the estimate of total floodplain area. Alluvium thickness was estimated by examining exposures on stream channel banks.

The distribution and mass of colluvial deposits is widespread in the watershed on footslopes and near the heads of zero-order basins. No accurate method for estimating this mass exists short of detailed field measurements. The total volume of landslide scars affords an upper-limit estimate of the total amount of colluvium generated if it is assumed that most colluvium postdates the oldest visible scars. This is a reasonable assumption based on observations in a nearby forested watershed with the same bedrock type where little colluvium exists on footslopes [*Larsen*, 1997; *Larsen and Torres Sánchez*, 1998]. An alternative approach provides a more conservative estimate of colluvial storage. Using GIS software and the 5-m DEM, a map of hillslopes with slopes of 10° or less was made. The low slope areas on the tops of ridges and saddles were eliminated as no colluvium is deposited in these locations. In addition, all watershed surface area inside the floodplain boundaries was eliminated. The resulting map provides an estimate of the total surface area of colluvial deposits; the surface area then was multiplied by the average thickness of colluvium, as determined by soil augering, and field observations at construction sites, road cuts and stream banks.

Fluvial Sediment and Runoff

Suspended and bed load sediment transport data were compiled from previous USGS studies to summarize annual yield [*Guzmán-Ríos*, 1989; *Larsen*, 1997; *Simon and Guzmán-Ríos*, 1990; *USGS*, 1999]. Annual runoff data were derived from a USGS data base for a surface-water gaging station, USGS ID 50051310, at the outlet of the Cayaguás watershed [*USGS*, 1999]. A continuous stage recorder has been maintained at this location since 1977.

RESULTS

Land use

Land-use characterization based on aerial photography indicates that by 1937, pasture was the dominant land cover for 84% of the surface area of the Cayaguás watershed (Fig. 4). Forest cover accounted for 12% of the area and the remainder was in crop and farm houses. Recent work indicates that as cultivated areas are abandoned, the land typically has reverted to pasture [*Aide et al.*, 1995]. This serves as a transitional, or buffer state, as it may be easily returned to cultivation after remaining fallow for several years. Alternatively, if cattle are kept off the land, secondary forest has been shown to recover in similar watersheds in Puerto Rico within about 20 years [*Aide et al.*, 1995]. Land-use practices in the watershed during the latter half of the 20th century diverged from the island average as pasture continued to be the primary land use at 66% of land cover in the Cayaguás watershed in 1995. This compares to pasture land cover of less than 19% for the island overall in 1987 (Fig. 2). During the second half of the 20th century, cropped areas decreased while forest cover and clusters of rural housing increased. Forest cover in 1937 was about 12% and decreased to less than 5% by 1950 [*Brockman*, 1952]. Forest recovery to 23% of the Cayaguás watershed had occurred by 1995.

Landslides

Most landslide scars mapped from aerial photographs predate the earliest available photographs (1937) and oral histories of local residents. Seventy percent of landslides mapped from 1951 and 1995 aerial photographs were present on the 1937 photographs. It is likely that accelerated rates of landsliding began in the early 19th century with the onset of

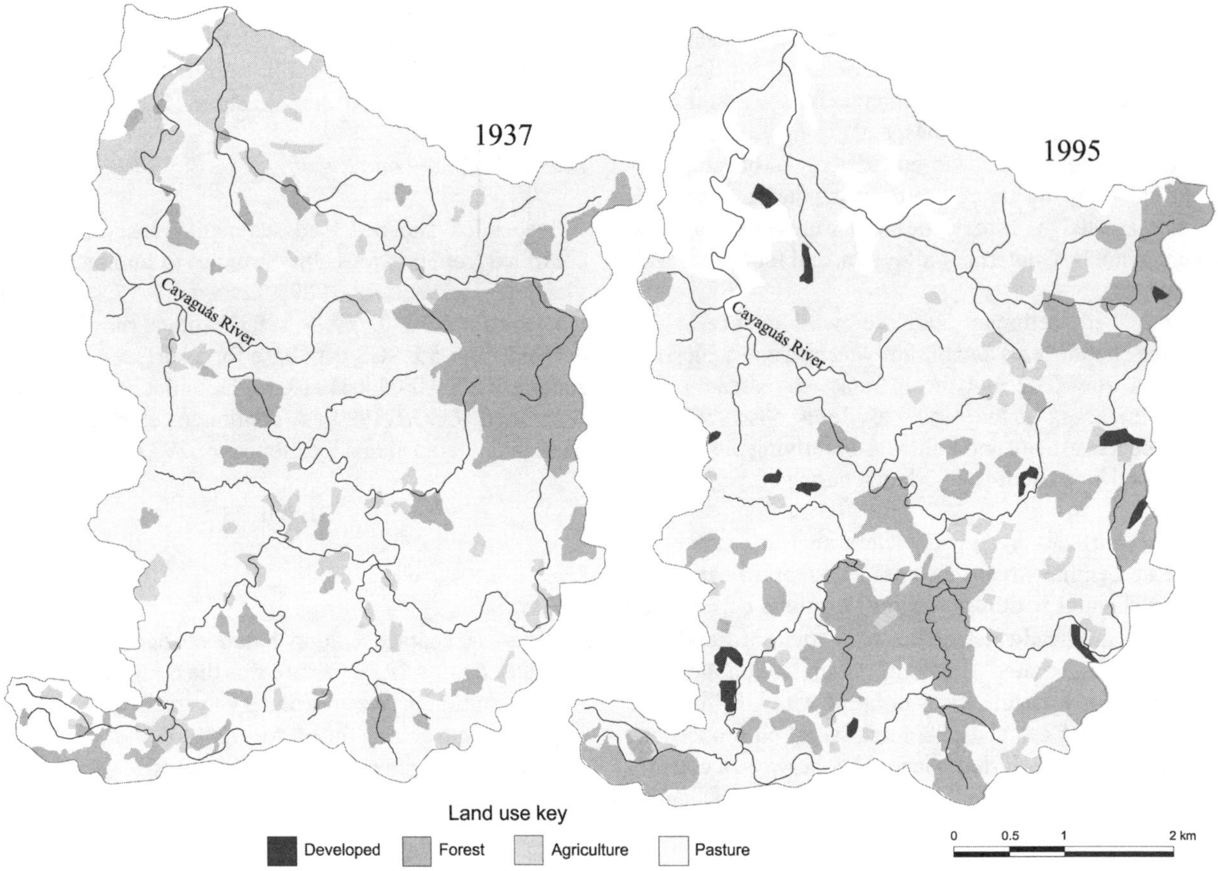

Figure 4. Generalized land use for 1937 and 1995 classified from black and white aerial photography, Cayaguás watershed, Puerto Rico.

forest clearing for agriculture. The causative storm events for landslide scars are therefore difficult to ascertain. The greatest rainfall intensities and widest distribution of storm rainfall in Puerto Rico usually are associated with tropical disturbances, which comprised 61% of the landslide-triggering storms recorded in the central mountains of Puerto Rico between 1960 and 1990 [*Larsen and Simon*, 1993]. Major hurricanes crossed the island in 1867, 1899, 1928, 1932, and 1960. These storms delivered precipitation totals of 300 to 800 mm in a 1 to 3 day period, and are the likely type of storm rainfall that triggered many of the landslide scars visible in the aerial photographs [*Scatena and Larsen*, 1991]. Conversations with watershed residents confirm that rainfall associated with hurricanes is commonly the trigger for landslides. This relation was observed following Hurricane Hortense (September 1996) and Hurricane Georges (September 1998) when dozens of debris flows, many extending from ridge top to the stream channel, were observed throughout the watershed (Fig. 5)[*Torres-Sierra*, 1997]. The rainfall totals from these two hurricanes had recurrence of intervals of 5 to 10 years in the Cayaguás watershed. The influx of landslide-derived and other sediment into channels resulted in 1.3 m of aggradation of the Cayaguás channel bed at the gage outlet following Hurricane Hortense [*Larsen*, 1997]. The channel bed surface was sustained at this elevation for three years before gradual degradation began in late 1999.

A total of 2,321 landslides were mapped from aerial photographs in selected portions of the watershed (Table 3). Landslide scars generally were small with a median length of 42 m, and affected a median surface area of 610 m^2, reflective, as noted above, of the high-intensity and short-duration nature of triggering rainstorms. The mean and median hillslope angles at landslide scars were both 25° (standard deviation: 6°, range: 10 to 38°, n: 159). Most landslides (77%) were on hillslopes with angles between 20 and 30°.

Landslide scar dimensions estimated from aerial photographs compared reasonably well to measurements made on the ground for length, but not as well for width and

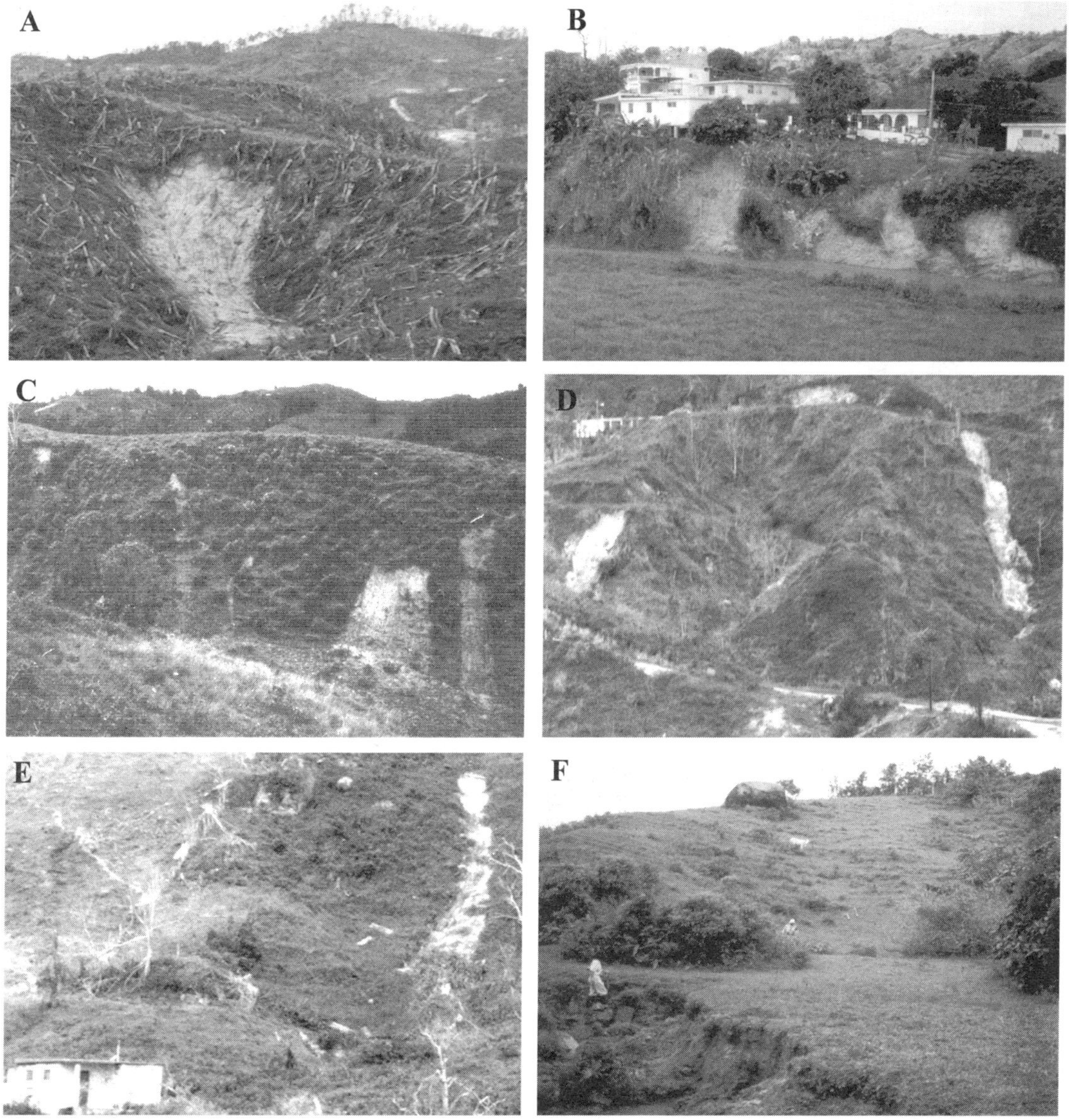

Figure 5. Views of landslide scars and stream channel headcutting associated with Hurricanes Hortense (September 1996) and Georges (September 1998) in the Cayaguás watershed, Puerto Rico. (A) Soil slip, 10 m long, on hillslope plantain field, September 1998. (B) Soils slips along Cayaguás tributary, September 1996. (C) Soil slips, 20 to 30 m long, September 1996. (D) Soil slip/debris flows, 10 and 30 m long, September 1998. Note larger deep landslide scar between the two recent ones. (E) Soil slip/debris flow, September 1998. (F) Headcut of first-order stream in colluvium at base of hillslope. Note large granodiorite corestone on ridgetop.

depth (Fig. 6). This result was expected as landslide scar length is relatively easy to determine while width and depth are more subjective, both from aerial photographs, as well as on the ground. The majority of scars have been modified by processes such as slopewash, trampling by cattle, plowing, and additional slumping on the side and head scarps [*Trimble and Mendel*, 1995; *Larsen et al.*, 1999]. This results in rounded head and side scarps and irregular scar depths that prevent precise measurements on the ground, as well as from the air (Fig. 7). The area, volume, and eroded mass of landslide scars discussed below are, therefore, approximations. The length, width, and depth of landslide scars measured on aerial photographs were adjusted using the equations shown in Figure 6.

Most landslide scars, 61%, were classified as debris flows (Table 2). Soil slips, 23%, were the second most common

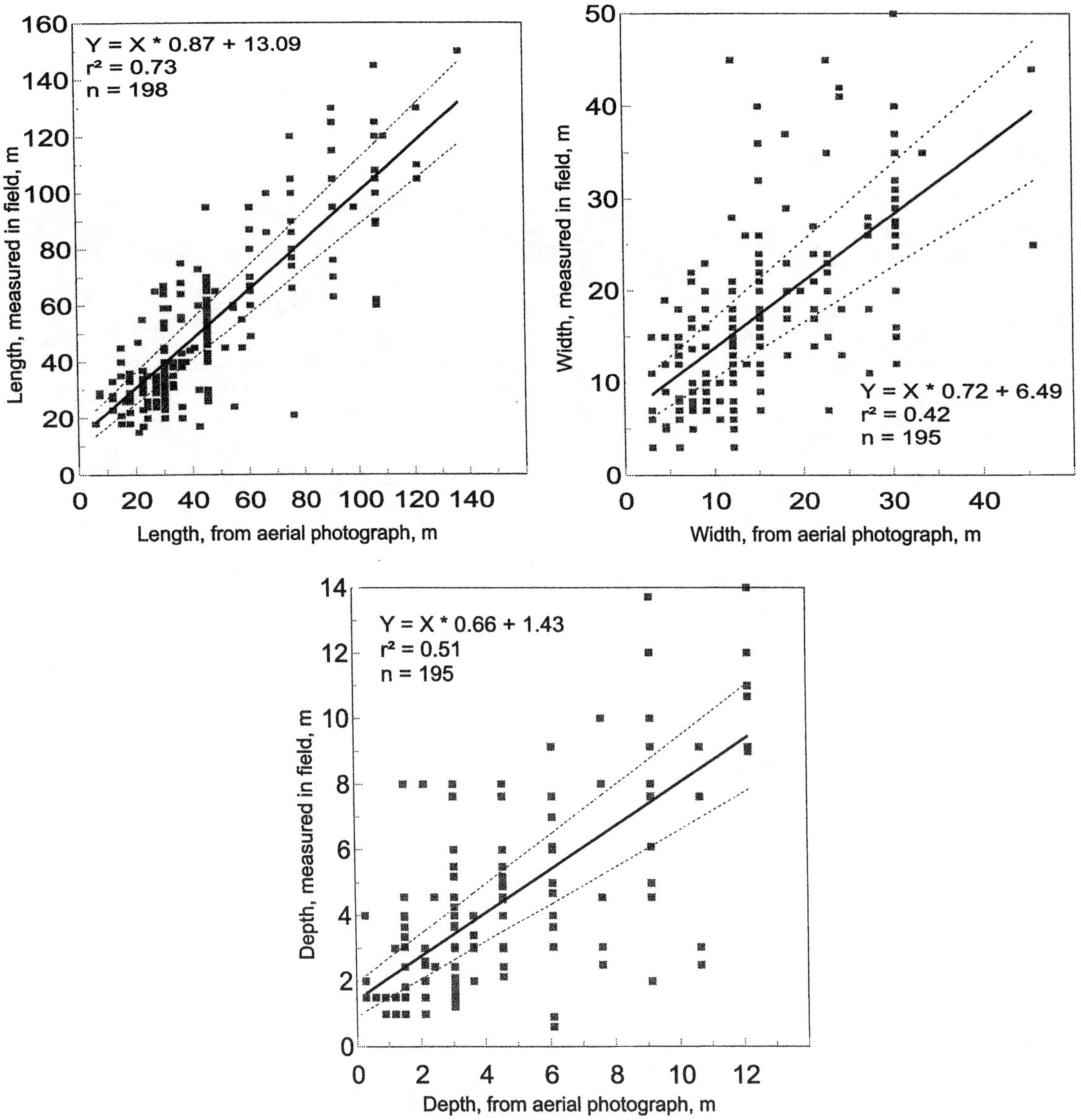

Figure 6. Relation between length, width, and depth of landslide scars measured in the field and measurements estimated from 1:20,000-scale aerial photographs, Cayaguás watershed. Simple linear regression models predict actual landslide scar dimension from that estimated using aerial photographs. Dashed lines show 95 percent confidence interval.

type of scar. Both of these landslide types are characteristic of high-intensity, short-duration storms in Puerto Rico in which 24-hour rainfall totals are on the order of 200 mm [*Larsen and Simon*, 1993]. Storms of this magnitude have about a 5-year recurrence interval [*U.S. Department of Commerce*, 1961]. The remaining landslide scars were classified mainly as slumps, debris avalanches, and complex. About 1% of the landslide scars could not be categorized because of poorly defined features.

Landslide frequency was 140 scars/km^2, which, for an estimated period of about 175 years, equals 80 landslides per km^2/100 y (Fig. 8). Mean soil density was determined to be 1.32 Mg/m^3. This value exceeds that measured in nearby undisturbed forest where the effects of compaction from grazing and agriculture are absent and also reflects the abundance of dense clay [*Larsen*, 1997; *Larsen et al.*, 1999; *Simon et al.*, 1990]. The combined total surface area of all landslide scars was 2.27 million m^2 and the total mass

Figure 7. Amphitheater-shaped landslide scars and cow terracettes (created by cattle moving on preferred paths parallel to the slope contour), Cayaguás watershed. Landslide scar morphology is subdued because of past agricultural activity and grazing. Much of the soil and saprolite eroded by the mass wasting in this watershed remains in storage in deposits 3 meters thick on footslopes and small (zero-order) valley bottoms such as that shown here.

equaled 14.4 million Mg (Table 3). Normalized to the watershed area, these values equal 140,000 m^2/km^2 and 870,000 Mg/km^2, respectively.

The volume of landslide scars totaled 660,000 m^3/km^2. If all colluvium was transported from the catchment, then the volume would be equivalent to a mean surface lowering of the entire watershed by 660 mm, or 3.8 mm/ y (Table 3). This is an extremely high rate of erosion, on the order of the accelerated rate of 8.5 mm/y documented at open-pit mines and construction sites in the United States [*Saunders and Young*, 1983; *Summerfield*, 1991]. The erosion rate of 3.8 mm/y is comparable to the anthropogenically induced erosion of 5 mm/y of soil and saprolite estimated for the Cayaguás watershed using the cosmogenic isotope of beryllium, ^{10}Be, documented by *Brown et al.* [1998]. Despite the high erosion rate, a large portion of the colluvium remains in the watershed.

Fluvial Sediment

The accurate determination of watershed sediment transport rates may require decades of sampling because of the importance of a few extreme events, as well as uncertainties due to changing land-use effects [*Dietrich* and *Dunne*, 1978; *Meade*, 1982; *Nordin*, 1985]. The data discussed below are, therefore, preliminary because of the short time period sampled.

Mean annual fluvial sediment yield, which includes suspended and bedload sediment for the 8 years with available data, was 2,570 $Mg/km^2/y$. Fluvial sediment yield was extremely low, 252 Mg/km^2, in 1991 when runoff was only 47% of the mean compared to a high of 8,570 Mg/km^2 in 1985, when runoff was 138% of the mean of 1,480 mm (Table 4). The largest storm of 1991 had a mean daily discharge of only 6.2 m^3/s. In comparison, there were 17 storms in 1985 with mean daily discharge that exceeded 6.2 m^3/s; the largest storm recorded in 1985 had a mean daily discharge of 53.5 m^3/s. This great interannual variability in discharges strongly influenced the mobilization and transport of fluvial sediment. The large difference (factor of 34) between sediment yield in 1985 and 1991 emphasizes the well known problem of characterizing mean annual fluvial

Table 2. Classification of 2,321 landslides by type in the Cayaguás watershed. Note, some totals may appear incorrect due to rounding.

Landslide type	Number	Percent
Debris flow	1,421	61.2
Soil slip	542	23.4
Slump	147	6.3
Debris avalanche	104	4.5
Complex	84	3.6
Undetermined	22	0.9
Earthflow	1	0.0

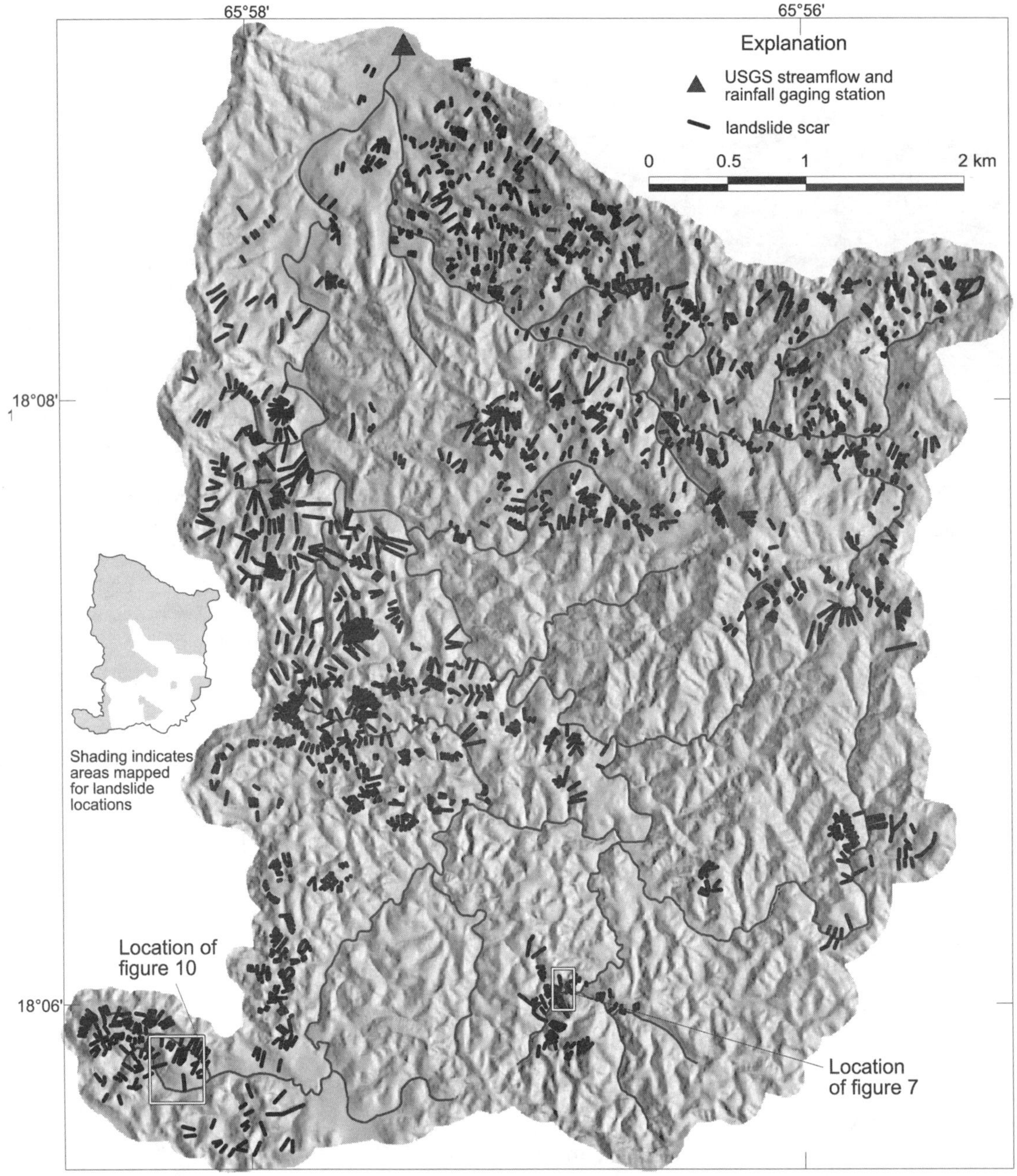

Figure 8. Shaded relief map of the Cayaguás watershed showing the location of 2,321 landslide scars mapped in a 16.6 square kilometer area (inset map shows area in which landslide scars were mapped) from 1937, 1951, and 1995 aerial photographs.

sediment yield with few years of data. The mean annual sediment yield for the 5 years of data representing the 1990's was only one-fourth that of the mean annual sediment yield for the 3 years of data collected in the 1980's. As is typical of many small watersheds, most of the annual sediment load was transported out of the watershed in a few days. For example, in 1985, the year with the greatest sediment yield, 63% of the annual yield occurred during a 5-day storm.

Sediment Storage

Channel widths were measured at 38 locations in the watershed and showed a pattern typical of fluvial systems where the increase in width with stream order defines a linear trend in semi-log space (Fig. 9; Tables 5,6). A total of 1.6 km of stream channel was surveyed to measure the dimensions of channel bars (Table 5). This length of channel represents a

Table 3. Summary of dimensions of 2,321 landslides mapped from aerial photographs (1937, 1951, and 1995) covering a 16.6 km^2 area of the Cayaguás watershed, Puerto Rico. The length, width and depth of landslide scars measured on aerial photographs and reported below were adjusted using linear regression models developed from comparison of ground-based and photograph measurements (Fig. 6). Range for total area, volume, and mass reflects the minimum and maximum estimates for length, width, and depth.

	Median	Mean	Standard error	Total
Length, m	42	53	0.66	
Width, m	14	16	0.18	
Depth, m	3.4	3.5	0.04	
Surface area, m^2	610	980	24.43	
Volume, m^3	1,800	4,700	213.63	
Mass, Mg	6,200	2,300		
Total surface area, million m^2 (range: 1.49 to 3.22)				2.27
Total volume, million m^3 (range: 5.57 to 18.8)				10.9
Total mass, million Mg (range: 7.35 to 24.9)				14.4
Landslide scars/km^2		140		
Landslide surface area, m^2/km^2		140,000		
Percent of basin surface area		14		
Landslide volume, m^3/km^2		660,000		
Total surface lowering averaged over study area, m		0.66		
Landslide mass, Mg/km^2		870,000		
Landslide mass, Mg/km^2/y (1820 to 1995)		5,000		
Landslides/km^2/100 y (1820 to 1995)		80		

channel surface area of 14,500 m^2. Mean channel bar thickness was 0.12 m and, using a mean density of 1.14 Mg/m^3 measured for sand bar deposits, the mean bar mass was 21 kg/m^2 of channel. A total of 5,300 Mg of sediment was calculated to be in active storage in channel bars (Table 6). Sediment in the other active compartment, the channel bed, was calculated to be 216,000 Mg. Using a 1.5-m mean thickess of alluvium and 3 m of colluvium thickness, the quantity of sediment stored in floodplains was estimated to be 1.79 million Mg; colluvial storage was about double this at 3.94 million Mg (Table 7). These two compartments are defined as semiactive. Inactive sediment, defined as that stored in approximately 4-m thick deposits of Quaternary alluvium, was calculated to be 0.68 million Mg. These data indicate that only a small portion of watershed sediment, 3.3%, exists in active storage compartments (Table 8). The majority of sediment, 87%, is in the semiactive compartments (i.e., floodplains and colluvium). The remainder, about 10%, is in Quaternary alluvium.

DISCUSSION

Landscape conversion in the Cayaguás watershed was carried to an extreme by 19th- and early 20th-century subsistence farmers who had removed at least 88% of the forest cover by 1937 (Fig. 3). It is probable that the farmers had eliminated even more forest cover prior to 1937, but aerial photographic documentation is not available. The removal of forest cover and subsequent intense year-round agriculture seems to have greatly accelerated hillslope erosion, as noted in other areas where human landscape disturbance is followed by very high erosion rates, at least

Table 4. Annual fluvial sediment discharge and runoff from the Cayaguás watershed, Puerto Rico *[Guzmán-Ríos*, 1989; *Larsen*, 1997; *Simon and Guzmán-Ríos*, 1990; *U.S. Geological Survey*, 1999). Mean annual runoff, 1977-95, is 1,480 mm.

Year	1984	1985	1986	1991	1992	1993	1994	1995	Mean, 1980's	Mean, 1990's	Mean, all years
Suspended sediment yield, Mg/km²	1,360	4,970	2,230	97	2,030	1,090	341	177	2,850	746	1,540
Bed-material yield, Mg/km²	984	3,600	1,620	154	497	785	382	268	2,070	417	1,040
Total fluvial sediment, Mg/km²	2,340	8,570	3,850	252	2,530	1,870	723	445	4,920	1,160	2,570
Runoff, mm/y	1,330	1,990	1,730	678	1,360	1,660	940	920	1,690	1,110	1,330
Annual runoff/mean	0.90	1.34	1.17	0.46	0.92	1.12	0.63	0.62	1.17	0.77	0.92

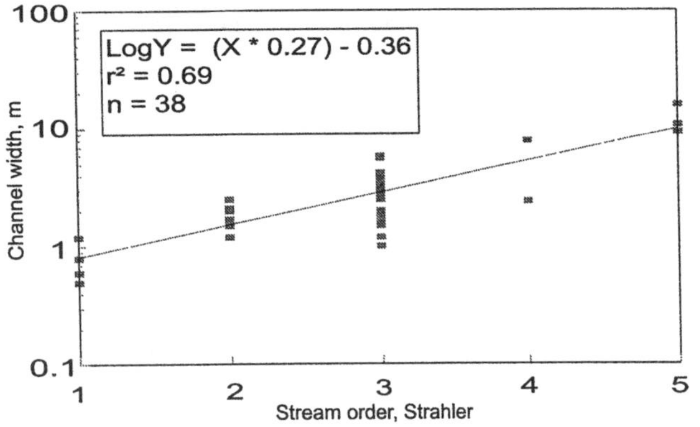

Figure 9. Relation between channel width and Strahler stream order in the Cayaguás watershed.

immediately after the disturbance [*Montgomery et al.*, 2000]. The nearby Icacos watershed, which is located in the Luquillo Experimental Forest and consists of primary forest with comparable bedrock geology (quartz diorite) and a mean slope (22% or 12.4°) has lower rates of landsliding (43 landslides/km^2/y), fluvial sediment yield (954 Mg/km^2/y), and sediment storage on footslopes [*Larsen*, 1997; *Larsen and Torres-Sánchez,* 1998; *Larsen et al.*, 1999]. This, in spite of more than double annual rainfall, 4,156 mm, in the Icacos watershed.

The number of landslide scars in the Cayaguás watershed, at 140 landslide scars per km^2, is the highest ever documented for Puerto Rico. Few studies elsewhere in the world have described landscapes of comparable area with greater densities of landslide scars. Scenes such as those shown in Figures 5, 7, and 10, are typical of the watershed: hillslopes are striated with numerous debris flow and soil slip scars and valley floors and footslopes are mantled with several meters of colluvium. No colluvium samples were dated but anecdotal evidence that colluvial material dates from the period of European forest clearing is suggested by the sharply angular form of wood and charcoal fragments present at several locations. The fragments appear to have been cut by a sharp-bladed instrument such as a machete, commonly used by farmers from the 19th century to the present.

Figure 10, which can be viewed in three dimensions with a stereoscope, highlights the abundance of landslide scars and their effect on hillslope morphology and land use. Landslide features dominate most of the hillslopes in the pair of photographs. Farmers have adapted to these conditions by planting plantains or grazing cattle on hillslopes, in spite of the presence of landslide scars. Landsliding, slopewash, and soil creep have shifted much of the nutrient-rich topsoil from the hillslope to the valley bottom. This large-scale, relatively rapid movement of sediment to valley floors which began in the 19th century, seems to have inhibited the pre-disturbance relatively rapid channel development here and elsewhere in the watershed, and has created a boggy, poorly drained substrate. On the colluvium- and alluvium-filled valley floors that result from these processes, farmers raise *Colocasia esculenta*, (malanga) the crop, noted above, that tolerates saturated soil conditions (Fig. 10). The most profound aspect of the rate of mass wasting in this watershed is the total volume of material shifted from hillslopes and moved to footslopes and river channels. This volume, 10.9 million m^3, indicates that, on average, the ground surface in this watershed was lowered by 660 mm in approximately 175 years, attesting to the remarkable effectiveness of humans as geomorphic agents (Table 3) [*Osterkamp*, 2000].

A conceptual geomorphic model of landform development for the Greater Antilles proposed by *Ahmad et al.* [1993] postulates that landsliding is the major contributor of sediment to stream channels. This concept has also been proposed for the Lesser Antillean island of Dominica by *Reading* [1986], who stated that landslides are the most important mechanism by which sediment is supplied to rivers. These models are supported by the present work in the Cayaguás watershed. Erosion of hillslopes is dominated by episodic mass-wasting processes, mainly landsliding, which, at decadal or greater time scales, deliver orders of magnitude more sediment to stream channels than all other hillslope-

Table 5. Summary of field survey of channel bar dimensions and distribution in the Cayaguás watershed. Density of sediment is 1.14 Mg/m^3.

Channel bar thickness, mean, m (n=134)	0.12
Surveyed channel bar surface area, total, m^2	2,500
Surveyed channel length, total, m	1,580
Channel bar area, mean, m^2/m of channel length	1.44
Channel area surveyed, m^2	14,500
Bar sediment, mean, kg/m^2 of channel	21

erosion processes combined [*Larsen*, 1997, *Larsen et al.*, 1999, *Gellis et al.*, 1999].

GIS analysis indicates that 1.15 million m^2 of the Cayaguás watershed has hillslopes of 10° or less, if ridgetops, floodplains, and river channels are excluded. If 3 m of colluvium are evenly distributed on these low-slope surfaces, a volume of 3.46 million m^3 is present in the watershed (Table 7). By subtracting this volume from the total volume of historic landslide scars in the Cayaguás watershed (10.9 million m^3), the difference, 7.44 million m^3 (282,000 m^3/km^2), provides an estimate of the amount of fluvial sediment export that may have occurred since deforestation began in 1820. Converting the difference to mass, 0.32 million Mg/km^2 of sediment may have passed the watershed outlet, equivalent to an average annual yield of about 1,800 Mg/km^2. This estimate may be low, but is reasonable, as the sediment yield during the years sampled in the 1980's and 1990's averaged from 1,160 $Mg/km^2/y$ (1990's) to 4,900 $Mg/km^2/y$ (1980's) (Table 4). It is assumed that during peak agricultural use of the watershed sediment yield was higher than the 1990's rate. Other evidence suggesting that 1,800 Mg/km^2 is reasonable, but may be a conservative approximation for the period of peak agricultural activity in the Cayaguás watershed, is obtained from a comparable site in the Greater Antilles. Estimated annual soil erosion in the humid central mountains of the Dominican Republic, where shifting cultivation and monoculture is still practiced, ranges from 5,930 to 6,930 $Mg/km^2/y$ [*Altieri*, 1990].

The mean landslide rate, 0.8/km^2/y, is equal to 5,000 Mg km^2/y (median landslide mass of 6,200 Mg times 0.8), which could explain more than 100% of the annual fluvial sediment yield (Tables 3, 4). Another source of fluvial sediment observed, but not quantified in this study, is the erosion of channel banks and flood plains caused by high flows. The 0.8/km^2/y landslide rate exceeds the contribution to sediment yield from landslides during recent years, especially when considered with other sources of fluvial sediment. Nonetheless, new landslides occurred during the period of study and many old scars are observed to be actively eroding in the Cayaguás watershed. Furthermore, hillslope lengths are short and many scars extend from ridgetop to stream channel, so landslide-derived sediment can be delivered directly to the fluvial system (Fig. 5B,10). Slopewash of 10 to 86 $Mg/km^2/y$ account for only a small portion of overall fluvial sediment yield. Although the contribution was not determined in this study, gullying provides a minor portion

Table 6. Stream channel dimensions and estimates of sediment in active storage compartments in the Cayaguás watershed. Channel lengths determined using GIS software analysis of digital elevation model. Channel widths derived from field-based model (see fig. 7). Estimates of the quantity of sediment in channel bars and bed derived from field surveys. Note, some totals may appear incorrect due to rounding.

Stream order	Length, m	Mean width, m	Area, m^2	Estimated mean bed thickness, m	Total bed sediment, Mg	Total bar sediment, Mg
1	83,900	0.82	68,800	0.02	1,600	450
2	44,500	1.53	68,100	0.05	3,900	890
3	23,500	2.85	67,000	0.5	38,200	1,750
4	9,600	5.32	50,900	1	58,000	1,560
5	10,100	9.93	101,000	1	115,000	630
Total	172,000		355,000		216,000	5,300

Table 7. Estimated sediment in semiactive (floodplain) and inactive (Quaternary alluvum) compartments, Cayaguas watershed. Mean floodplain thickness is 1.5 m, Quaternary alluvium thickness estimated at 4 m, and mean colluvium thickness is 3 m. Mass was calculated using a density of 1.14 Mg/m^3. Note, some totals may appear incorrect due to rounding.

	Area, m^2	Volume, m^3	Mass, Mg	Mass, Mg/km^2
Floodplains (semiactive)	1,040,000	1,570,000	1,790,000	67,600
Quaternary alluvium (inactive)	150,000	598,000	682,000	25,800
Colluvium (semiactive)	1,150,000	3,460,000	3,940,000	149,000

of fluvial sediment yield. The abundance of landslide scars and the degree of channel aggradation following Hurricane Hortense are dramatic evidence of the substantial role of episodic landslide-induced erosion in this watershed.

The storage of colluvium is extensive in the Cayaguás watershed. Storage of sediment is noted even in the headwaters of the Cayaguás watershed where contributing areas apparently limit runoff to below the competence required to erode valley-bottom deposits, except during infrequent events (Figs. 7,10). Another factor limiting the transport of colluvial sediment and channel development in some headwater areas is the effect of cattle on the bottomland soil surface. The combination of high annual rainfall and random grazing patterns on these relatively low gradient surfaces has resulted in saturated, poorly drained soil with very limited channel development. Standing water with depths of several 10's of cm occurs in pools created by cattle hoof prints and is widespread during most of the year at the heads of drainages. Nonetheless, field observations at dozens of sites immediately down-gradient of such areas indicate that recent headcutting into colluvium is active in the Cayaguás watershed (Fig. 5F). This suggests that at least some of the fluvial sediment now being transported out of the watershed is derived from sediment stored in semiactive compartments (floodplains and colluvium) deposited perhaps as long ago as the 1800's. The estimated amount of colluvium remaining in the watershed, 3.94 million Mg, is a 30- to 129-year supply of sediment at the 1980's and 1990's transport rates, respectively (Table 7). Although these calculations are based on a series of approximations, they provide a means to evaluate historic transport rates, estimate colluvial storage, and predict the potential mass of material available for fluvial transport out of the watershed.

The determination of the mass of sediment available for transport in the Cayaguás watershed (and other Río Grande de Loíza upland watersheds within the boundaries of the San Lorenzo batholith) is of more than theoretical interest. The Loíza Reservoir is downstream of the Cayaguás River (Fig. 1). This 267-ha reservoir is the source of about 50% of the water supplied to the 1.6 million inhabitants of metropolitan San Juan, at an average daily rate of 302,000 m^3 [*Webb and Soler-Lopéz*, 1997]. Sediment has accumulated in the reservoir at annual rates ranging from 110,000 m^3 to 452,000 m^3 since the first bathymetric survey in 1963. These volumes

Table 8. Summary of sediment stored in active, semiactive, and inactive compartments in the Cayaguas watershed, Puerto Rico. Note, some totals may appear incorrect due to rounding.

	Mass, Mg	Mass, Mg/km^2	Percent
Channel bars (active)	5,300	200	0.1
Channel bed (active)	216,000	8,200	3.3
Floodplains (semiactive)	1,790,000	67,600	26.9
Colluvium (semiactive)	3,940,000	149,000	59.5
Quaternary alluvium (inactive)	682,000	25,800	10.3
Total	6,630,000	251,000	

Figure 10. Stereo aerial photographs of southeastern Cayaguás watershed, Puerto Rico, 1951, showing 15 landslide scars of various ages, and sediment stored as colluvial/alluvial fill on valley floor. Sunlight from right (east) side of photographs. Second-order stream flows towards southeast.

are equivalent to an annual capacity loss of 1.1 to 1.5%, and indicate that average sediment yield from the basin has fluctuated between 750 and 900 Mg/km^2, assuming an average trap efficiency of 75% [*Webb and Soler-Lopéz*, 1997]. Current infilling is estimated to be 242,000 Mg/y. The current 1,160- to 4,920-Mg/ km^2/y annual sediment yield from the Cayaguás watershed, (which represents only 5% of the 538 km^2 contributing area of the reservoir) is 31,000 to 130,000 Mg, or an impressive 13 to 54% of the total annual reservoir sediment input. In addition, the estimated volume of colluvial storage in the Cayaguás watershed was approximated above to be 3.5 million m^3. The current capacity of the Loíza reservoir has been computed to be 14.2 million m^3. A high-magnitude storm, if able to mobilize a large portion of this stored sediment, could have a severe impact on reservoir storage and the San Juan water supply. Mitigation is expensive: in 1998-1999, 5 million m^3 of sediment was dredged from the reservoir at the cost of \$12/m^3 [*Vega and Terrasa-Soler*, 1998].

The Cayaguás mass-wasting erosion and fluvial-sediment tranport rates described above are high for the island of Puerto Rico. Even on geologic time scales, landscape denudation in the Cayaguás watershed seems to have been extreme. This is observed on topographic maps where the region underlain by the intrusive bedrock in southeastern Puerto Rico is anomalously lower than surrounding non-intrusive bedrock terrain. The San Lorenzo batholith appears as an extensive area of highly weathered rock with reduced relief compared to the ridges bounding the batholith along the western and southern margins (Fig. 1).

CONCLUSIONS

Landslide frequency in the Cayaguás watershed is the highest documented for the island of Puerto Rico, and is high compared to rates published for watersheds of comparable size elsewhere [*Summerfield*, 1991]. Mass wasting enhanced by human activity during the last two centuries has, on average, lowered the land surface of the watershed at the rate of 3.8 mm/y. Based on landslide frequency, the storage of colluvium, and annual fluvial sediment yields, hillslope processes that were triggered as long ago as 1820 by the onset of forest clearing, maintained high sediment yield during the 20th century and will probably affect fluvial sediment yield for most of the 21st century. This is sobering news to governments attempting to manage landscapes and watersheds in a sustained manner.

Land-use conversion in the Mediterranean-European region required several millennia and occurred in North America during a few centuries. In the humid tropics, it is occuring at a far more rapid rate, measureable in decades. In the context of inexorable population growth and increasing resource consumption, the impacts of development on natural systems in the low-latitude regions have, therefore, never been more intense than at the present time. The adverse effects of forest cutting, agricultural expansion, and urbanization are recognized at the global scale and are international concerns [*Osterkamp*, 2000].

What does the future hold? A rational appreciation of development processes in the humid tropics necessitates educated debate and compromise on how landscapes are to be converted or managed in the near and long term. Sustainable development of natural resources requires accumulation of extensive knowledge of geomorphic processes and sediment budgets in undeveloped and developing regions. The data base in the tropics is limited. One objective of this study was to contribute to that knowledge. Puerto Rico is uniquely positioned to provide a glimpse into the future for humid-tropical environments in that removal of forest and conversion of landscapes to agriculture peaked about 50 years ago. Subsequent industrial growth has allowed forests to recover as the agricultural sector has contracted. This setting is comparable to what might be expected during the 21st century in the impacted areas of Africa, Asia, and Amazonia, and therefore merits close scrutiny.

Acknowledgments. This work was funded though the U. S. Geological Survey, Global Change, Water, Energy, and Biogeochemical Budgets (WEBB) Program [*Lins*, 1994; *Larsen et al.*, 1993]. Nel Caine, University of Colorado, merits special thanks because of his advice and suggestions for supervising the dissertation from which data were extracted. The manuscript benefited from reviews by Kevin M. Schmidt, Mary Ann Madej, Waite Osterkamp, Heriberto Torres-Sierra, USGS, Mitchell Aide, University of Puerto Rico, and David R. Montgomery, University of Washington. Unpublished data on land cover in Puerto Rico were provided by Tania Del Mar López, U.S. Forest Service. Thanks are owed to Isaac Santiago Román, Vega Baja, Puerto Rico, Rebecca Kyle, Antioch College, Rachel Sears, Shepherd College, West Virginia, and Tim Doherty, Steve Manning, Chris Seveney, Worcester Polytechnic Institute, for assistance in field mapping of stream channels and landslide scars.

REFERENCES

Ahmad, R., Scatena, F.N., and Gupta, A., Morphology and sedimentation in Caribbean montane streams: examples from Jamaica and Puerto Rico, *Sedimentary Geology* 85, 157-169, 1993.

Aide, T.M., Zimmerman, J.K., Herrera, L., Rosario, M., and Serrano, M., Forest recovery in abandoned tropical pastures in Puerto Rico *Forest Ecology and Management* 77, 77-86, 1995.

Alberts, H.W., Pastures in Puerto Rico: Soil Conservation, Our work in Puerto Rico, *U.S. Department of Agriculture-Soil Conservation Service*, 6, no. 1, 15-18, 1940.

Altieri, M.A., Stabilizing hillside farming systems in the Sierra of the Dominican Republic *IAHS Publication 192, Research needs and applications to reduce erosion and sedimentation in tropical steeplands*, 355-363, 1990.

Beishlag, G., Trends in land use in southeastern Puerto Rico *in* Jones, C.F., and Picó, R., *Geography of Puerto Rico*, University of Puerto Rico Press, Río Piedras, PR, 269-296, 1955.

Birdsey, R. A., and Weaver, P. L., Forest area trends in Puerto Rico, *U.S. Department of Agriculture-Forest Service Research Note* SO-331, 5 pp. 1987.

Brockman, V., Land utilization in Caguas-San Lorenzo; in *Rural Land Classification Program of Puerto Rico*, Northwestern University Studies in Geography, no. 1, Northwestern University, Evanston, Illinois, 1952.

Broedel, C.H., Preliminary Geologic Map Showing Iron and Copper prospects in the Juncos Quadrangle, Puerto Rico, *U.S. Geological Survey Miscellaneous Geologic Investigations Map* I-326, scale 1:20,000, 1961.

Brown, E. T., Stallard, R. F., Larsen, M. C., Bourlès, D. L., Raisbeck, G. M., and Yiou, F., Determination of predevelopment denudation rates of an agricultural watershed (Cayaguás River, Puerto Rico) using in situ-produced ^{10}Be in river-borne quartz, *Earth and Planetary Sciences Letters,* 160, 723-728, 1998.

Caine, N., and Swanson, F.J., Geomorphic coupling of hillslope and channel systems in two small mountain basins, *Zeitschrift für Geomorphologie N.F.*, 33, no. 2, 189-203, 1989.

Campbell, R.H., Soil slips, debris flows and rainstorms in the Santa Monica Mountains and vicinity, southern California, *U.S. Geological Survey Professional Paper* 851, 51p, 1975.

Church, M., and Slaymaker, O., Disequilibrium of Holocene sediment yield in glaciated British Columbia, *Nature*, 337, 452-454, 1989.

Costa, J.E., Effects of agriculture on erosion and sedimentation in the Piedmont Province, Maryland, *Geological Society of America Bulletin*, 86, 1281-1286, 1975.

Cruden, D.M., and Varnes, D.J., Landslide types and processes, in *Landslides, investigation and mitigation*, edited by Turner, A.K., and Schuster, R.L, Special Report 247, Transportation Research Board, National Academy Press, Washington, D.C. pp. 36-75, 1996.

Dietrich, W., and Dunne, T., Sediment budget for a small catchment in mountainous terrain, *Zeitchrift für geomorphologie N.F.*, Supplementband 29, 191-206, 1978.

ESRI, *Understanding GIS--The ARC/INFO method*, Environmental Systems Research Institute, Inc., J. Wiley and Sons, New York, 535 p, 1993.

Franco, P. A., Weaver, P. L., and Eggen-McIntosh, S., Forest resources of Puerto Rico, 1990, *U.S. Department of Agriculture-Forest Service Resource Bulletin* SRS-22, 45 pp., 1997.

Garwood, N.C., Janos, D.P. and Brokaw, N., Earthquake-caused landslides: a major disturbance to tropical forests, *Science*, 205, no.4410, 997-999, 1979.

Gellis, A.C., Webb, R.M.T., Wolfe, W.J., and McIntyre, S.C.I., Effects of land use on upland erosion, sediment transport and reservoir sedimentation, Lago Loíza watershed, Puerto Rico, *U.S. Geological Survey Water Resources Investigation Report* 99-4010, 60 p, 1999.

Gerrard, J., Weathering of granitic rocks--Environment and clay mineral formation, in, Robinson, D.A., and Williams, R.B.G., eds., *Rock Weathering and Landform Evolution*, John Wiley and Sons Ltd., 3-20, 1994.

Graf, W., Late Holocene sediment storage in canyons of the Colorado Plateau, *Geological Society of America Bulletin*, 99, 261-271, 1987.

Gupta, A.V., Large floods as geomorphic events in the humid tropics in *Flood Geomorphology*, edited by Baker, V.R., Kochel, R.C., and Patton, P.C., J. Wiley and Sons, New York, pp. 301-315, 1988.

Gupta, A.V., Magnitude, frequency, and special factors affecting channel form and processes in the seasonal tropics *in Natural and Anthropogenic Influences in Fluvial Geomorphology*, edited by Costa, J.E., Miller, A.J., Potter, K.W., and Wilcock, P., American Geophysical Union, Geophysical Monograph 89, pp. 125-136, 1995.

Guzmán-Ríos, S., Suspended-sediment data in the upper Río Grande de Loíza basin, Puerto Rico, *U.S. Geological Survey Open File Report* 88-324, 42 p., 1989.

Haggett, P., Land use and sediment yield in an old plantation tract of the Serra do Mar, Brazil, *Geographical Journal*, 27, 50-59, 1961.

Jacobson, R.B., and Coleman, D.J., Stratigraphy and recent evolution of Maryland piedmont flood plains, *American Journal of Science* 286, 617-637, 1986.

Jones, A., Landslides of Río de Janeiro and the Serra da Araras escarpment, Brazil, *U.S. Geological Survey Professional Paper* 697, 42 p, 1973.

Kelsey, H.M., A sediment budget and an analysis of geomorphic process in the Van Duzen River basin, north coastal California, 1941-1975: Summary, *Geological Society of America Bulletin*, 91, 190-195, 1980.

Kelsey, H.M., Lamberson, Roland, and Madej, M.A., Stochastic model for the long-term transport of stored sediment in a river channel, *Water Resources Research*, 23, no. 9, 1738-1750, 1987.

Larsen, M.C. Tropical geomorphology and geomorphic work: A study of geomorphic processes and sediment and water budgets in montane humid-tropical forested and developed watersheds, Puerto Rico, *PhD dissertation, University of Colorado*, 341 pp, 1997.

Larsen, M.C. and Parks, J.E., How wide is a road? The association of roads and mass-wasting disturbance in a forested montane environment, *Earth Surface Processes and Landforms*, 22, 835-848, 1997.

Larsen, M.C., and Simon, A. Rainfall-threshold conditions for landslides in a humid-tropical system, Puerto Rico, *Geografiska Annaler*, 75A, (1-2) 13-23, 1993.

Larsen, M.C., Collar, P.D., and Stallard, R.F., Research plan for the investigation of water, energy, and biogeochemical budgets in the Luquillo mountains, Puerto Rico, *U.S. Geological Survey Open File Report* 92-150, 19 pp, 1993.

Larsen, M.C., and Torres-Sánchez, A.J., The frequency and distribution of recent landslides in three montane tropical regions of Puerto Rico, *Geomorphology*, 24, (4) 309-331, 1998.

Larsen, M.C., Torres-Sánchez, A.J., and Concepción, I.M., Slopewash, surface runoff, and fine-litter transport in forest and landslide scars in humid-tropical steeplands, Luquillo Experimental Forest, Puerto Rico, *Earth Surface Processes and Landforms*, 24, 481-502, 1999.

Lewis, L. A., Slow movements of earth under tropical rainforest conditions, *Geology*, 2, p. 9-10, 1974.

Lins, H.F., Recent directions taken in water, energy, and biogeochemical budgets research, *EOS, Transaction of the American Geophysical Union*, 75, (38) 433-439, 1994.

Madej, M.A., Changes in channel-stored sediment, Redwood Creek basin, northwestern California, 1947-1980, *U.S. Geological Survey Professional Paper* 1454-O1-O27, 1995.

Madej, M.A., and Ozaki V., Channel response to sediment wave propagation and movement, Redwood Creek, California, USA, *Earth Surface Processes and Landforms*, 21, no. 10, 911-928, 1996.

Maharaj, R.J., Landslide processes and landslide susceptibility analysis from an upland watershed: a case study from St. Andrew, Jamaica, West Indies, *Engineering Geology*, 34, 53-79, 1993.

Meade, R.H., Sources, sinks, and storage of river sediment in the Atlantic drainage of the United States, *Journal of Geology*, 90, 235-252, 1982.

Monroe, W.H., Geology of the middle Tertiary Formations of Puerto Rico, *U.S. Geological Survey Professional Paper* 953, 93 p., 1980.

Montgomery, D. R. and Dietrich, W.E., Source areas, drainage density, and channel initiation, *Water Resources Research*, 25 (8) 1907-1918, 1989.

Montgomery, D. R. Schmidt, K.M., Greenberg, H.M., and Dietrich, W.E., Forest clearing and regional landsliding, *Geology*, 28(4)311-314, 2000.

Morales-Munoz, G.E., *Orígenes historicos de San Miguel de Hato Grande- actual pueblo de San Lorenzo*, Esmaco Printers Corp., Hato Rey, PR, 333 p., 1943.

Nordin, C.F. Jr., The sediment loads of rivers, in *Facets of Hydrology*, edited by J.C. Rodda, vol. 2, John Wiley and Sons, New York, 183-204, 1985.

Osterkamp, W. R., 2000, Earth surface processes, materials use, and urban development--A case study of the San Juan metropolitan area, northeastern Puerto Rico, in Puerto Rico marine sediments, terrestrial and seafloor imagery, and tectonic interpretations, edited by K.M. Scanlon and P.R. Briere, *U.S. Geological Survey Open-File Report* 00-006.

Phillips, J.D., Sediment storage, sediment yield, and time scales in landscape denudation studies, *Geographical Analysis*, 18, no. 2, 161-167, 1986.

Reading, A.J., Landslides, heavy rainfalls, and hurricanes in Dominica, West Indies, *Ph.D. Thesis, University of Wales*, 1986.

Reading, A.J., Thompson, R.D., and Millington, A.C. *Humid tropical environments*, Blackwell Publishers, Cambridge, MA, 429 pp, 1995.

Rogers, C. L., Cram, C. M., Pease, Jr., M. H., and Tischler, M. S., Geologic Map of the Yabucoa and Punta Tuna Quadrangles, Puerto Rico, *U.S. Geological Survey Miscellaneous Geologic Investigations Map* I-1086, scale 1:20,000, 1979.

Saunders, I. and Young, A., Rates of surface processes on slopes, slope retreat and denudation, *Earth Surface Processes and Landforms*, 8, 473-501, 1983.

Scatena, F. N., Relative scales of time and effectiveness of watershed processes in a tropical montane rain forest of Puerto Rico, in *Natural and Anthropogenic Influences in Fluvial Geomorphology*, edited by Costa, J.E., Miller, A.J., Potter, K.W., and Wilcock, P., American Geophysical Union, Geophysical Monograph 89, 103-111, 1995.

Scatena, F. N., and Larsen, M. C., Physical aspects of Hurricane Hugo in Puerto Rico, *Biotropica* 23, (4A) 317-323, 1991.

Simon, A., and Guzmán-Ríos, S., Sediment discharge from a montane basin, Puerto Rico: Implications of erosion processes and rates in the humid tropics, in *Research Needs and Applications to Reduce Erosion and Sedimentation in Tropical Steeplands*, IAHS Publication 192, 35-47, 1990.

Simon, A., Larsen, M.C., and Hupp, C.R., The role of soil processes in determining mechanisms of slope failure and hillslope development in a humid-tropical forest: eastern Puerto Rico, in *Soils and Landscape evolution*, edited by Kneuper, P.L.K., and McFadden, L.D., Geomorphology, 3, 263-286, 1990.

Simonett, D.S., Landslide distribution and earthquakes in the Bewani and Torricelli Mountains, New Guinea, in *Landform Studies from Australia to New Guinea*, edited by Jennings, J. N., and Mabbutt, J.A., Cambridge Univ. Press, 64-84, 1967.

Strahler, A.N., Quantitative slope analysis, *Geological Society of America Bulletin*, 67, p. 571-597, 1956.

Summerfield, M.A., *Global Geomorphology*, Longman Scientific and Technical, Essex, England, 537 p, 1991.

Sutherland, R. A., and Bryan R. B., Sediment budgeting: A case study in the Katiorin drainage basin, Kenya, *Earth Surface Processes and Landforms*, 16, p. 383-398, 1991.

Tarboton, D. G., Bras, R. L. Rodriguez-Iturbe, I., On the extraction of channel networks from digital elevation data, *Hydrological Processes*, 5, 81-100, 1991.

Torres-Sierra, H., Hurricane Hortense impact on surface water in Puerto Rico, *U.S. Geological Survey Fact Sheet* FS-014-97, 4 p, 1997.

Trimble, S.W., The fallacy of stream equilibrium in contemporary denudation studies, *American Journal of Science*, 277, 876-887, 1977.

Trimble, S.W., Changes in sediment storage in the Coon Creek basin, Driftless Area, Wisconsin, 1853-1975, *Science*, 214, 181-183, 1981.

Trimble, S.W., and Mendel, A.C., The cow as a geomorphic agent--a critical review, *Geomorphology*, 13, 233-253, 1995.

U.S. Department of Agriculture, Soil Survey of the Humacao Area of Eastern Puerto Rico, *Soil Conservation Service*, 103 pp, 1977.

U.S. Department of Commerce, Generalized Estimates of Probable Maximum Precipitation and Rainfall-Frequency Data for Puerto Rico and the Virgin Islands, *U.S. Weather Bureau Technical Report* 52, 94 p, 1961.

U.S. Department of Commerce, Census (1990) of population and housing: Puerto Rico, *U.S. Department of Commerce*, Bureau of Census, CPH 1-53, 199 p., 1991.

U.S. Department of Commerce, Climatological Data, Annual Summary, 1998, Puerto Rico and Virgin Islands, *National Oceanic and Atmospheric Administration,* 44, no. 13, 16 p, 1999.

U.S. Geological Survey, 1999, Water resources data-- Puerto Rico and the U.S. Virgin Islands, Water year 1998, *U.S. Geological Survey Water-Data Report* PR-98-1, 641 p.

Vega, T., and Terrasa-Soler, J.J., Dredging of Carraízo reservoir: restoration of a critical resource for San Juan in *Proceedings, Tropical Hydrology and Caribbean Water Resources*, edited by Segarra-García, R.I., American Water Resources Association, San Juan, Puerto Rico, July 13-16, 1998, 161-166, 1998.

Wadsworth, F. H. Notes on the climax forests of Puerto Rico and their destruction and conservation prior to 1900, *The Caribbean Forester,* 11, (1) 38-46, 1950.

Webb, R.M.T., and Soler-López, Luis, R., Sedimentation history of Lago Loíza, Puerto Rico, 1953-1994, *U.S. Geological Survey Water Resources Investigation Report* 97-4108, 18 p, 1997.

Wieczorek, G.F., Preparing a detailed landslide-inventory map for hazard evaluation and reduction, *Bulletin of the Association of Engineering Geologists*, 21, 337-342, 1984.

Matthew C. Larsen, U.S. Geological Survey, 651 Federal Drive Guaynabo, Puerto Rico, 00965-5703 USA; mclarsen@usgs.gov

Abigail Santiago Román, Geology Department, Binghampton University, Binghampton, New York, 13902 USA abssantiago@hotmail.com

Animal Response to River Evolution in the Tidal Freshwater Zone

Gregory B. Pasternack

Department of Land, Air, and Water Resources, University of California, Davis, California

Landscape and river channel change caused by historic land clearance and intensive agriculture in coastal areas of the eastern United States has resulted in significant changes in the distribution, volume, and quality of riverine habitats. In this study of biogeomorphic dynamics in a tidal freshwater river, the extent, effectiveness, and mechanisms by which mammalian populations counteract anthropogenically-accelerated landscape evolution in the tidal freshwater zone are identified. Mapping and monitoring of vegetation, animal activity, substrate conditions, delta geomorphology, and sedimentation rates were carried out April 1995 to November 1997 in the Otter Point Creek tidal freshwater zone of Bush River in Maryland. Animal activities that were observed include plant uprooting, soil surface mixing, channelization, dam building, and lodge building by beavers and muskrats. These activities decreased soil surface elevation, increased daily flooding depth and duration, and locally redistributed sediment. Time series of net sedimentation show that animal-impacted sites exhibit unique local sedimentation pulses and erosional phases that are distinguishable from non-impacted sites within the same habitat type. The primary consequence of animal activity was a reverse succession of plant associations and a predictable change in related geomorphic site variables. Even though changes are localized, aerial photos show large interdistributary regions that are two or more steps behind in plant succession relative to surrounding areas, and these lagging sites are likely the result of historic animal activity. Thus, even though animals cannot stop human-accelerated river evolution, they can change the distribution of habitats and rates of vertical accretion across a delta.

1. INTRODUCTION

Human geomorphic activity in the United States has redistributed vast quantities of sediment across the landscape, particularly in eastern states [*Hooke*, 1999]. Post-European-settlement deforestation, agriculture, and urbanization are responsible for much of the historic landscape changes that have occurred [*Trimble*, 1974; *Costa*, 1975; *Jacobson and Coleman*, 1986; *Whitney*, 1994]. An important consequence of landscape change has been significant alteration of the distribution, volume, and quality of riverine habitats in response to sediment loading.

Where streams grade into coastal tidal freshwater rivers, historic sediment loadings caused widespread delta progradation and subtidal sedimentation [*Gottschalk*, 1945; *Brush*, 1984], and this has fundamentally transformed aquatic communities [*Brush and Davis*, 1984]. *Froomer* [1980] analyzed stratigraphic cross-sections of four deltas in Chesapeake Bay tributaries and found that modern alluvium overstepped tidal marshes, completely destroying

Geomorphic Processes and Riverine Habitat
Water Science and Application Volume 4, pages 139-157

them. *Davis* [1985] and *Hilgartner* [1995] reconstructed paleoecological conditions in upper Chesapeake Bay tributaries and found local species extinctions of submerged aquatic vegetation coincident with initial deforestation and land use.

Because many tidal freshwater wetland plants have a competitive advantage in a narrow range of environmental conditions [*Pasternack et al.*, 2000], the initial burial of pre-existing aquatic habitats by watershed sediments was followed by a sequence of allogenic succession driven by continued geomorphic change. At first, unvegetated, frequently flooded mudflats were colonized by floating-leaf plants that survive because of a biological adaptation known as aerenchyma-- air spaces in roots and stems which allow diffusion of oxygen from aerial plant parts into roots [*Mitsch and Gosselink*, 1993]. As river bed elevation rose through time, plant species populations underwent a directional and continuous pattern of colonization and extinction, as demonstrated by paleoecological studies and space-for-time analyses [*Khan and Brush*, 1994; *Hilgartner*, 1995]. A similar process has been reported for saline systems [*Ranwell*, 1974].

In the face of widespread geomorphic evolution and allogenic plant succession in the tidal freshwater zone, historic wildlife populations must have struggled as their habitats were destroyed. Although it is difficult to determine the pre-settlement population size of animals that lived in the tidal freshwater zone of rivers, records of the early fur-trading industry demonstrate that large numbers of beavers were present on the east coast of the United States at the time of colonization [*Whitney*, 1994]. Other species such as otters and muskrats were likely to be similarly numerous. Most species have no ability to counteract geomorphic change, but some, notably beaver, may have been able to engineer intertidal and riparian regions to create and maintain preferred habitat conditions. Beavers have long been known to alter riverine habitat through channelization [*Berry*, 1922], dam building [*Ives*, 1942; *Ruedemann and Schoonmaker*, 1938; and *Gurnell*, 1998], and tree cutting [*Jenkins*, 1980]. These activities have been reported to have significant ecological and geomorphic consequences, such as development of river floodplains [*Ruedemann and Schoonmaker*, 1938], increased habitat diversity via patch creation [*Johnston and Naiman*, 1990; *Snodgrass*, 1997], and retention of sediment and biogeochemical loads [*Naiman et al.*, 1994; *Butler and Malanson*, 1995].

Given that fluvial geomorphology, plant ecology, and wildlife populations are all coupled, the question posed in this paper is how did animal populations respond to rapid habitat change in the tidal freshwater zone of rapidly changing tributaries to Chesapeake Bay and did their response effect the interactions between abiotic and biotic variables in the system? Based on the existing literature on beaver activity and the role of beavers in watersheds, it was hypothesized that animals did attempt to counteract habitat changes, so it was necessary to first determine the mechanisms by which such attempts would have been carried out in the tidal freshwater zone, as well as the consequences of such activity. Mechanisms and animal impacts were investigated by randomized field monitoring at the point scale. Once the current mechanisms by which animals alter habitat conditions were identified and their short-term consequences on biogeomorphology were quantified, a conceptual model for how they would affect geomorphic evolution over the long term was developed. Actual long-term consequences of observed animal activity at a study site were inferred by interpretation of the present geomorphic and ecological structure of the system in light of the conceptual model.

2. STUDY SITE

Otter Point Creek (OPC) is a research preserve located on the tidal freshwater delta at the head of Bush River in upper Chesapeake Bay (Fig. 1). The tidal freshwater zone where OPC is located has undergone major changes as a result of post-European-settlement land use activity in the Winters Run basin beginning in the mid-17th century. Maps of the delta exist for the years 1836, 1902, 1915, 1922, and 1951. These maps along with the oral history of the region document historic geomorphic changes [*Pasternack et al.*, 2001]. Prior to settlement, the river was deep and clear, with diverse beds of submerged aquatic vegetation. Paleoecological analysis of sediment cores from the river have shown that OPC had a stable ecosystem (i.e. consistently low sedimentation rates, no lasting record of major disturbance, and low turnover of plant species) for more than 1,000 years, even in the face of regular hurricanes, fires, and natural climate variations [*Hilgartner*, 1995; *Pasternack et al.*, 2001]. Once land clearance began, soil erosion and sediment transport accelerated the formation of a delta. The delta has since rapidly prograded in response to increasing land clearance and intensive land use. By the beginning of the 20th century, 80% of the watershed was deforested. Between 1836 and 1951 the delta grew by a factor of 7.5 [*Pasternack et al.*, 2001]. Presently, the basin is 48% forest and the delta continues to prograde, impacting downstream marinas.

The modern hydrology of the OPC delta is divided into two distinct regimes. Flow in the 54.4-ha riparian forest on the delta plain derives from the 150-km^2 Winters Run basin and is largely contained in two major distributary channels, except during floods, when it occupies a complex network

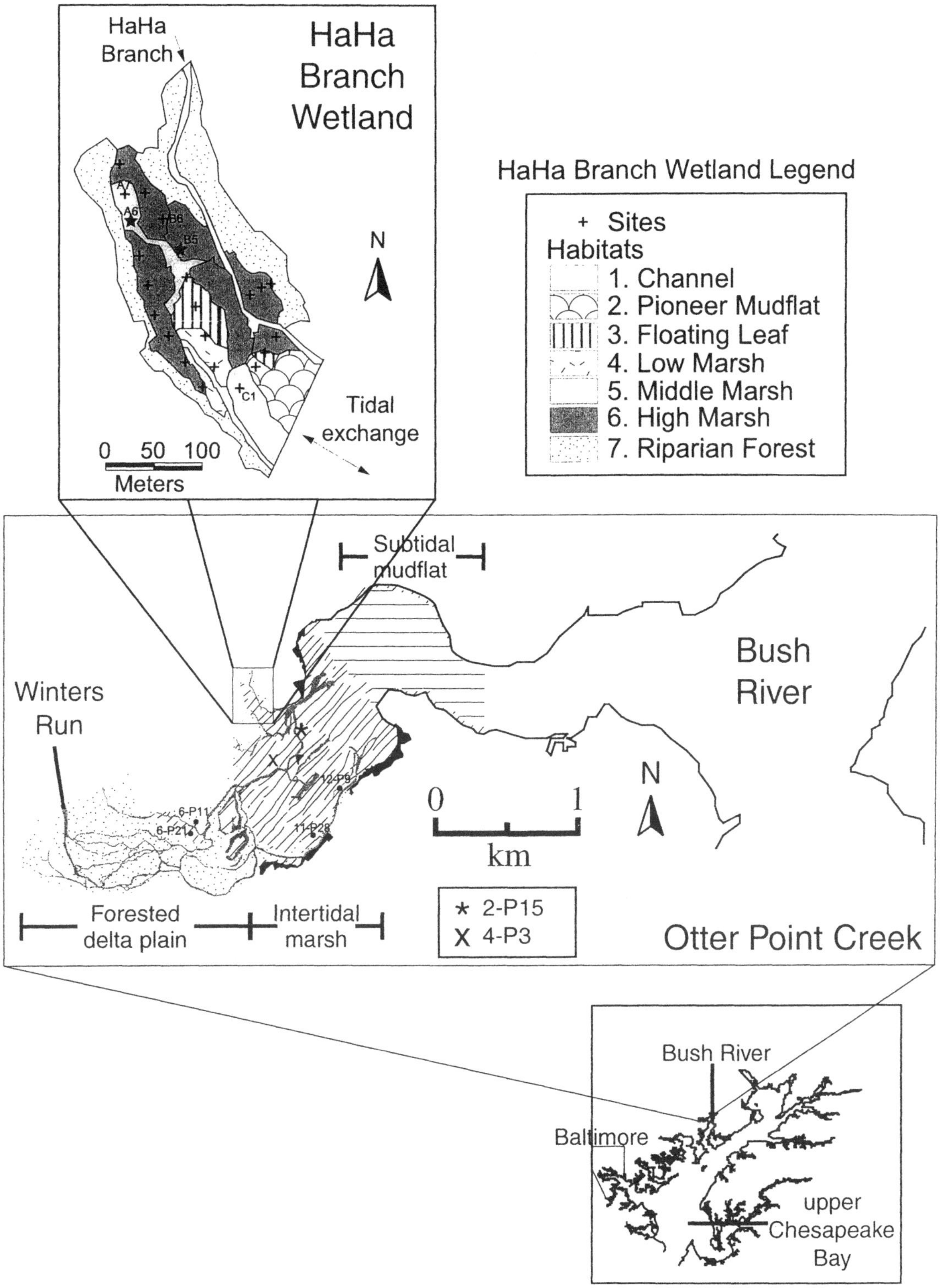

Figure 1. Maps showing the location and delta zonation of Otter Point Creek at the head of Bush River as well as the spatial pattern of habitats in HaHa Branch Wetland. Locations of sites mentioned in text are labeled; animal-impacted sites are shown using stars.

of ephemeral channels. Changes to the channel network are primarily controlled by large woody debris, which blocks channels during storms. For example, during Hurricane Agnes in June 1972, two logjams that are still present were formed and they blocked the southern distributary channel. In contrast to that of the forested delta plain, the hydrology of the 84-ha intertidal zone is primarily controlled by wind-enhanced tides [*Pasternack*, 1998].

Vegetation at OPC has been mapped using aerial photos and extensive ground-truthing field surveys. *Pasternack et al.* [2000] used multivariate statistical methods including cluster analysis, direct gradient analysis, and multiple regression analysis to show that the nine "plant associations" (i.e. groups of species commonly found together) present at OPC are aligned along an environmental gradient in abiotic and biotic conditions. Because of the presence of a predictable gradient, they were able to derive a simple algorithm that computes a "habitat index" that succinctly represents the ecological state of the site based solely on its plant species abundance data from 1-m^2 quadrat surveys (field method described below). The equation for the habitat index is

$$HI_x = \sum_{i=1}^{n} \left(c_{ix} \cdot PP_i\right) \qquad (1)$$

where HI_x is the habitat index for site x, n is the number of species at site x, c_{ix} is the relative percent cover of species i at site x, and PP_i is the preferred gradient position value for species i, which is obtained from the database published as Table 2 in *Pasternack et al.* [2000]. HI_x strongly correlates with site elevation, and together with elevation and distance from the tidal inlet, it accounts for ~90-95% of the spatial variation in sedimentation rates [*Pasternack et al.* 2000]. Thus, the habitat index is a useful tool for studying interdependent abiotic-biotic relations.

Wildlife observed at OPC include *Castor canadensis* (beavers), *Ondatra zibethicus* (muskrats), and *Lontra canadensis* (river otters). Signs of animal activity, including debarked trees, channel bank slides, animal lodges, and channelized marsh plains are widespread on the delta. No ecological studies of these species or others have been conducted at OPC.

3. METHODS

3.1 Field Surveys

Otter Point Creek has been under study since 1990 to better understand the interactions between physical and ecological processes in the tidal freshwater tributaries of upper Chesapeake Bay. Even though a study of animal activities was not initially planned, the extent of activity combined with the random siting of monitoring transects enabled sufficient observation to warrant this investigation. Mapping and monitoring of vegetation, sedimentation and erosion, substrate conditions, and animal activity were carried out April 1995 to November 1997 at OPC. During an 88-week period from April 1995 to March 1997, monitoring focused on the small HaHa Branch Wetland (HBW) on the north flank of OPC where the HaHa Branch stream terminates (Fig. 1). In the middle of that study, monitoring was expanded to sites throughout OPC that were observed for a 60-week period from September 1996 to November 1997. In all, 53 monitoring sites were used, and these were distributed along randomly located transects, as detailed by *Pasternack and Brush* [1998] and *Pasternack* [1998].

Percent cover of herbaceous plants and saplings was surveyed at each site during summer and autumn 1995-1997 using 1-m^2 quadrats. The methodology is identical to that used in previous work on the delta and reported in *Pasternack et al.* [2000]. It involves counting leaf cover for each species within each square decimeter of a quadrat [*Mueller-Dombois and Ellenberg*, 1974]. Taxonomy follows *Fernald* [1970]. Once plant species' abundances were known, equation 1 was used to convert vegetation data from each site into a habitat index.

An "anchored tile" approach suitable for monitoring sedimentation and erosion when it is extremely variable in time and space was used at the 53 sites on the OPC delta. This method is described by *Pasternack and Brush* [1998], with a detailed protocol available from the author upon request; it involves anchoring detachable ceramic tiles to the marsh surface and then collecting accumulated sediments biweekly for laboratory determination of net sedimentation (g cm^{-2} yr^{-1}), organic content (% loss-on-ignition), and geochemistry. Erosion was measured by multiplying the volume of sediment removed between a tile and the underlying marsh surface by local sediment bulk density. Because measurements are made so frequently, other potential sources of elevation change within the 1.22 m span of the anchor rod that might mimic erosion over long time scales, such as compaction due to respiration of organics or sediment consolidation, cannot be significant contributors to the observed elevation changes.

Substrate conditions at each site were periodically assessed for bulk density, grain size distributions, and post-animal activity disturbance characteristics. Bulk density (g ml^{-1}) was measured by coring soils with a 26-ml tube (1.29 cm in diameter by 5 cm in length), weighing the material dry, and dividing the weight by 26. Samples for grain size analysis were collected from the top 3 cm of the soil profile with a hand shovel, transferred to resealable plastic bags, and stored in a refrigerator at 4° C. Grain-size distributions were determined according to *Folk* [1974] after organics were removed using 30% H_2O_2 over a 24-hour period.

Animal activity in the vicinity of monitoring sites was observed and recorded on a biweekly basis during sediment collections. Qualitative observations included types of activities, general hydrogeomorphic impacts, and indicators of the species responsible for the activity. Quantitative measures included area impacted by activity and laboratory analyses of bulk density, organic content, and grain size of sediment samples collected from disturbed monitoring sites.

3.2 Data Analyses

Monitoring sites impacted by animal activity were compared with non-impacted sites in the same plant association to distinguish the biotic disturbance signal from background abiotic processes. One set of comparisons focused on time series of net sedimentation rates and percent organic content of deposited material in relation to animal activity at individual sites. A second set of comparisons focused on time-averaged process-morphology feedbacks among all study sites as evident in relations between sediment response variables (for example, sedimentation rate) and biogeomorphic controls (for example, habitat index and elevation). All analyses were combined to obtain a conceptual model of the impacts of animal activity on biweekly to interannual time scales.

Once the conceptual model was developed, implications of animal activity for delta evolution over longer time scales were assessed. Indicators of the actual long-term impacts of animal activity were identified by studying the structure of delta landforms as well as by considering distributions of plant and wildlife habitats. Broad scale spatial patterns were investigated using field observations coupled with Geographical Information System (GIS) data. GIS data included a 1.22 m resolution 1994 color infrared digital orthophoto of the delta as well as plant association, topographic, and sedimentation pattern data layers.

4. RESULTS

4.1 Monitoring Time Series

Four out of the fifty-three monitoring sites were directly impacted by animal activity. The animals responsible for the impacts were beavers and muskrats, with each causing a different effect and working during a different time of the year. Both animal species uprooted plants, mixed the surface sediment layer, dug new marsh plain channels, maintained old marsh plain channels, made mud piles, built new lodges, and maintained old lodges. Whereas beaver impacts were initially rapid and spanned a large radius, muskrats continuously worked over time and focused cutting along their channels. Also, beaver activity occurred in autumn while muskrat activity occurred in spring and early summer.

In HBW, beavers also cut trees in the riparian forest on the eastern side of the marsh and used those trees in an attempt to dam the HaHa Branch stream just upstream of an abandoned paleo-distributary channel (Fig. 2). The riparian forest that was the source of the wood for the dam had 313 trees, which were fortuitously all mapped, identified, and measured for diameter at breast height six months before dam building commenced. Arboreal species found in the forest included *Prunus serotina* (black cherry), *Acer rubrum* (red maple), *Liquidambar styraciflua* (sweet gum), *Ilex opaca* (American holly), *Betula nigra* (river birch), *Salix nigra* (black willow), and *Prunus virginiana* (choke cherry). *Acer rubrum* was the most prevalent species, and it was the primary one used for the dam. This population was very young, with diameters ranging from 2-4 cm in most cases. The dam failed during spring high flows in 1996 and was never rebuilt. Instead, beavers used the remaining debris to make a large lodge in the middle marsh habitat on the western side of the channel adjacent to the paleo-channel (Fig. 3).

Site A6 was in an HBW area containing middle marsh and some high marsh plants that were all herbaceous. Beavers altered it in autumn 1995 by cutting down most plants in a 180-m^2 region and re-excavating a channel network that was barely visible prior to their effort due to sediment deposition during the previous year. Where they passed through the high marsh and middle marsh near A6 (Fig. 1), the rehabilitated beaver channels were rectangular and 30 cm wide x 20 cm deep (Fig. 4), with dredged sediment piled on top of channel banks. The main stem of the excavated network that ran through the low marsh was not rectangular because the sediment was not cohesive enough to sustain straight-sided banks, and there were no plant roots on the banks to aid cohesion. Field observations during biweekly sampling trips indicated that the ground surface elevation in the impacted area was decreased relative to the surrounding land, though no repeat elevation surveys were possible. Associated with the decreased elevation was a clearly visible, increased flooding depth and duration relative to the surrounding undisturbed area.

According to monitoring data from site A6, beaver activity locally increased sedimentation for a 2-month period (Fig. 5a). This increase was a direct result of lowered elevation and increased flooding, which enabled more sediment transport. No measurable variation in organic content of accumulating sediment was observed at A6 during the

Figure 2. Beaver dam on HaHa Branch adjacent to wetland during winter 1996. The dam was blown out during subsequent high spring flows. For scale, the stream is 7 m wide.

beaver activity (Fig. 5b), because organic content was low in these sediments and organic matter has very low mass. Nevertheless, the sediment was quite distinguishable in appearance, looking like muddy hay, with many fragments of shredded plant stalks and roots. Sedimentation rates at A6 dropped down to background values in 1996, though vegetation underwent reverse allogenic succession to low marsh species, particularly *Peltandra virginica* (arrow arum), providing further evidence for lower elevations and increased flooding. However, in the mild winter of 1997 the site was significantly eroded as were other low marsh sites reported by *Pasternack and Brush* [1998] and unlike other middle marsh sites (Fig. 5a), illustrating long term consequences of animal disturbance. Pulses of organic deposition occurred in autumn 1996, and were caused by foliage dropped from the adjacent riparian forest buffer zone (Fig. 5b).

A comparison of time series among middle marsh sites shows that the signal of beaver disturbance at A6 is significantly different from the background abiotic signal. Sites C1 and A7 are the other middle marsh sites monitored at HBW. Because winter 1997 erosion at A6 counterbalanced the earlier deposition, the mean net sedimentation of the three sites is similar. However, the standard deviation in sedimentation was much higher at A6 than at the other sites (Table 1A). Neither A7 nor C1 showed a significant spike in sedimentation in October 1995, demonstrating the local nature of the disturbance, even though A7 was only 30 m away from A6 (Fig. 5a). Toward the end of the beaver activity, some sediment was carried further into the marsh where it deposited at site A7, causing the only significant temporal variation in sedimentation rate at that site over the whole study. In terms of organic content, A7 showed consistently higher values-- twice those of the other two sites (Table 1A)-- due primarily to its inability to flush autumn arboreal debris from the fringing, forested buffer zone except in early spring when no vegetation was present on the marsh surface. Beaver activity in the vicinity of A6 caused so much surface sediment mixing in autumn 1995 that arboreal debris could not accumulate there. Also, in the fall of 1996, the beaver-created channel near A6 was still effective enough to continue to export arboreal debris. No other site along the A-transect was able to export accumulating debris, so regardless of which mechanism-- surface mixing or channel export-- was more important, beaver activity was an important factor in the fate of a major source of organic deposition.

Site B5 was in a *Phragmites australis* (common reed) patch in the high marsh, and this patch occurs on the edge of a sand splay caused by overbank flooding of the HaHa

Figure 3. Beaver lodge constructed out of Red Maple and Black Willow branches after dam failure.

Branch stream. An actively used muskrat lodge was present at the site throughout the study. B5 showed a high average sedimentation rate and a low average organic content (Table 1B). A spike in organic content at B5 in May 1996 was an artifact of extremely low sedimentation. Under this condition, accumulating material was predominantly composed of in situ plant leaves whose deposition is capricious. Significant over-winter deposition and frequent large variations in sedimentation rate distinguished the muskrat-impacted site from comparable sites without the influence (Fig. 6). For example, B6 was a high marsh site 30 m further upslope in the marsh that was not in the stand of reeds. That site showed little temporal variation in sedimentation rate except for a single spike of inorganic sedimentation in the first half of April 1996. The cause of the spike at B6 was unknown.

Site 2-P15 was in a middle marsh area dominated by *Typha angustifolia* (cattails). In late summer and autumn 1996, beavers cut down all cattails in a 40 m^2 area near 2-P15 and used them to build a lodge (Fig. 7). As at A6, the impacted area was visibly depressed and was flooded for a greater duration. The result of this change was a high average sediment-deposition rate and a low average organic content (Table 1C). Sedimentation rates in late summer and autumn 1996 were moderate, while organic contents were low (Fig. 8). On December 25, 1996 a dead beaver was found on the western side of the cleared area. During winter 1997, the area around the lodge, including site 2-P15, experienced significant net erosion, which was typical of low marsh habitat rather than the middle marsh habitat that was present prior to beaver activity. During the following spring, muskrats took over the abandoned lodge and dug small channels throughout the area. These muskrat channels connected the site to the marsh front 10 m away and a major distributary channel 25 m to the east (Fig. 1). Meanwhile, few new middle marsh plants sprouted in the impacted zone, though they grew in the adjacent area. Instead, the low-marsh plant *Peltandra virginica* colonized

Figure 4. Example of a channel cut by animals at HaHa Branch Wetland. This channel begins in the tidal inlet and cuts NE all the way to the natural levee of HaHa Branch. It is 30 cm wide and 20 cm deep. No abiotic channels exist in HBW.

the disturbed ground. Since this plant decomposes quickly upon death, it is unsuitable for constructing lodges. Consequently, very little material was available for lodge maintenance, and the lodge was observed to deteriorate rapidly during summer 1997.

Concomitant with the muskrat activity, extreme swings in sedimentation rate followed (Fig. 8a). The peak sedimentation rate was 40 g cm^{-2} yr^{-1} while the minimum was 0.12 g cm^{-2} yr^{-1}. Unfortunately, the data cannot distinguish whether the high loads and fluctuations were due to changes in the rate of sediment import or changes in local redistribution, but both mechanisms are directly caused by animals in this instance. By comparison, non-animal impacted sites such as 12-P9 had almost no measurable deposition or erosion throughout the study. Under background conditions, OPC middle marsh areas receive less than 0.07

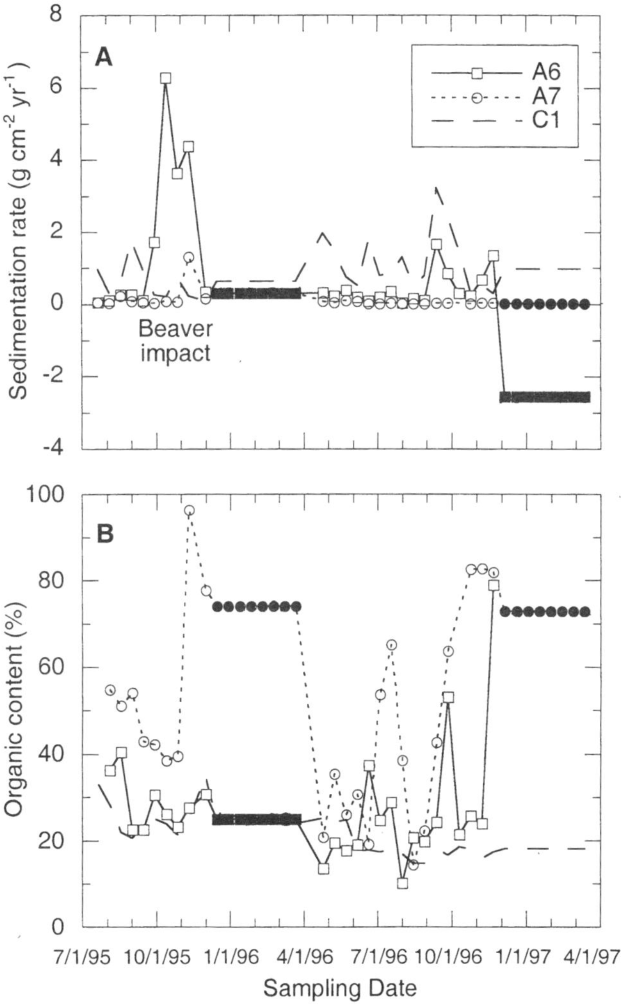

Figure 5. Comparison of (a) sedimentation and erosion and (b) organic content between a beaver-impacted site (A6) and two non-impacted sites in the same habitat type. Filled symbols indicate data that is averaged over the winter when anchored tiles were frozen into the marsh and could not be sampled.

g cm^{-2} yr^{-1}. Site 2-P15 experienced more than an order of magnitude more sedimentation than that on average (Table 1C).

Site 4-P3 was in a middle marsh locale with a mixture of arrow arum and cattails. Surrounding the disturbed area was a densely populated cattail field. The monitoring site showed many signs of animal activity in recent years, including in-filled animal channels, cut cattail stalks, and a dilapidated lodge. Muskrats began working the site in spring 1997. During the following summer, they intensively channelized the area and built up a former beaver lodge with stalks from surrounding cattails. In contrast to beaver activity observed at A6 and 2-P15, muskrats did not graze all vegetation indiscriminately, just plants where channels would go. As a result, the site developed pockets of arrow arum mixed in with remaining cattail stands. Sedimentation and erosion dynamics at 4-P3 reflect the timing of muskrat activity. During autumn for both years, the sedimentation rate was a moderate 1-2 g cm^{-2} yr^{-1}. During winter 1997, the site experienced substantial erosion, while during summer 1997 it experienced a large pulse of deposition (Fig. 9). The average sedimentation rate over the study was the highest at the disturbed site, at 1.18 g cm^{-2} yr^{-1}, and had the highest standard deviation (Table 1D). Meanwhile, organic content in the disturbed area was the lowest, though there was a large pulse of organics in autumn 1997 when cattail debris was deposited.

4.2 Time-averaged Biogeomorphic Relations

When sedimentation rates are plotted as a function of habitat index, disturbed sites show a predictable process-morphology feedback in relation to the delta's biogeomorphic evolution line. The autumn 1995 time-averaged sedimentation rates for non-impacted HBW sites show a strong response to habitat index (Fig. 10a). The relation suggests that biogeomorphic parameters such as elevation and plant associations control sedimentation processes at the seasonal time scale. As discussed by *Pasternack et al.* [2000], there is a dynamic interdependence between abiotic and biotic processes, such that plant associations and ground elevation are in turn changed by sedimentation over annual to interannual time scales. When a space-for-time transformation is applied to the seasonal relation in Figure 10, it appears that as a site evolves into higher and drier habitats, sedimentation rates decrease exponentially under normal conditions.

In contrast with non-impacted sites, the beaver-impacted site (A6) plotted far off the trend line during the time of beaver activity, while the muskrat-impacted site (B5) showed only a small deviation (Fig. 10a). Arrows in Figure 10a graphically represent site A6's evolution. The decreased elevation and increased inundation induced by beavers at A6 caused an immediate increase in sedimentation (vertical arrow) followed by a reverse succession in the plant association toward low marsh plants during the next growing season (horizontal arrow), as described earlier. Thus, the data from A6 shows that when plant species and sedimentation rates are not mutually adjusted due to a rapid disturbance, there is a natural process response that returns the system to the equilibrium balance between abiotic and biotic variables defined by the line in Figure 10a. Lower elevation, more flooding, and higher localized

Table 1. Time-averaged values comparing animal-impacted and non-impacted sites.

Site	Animal Activity	Deposition ($g\ cm^{-2}\ yr^{-1}$)			Organic Content (%)		
		Mean	Standard Devation	N	Mean	Standard Devation	N
A)							
A6	beaver	0.14	1.81	42	27.11	12.13	33
A7	none	0.12	0.22	41	58.76	21.38	40
C1	none	0.91	0.62	42	21.11	4.53	42
B)							
B5	muskrat	0.43	0.28	42	25.14	11.25	42
B6	none	0.13	0.29	41	37.05	15.08	41
C)							
2-P15	beaver & muskrat	0.80	9.88	30	19.75	9.79	30
12-P9	none	0.08	0.08	27	47.18	18.30	18
D)							
4-P3	muskrat	1.18	12.54	30	16.75	15.94	28
11-P28	none	0.05	0.06	27	78.33	12.25	24

sedimentation requires a shift to more inundation- and sedimentation-tolerant species.

Once equilibrium is re-established, the higher sedimentation rate of the new regime causes elevation to increase faster, thereby undoing the animals' work. Ultimately the site should return to its pre-disturbance site should return to its pre-disturbance habitat state over time (diagonal arrow), requiring animals to periodically maintain their channels and cleared areas. Overall, the system has a negative feedback mechanism that counteracts local animal activity. Because biweekly sedimentation data is not suitable for making accurate long term calculations, it is not possible to accurately estimate how long it would take the site to revert back to middle marsh, assuming that animals allow such a reversion to occur. Most likely this process takes several years, depending on how much winter erosion occurs under the imposed low marsh regime.

A similar feedback process as reported for HBW is evident in the data from OPC. Summer 1997 time-averaged sedimentation rates for sites 4-P3 and 2-P15 show that both underwent reverse succession from middle to low marsh (Fig. 10b). Whereas the autumn 1995 data from A6 captured a system out of equilibrium, the summer 1997 data for 4-P3 and 2-P15 represent a system after the vegetative response to changed abiotic conditions, as both sites plot very close to the biogeomorphic evolution line in the vicinity of expected low marsh conditions. Neither site has any characteristics resembling the middle marsh that surrounds them.

Because the data presented thus far represent short-term snapshots of processes occurring at OPC, a more integrated measure of longer-term hydrogeomorphic dynamics of the system might provide a broader context for understanding the impacts of animal activities. The spatial distribution of inorganic sediment grain sizes in HBW provides just such a measure, as it characterizes the energy distribution in the system and identifies the primary direction of sediment transport. A ternary plot of bed material sizes from HBW monitoring sites shows two distinct hydraulic sorting lines (Fig. 11). The first line represents sorting of sandy loam as it is transported through the tidal inlet into HBW or over the banks of the HaHa Branch stream (Fig. 1). The second line represents sorting of the resulting clayey silt as it is transported into the most distal regions of HBW, where only clay remains in transport. Neither animal-impacted site shows a deviation from its expected position along the sorting pathway. This suggests that the biogeomorphic process response that restores the biotic-abiotic balance at an animal-impacted site is localized and does not affect the system-wide hydrogeomorphic mechanism.

4.3 Delta Features

The general geomorphic structure of the OPC delta is typical of shallow, semi-enclosed, fluvial-dominated receiving basins. Beyond the easily identifiable forested, intertidal, and subtidal zones (Fig. 1), there are characteris-

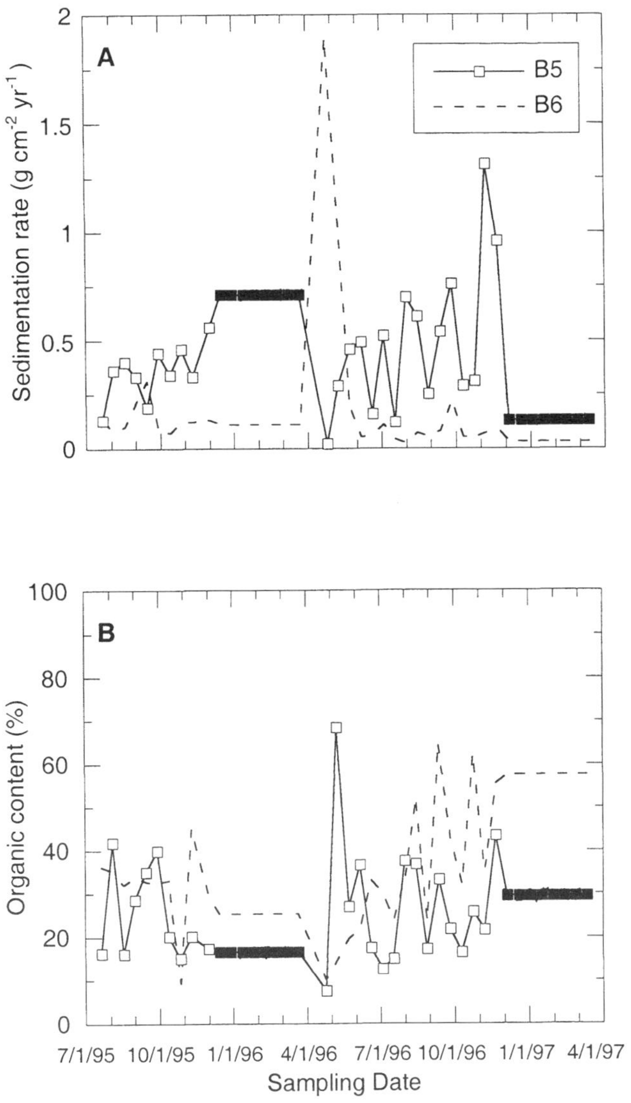

Figure 6. Comparison of (a) sedimentation and erosion and (b) organic content between a muskrat-impacted site (B5) and a non-impacted site 30 m away (B6). Filled symbols indicate data that is averaged over the winter when anchored tiles were frozen into the marsh and could not be sampled.

tic channel features such as multiple distributary channels, abandoned channels, and "middle ground" bars where channels end (Fig. 12). Also, distributary channels have well-defined sandy levees. Interdistributary areas away from wind-waves contain quiescent marshes that slowly accrete over time.

However, there are four regions on the delta that defy the standard geomorphic model. One such area was once the site of two sewage lagoons built 1966-1968 and used until 1971 while a pipeline and sewage-pumping stations were constructed (Fig. 13). Since abandonment the lagoons have evolved from shallow open-water all the way to middle and high marsh habitats in 30 years. Presently the pipeline is preventing the largest distributary channel from meandering (Fig. 13). A second area impacted by a known disturbance is the large shrub marsh located at the western end of the riparian forest (Fig. 14). This marsh was formed after June 1972 when two large woody-debris jams were deposited during Hurricane Agnes. The jams backed up flow, flooding a large section of forest and converting it to shrub marsh dominated by *Salix nigra* (black willow), *Typha latifolia* (broad-leaved cattails), *Leersia oryzoides* (rice cutgrass), and several species of *Polygonum sp.* (smartweeds). Dead trees rooted in the shrub marsh provide high quality wood duck habitat and are a key indicator of this disturbance mechanism. Distributary channels that transport sediment through the forested delta plain currently bypass this isolated shrub marsh, so there is very little sediment supply for marsh growth in this zone, except for in situ accumulation of organics, which occurs at a rate of 0.01 – 0.05 g cm^{-2} yr^{-1} [Pasternack, 1998]

Two areas on the delta are very distinct from their surroundings and do not have a definitive explanation. The first is Snake Island, which is an elevated tract of upland forest in the middle of the intertidal zone that is prime Bald Eagle habitat (Fig. 12). Soils on the island are well developed and well drained. The island does not appear on any historic maps, but is visible in a 1957 aerial photo. It is possible that the island was simply overlooked in map production.

The second distinct feature is a large tract of high marsh cut by a tributary to OPC in the riparian forest (Fig. 13). This marsh tract has many signs of animal activity including well-maintained marsh channels and channel-bank slides. Two monitoring stations, 6-P11 and 6-P21, were located within the area. Site 6-P21 was in a stand of *Acorus calamus* (sweetflag), the dominant high marsh species, while 6-P11 was located on the eastern levee of the tributary. These sites accumulated 0.21 and 0.13 g cm^{-2} yr^{-1} of sediment, respectively. While 6-P21 received more sediment on average, 6-P11 experienced significant erosion during winter 1997 followed by a single large deposition event cause by a spring flood (Fig. 15). The flood deposited 765 g of sediment at 6-P11. Given the high rate of overbank deposition on the tract as well as the tract's position far upstream from the main intertidal zone, it seems very unlikely that the tract should naturally be high marsh at all without some unknown prior disturbance. Based on the processes observed on the delta, it is possible that historic animal activity or unreported human activity could be responsible for this region that lags behind surrounding areas in its evolution. Since debris jams have been seen to form shrub marshes with relict riparian tree species, they cannot be responsible for the state of the high marsh tract.

Figure 7. Beaver lodge in the vicinity of site 2-P15 during winter 1997. Note the complete cutting of vegetation in a uniform radius around the lodge.

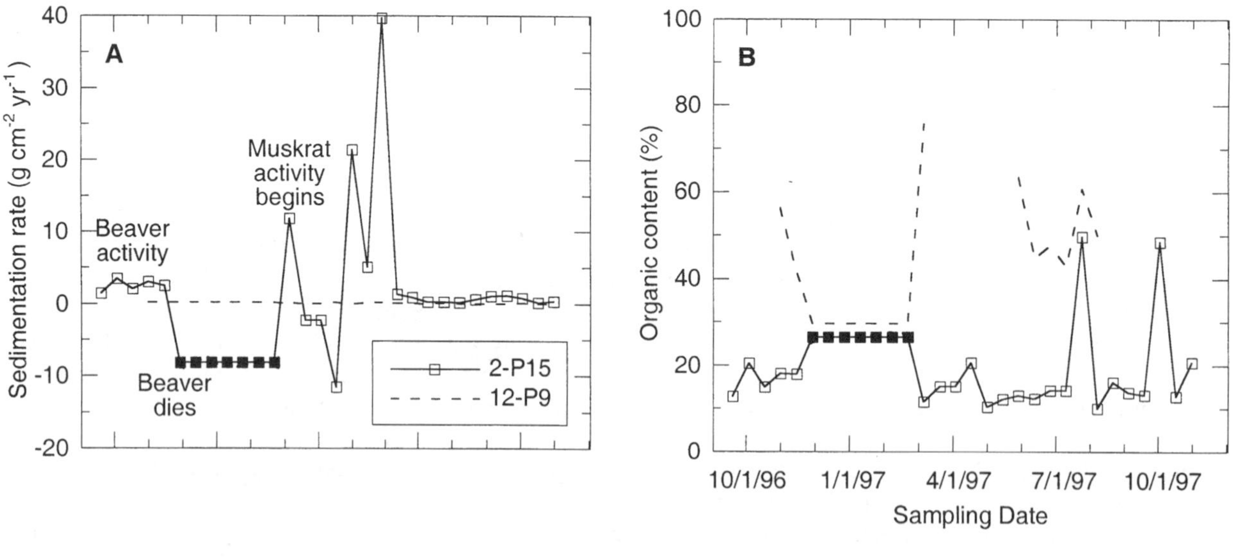

Figure 8. Comparison of (a) sedimentation and erosion and (b) organic content between an animal -impacted site (2-P15) and a non-impacted site in the same habitat type (12-P9). Filled symbols indicate data that is averaged over the winter when anchored tiles were frozen into the marsh and could not be sampled.

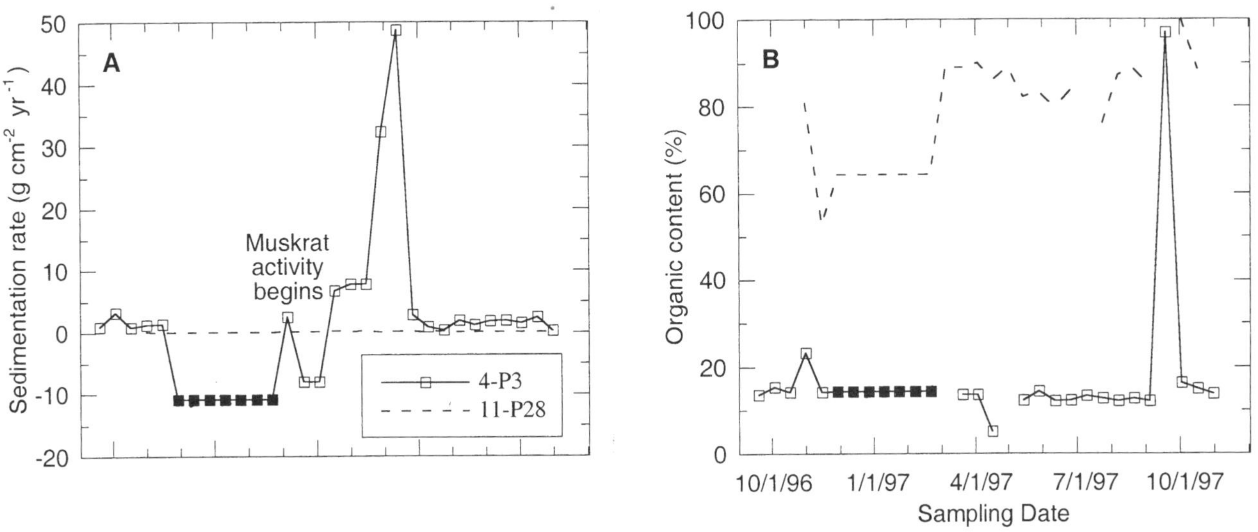

Figure 9. Comparison of (a) sedimentation and erosion and (b) organic content between an animal -impacted site (4-P3) and a non-impacted site in the same habitat type (11-P28). Filled symbols indicate data that is averaged over the winter when anchored tiles were frozen into the marsh and could not be sampled.

5. DISCUSSION

Time series data of sedimentation and erosion on the OPC delta show that animals have a measurable impact on geomorphic and ecologic dynamics in the tidal freshwater zone. Animal-impacted sites have higher sedimentation rates and greater temporal variation in sedimentation and erosion than non-impacted sites. They also have lower organic content and lower temporal variation in organic content. Field observations along with analyses of substrate characteristics such as grain sizes suggest that these differences were largely caused by a redistribution of local bed material within the disturbed area. Increased sedimentation on a tile must therefore be offset by erosion elsewhere, perhaps in the newly formed channels. Even though the redistribution can be accomplished in a single season, the

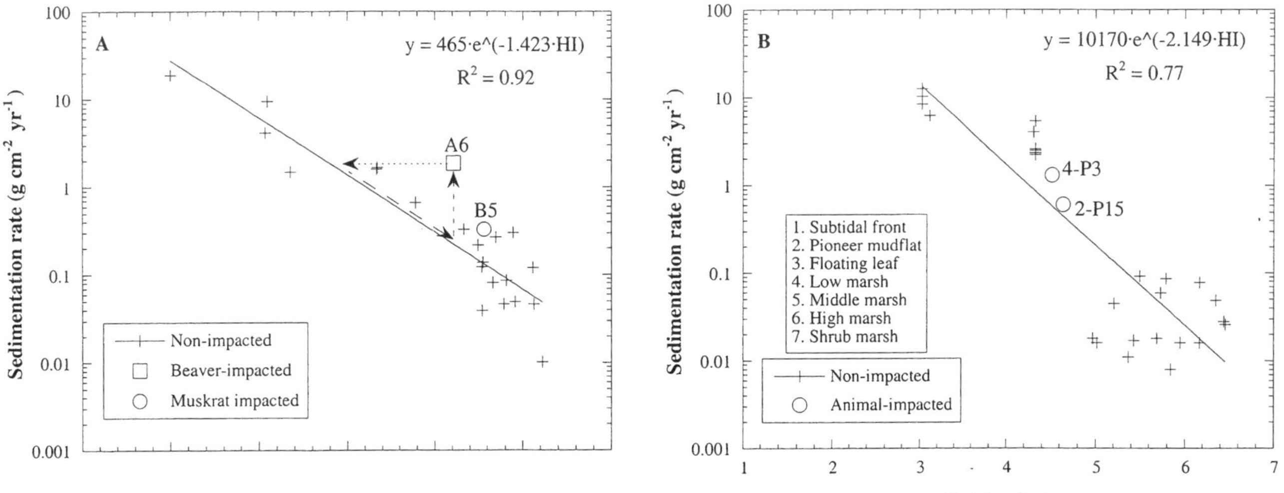

Figure 10. Relationship between sedimentation rate and a habitat index derived from plant species abundances for (a) autumn 1995 HBW data and (b) summer 1997 OPC data.

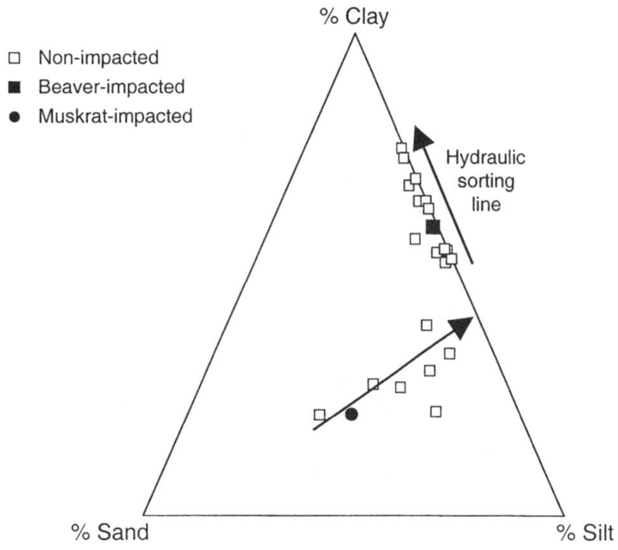

Figure 11. Ternary diagram of grain size distributions for HBW sites showing two hydraulic sorting lines that confirm the presence of a hydrogeomorphic gradient in the delta. Disturbed sites plot on the sorting lines, suggesting that animal impacts are localized and do not disturb the system-wide geomorphic dynamics.

disturbed area does not recover to its pre-impact condition for many years. In comparison to HBW sites, animal-impacted sites at OPC had order-of-magnitude larger pulses in sedimentation. These extreme fluctuations are thought to stem from the proximity of OPC sites to major distributary channels that are conduits for sediment, whereas the HBW sites are isolated, with the only channels present being the ones made by animals.

Both beavers and muskrats were observed to alter habitat conditions on the delta. Beavers worked for 3-6 weeks each autumn when marsh plants were dying back, whereas muskrats worked for 9-13 weeks as plants grew in spring and summer. Also, beavers channelized and devegetated more area than muskrats at any one time, but were less likely to return to the same spot year after year. Because beaver activities took place during the time when vegetation was decreasing, they exacerbated an already significant susceptibility to sedimentation and erosion. This timing led to higher sedimentation pulses at the beaver-impacted site (A6) than the muskrat-impacted site (B5) at HBW. However, muskrat activity succeeded over time in exposing the marsh to greater sedimentation and erosion on average between those same two locations.

Based on the field data, it is possible to construct a conceptual model of how animal activity impacts delta evolution. When no animals are present, spatio-temporal distributions of sedimentation and erosion in the intertidal zone of the delta are controlled by biogeomorphic boundary conditions such as plant association and elevation as well as marsh geometry in terms of distance from the subtidal front and pioneer mudflat where suspended sediment is initially entrained. As sites grow vertically, the rate of their further growth diminishes exponentially. In the face of this long-term delta evolution driven by sediment supplied from the upstream watershed, animals work hard to maintain their habitat, which was observed to be middle marsh, and preferably areas that are already somewhat disturbed. Middle marsh is preferred because it is the only intertidal habitat with plants suitable for building lodges. The geomorphically significant changes induced by animals include plant cutting, elevation reduction, and channelization. These changes cause higher sedimentation rates and reverse succession of an area from middle marsh to low marsh. It takes years for the system to aggrade back to its pre-disturbance state, partly because of the improved drainage of disturbed sites and partly due to increased winter erosion characteristic of the low marsh.

Given that animals can significantly change local habitat conditions, the question remains as to whether the observed animal-induced reverse succession is widespread enough to fundamentally affect geomorphic evolution at the scale of the whole delta. Presently there are too few animals to have such an impact, but historically lowland animal populations in Maryland were very large. How would such a population respond to the anthropogenically induced speed-up in delta growth? It is hypothesized that once the tidal freshwater zone evolved from open water submerged aquatic vegetation habitat to intertidal marsh, a large area of suitable habitat became available to the animals. However, as sedimentation continued, animal populations must have struggled to maintain their preferred conditions. This struggle would have been more successful in areas that are isolated from the major distributary channels carrying anthropogenic sediment through the delta. Consideration of the delta's biogeomorphic structure points to both HaHa Branch Wetland (Fig. 1) and the high marsh tract surrounding the small tributary in the riparian forest (Fig. 13) as examples of just such isolated areas. Both had a diverse array of intertidal and riparian habitats that could provide all the resources animals needed and both are stunted in their long-term geomorphic evolution and plant succession relative to their landscape position. Therefore, it is surmised that these sites lag the rest of the delta plain because of animal activity. Once local animal populations were reduced to a negligible size, these areas would have evolved from low and middle marsh to their current states.

Using a combination of field monitoring and landscape interpretation it is deduced that animals in the Chesapeake Bay region cannot prevent human-induced rapid delta pro-

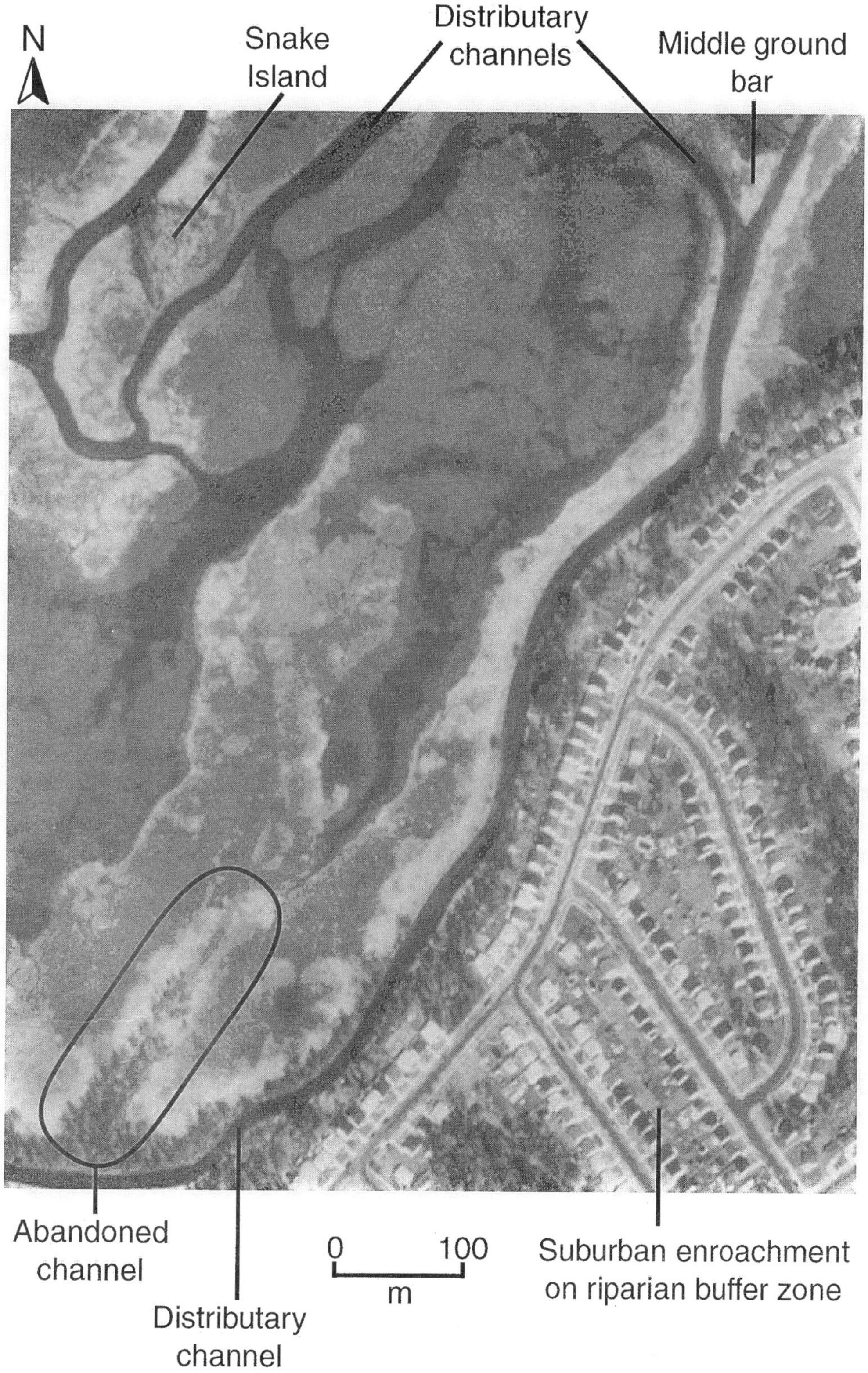

Figure 12. Aerial photo of the southeast portion of the OPC intertidal zone showing typical delta features including bars and distributary channels as well as unusual features such as Snake Island.

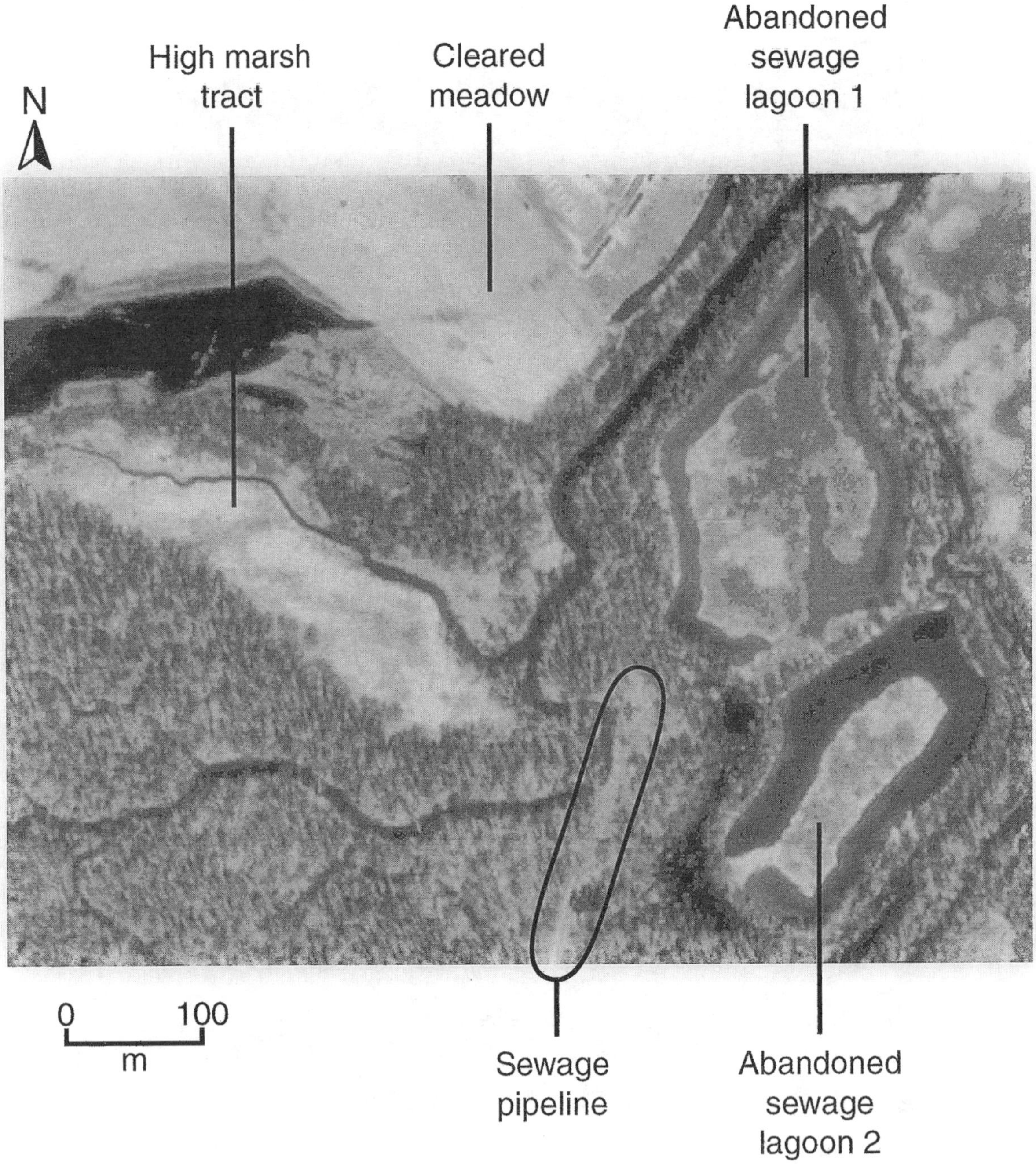

Figure 13. Aerial photo of the central portion of the OPC delta showing abandoned sewage lagoons that have evolved into middle and high marsh habitats as well as a very unusual high marsh tract along a tributary to OPC in an area isolated from the primary sediment transport conduits. This tract is thought to be two stages behind in plant succession relative to surrounding areas due to historic animal activity.

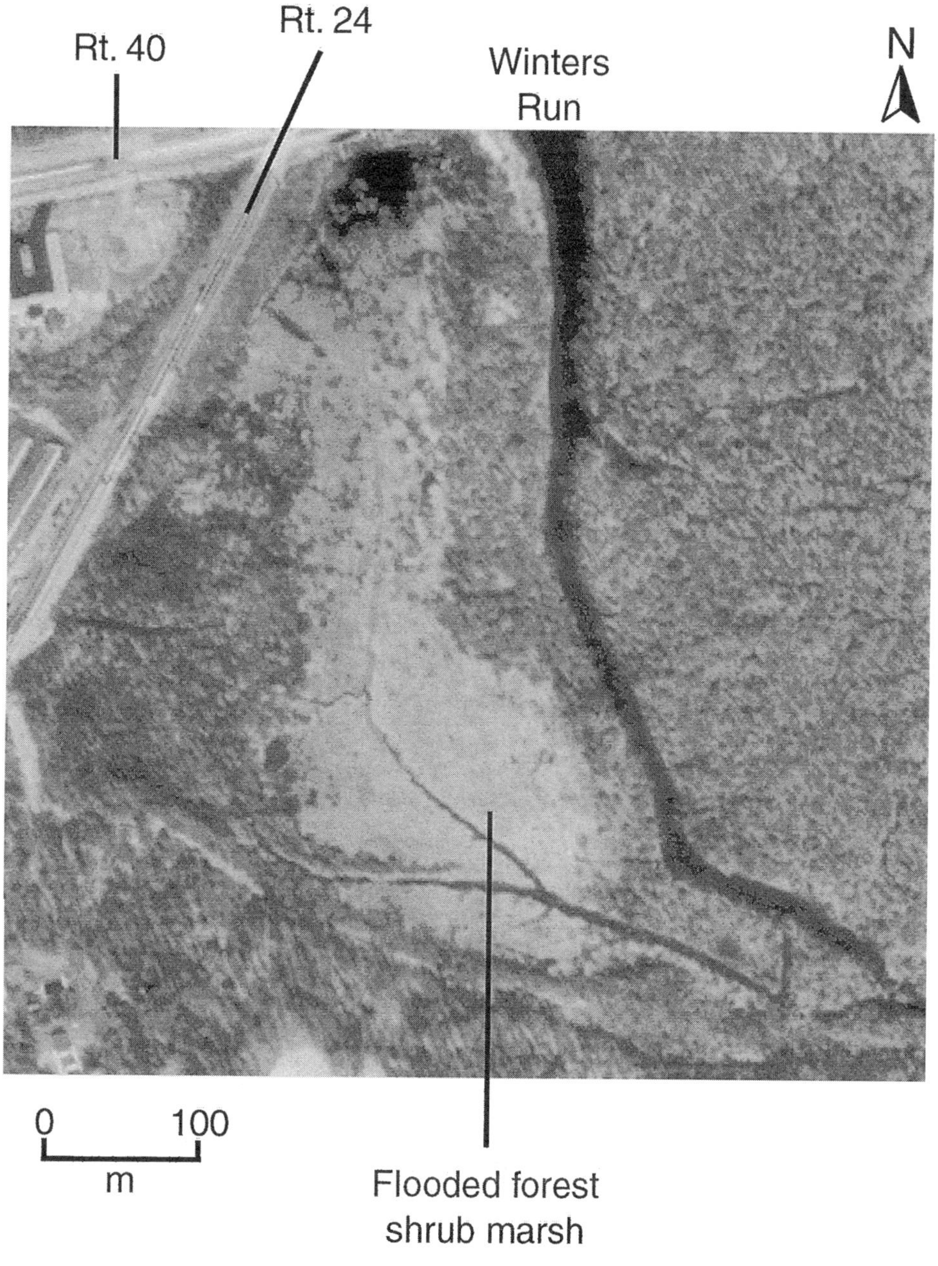

Figure 14. Aerial photo of the westernmost portion of the OPC delta plain showing a forested region flooded by backwater effects after the river was blocked by logjams. The disturbed area is now primarily shrub marsh, with relict dead trees that indicate the backwater mechanism.

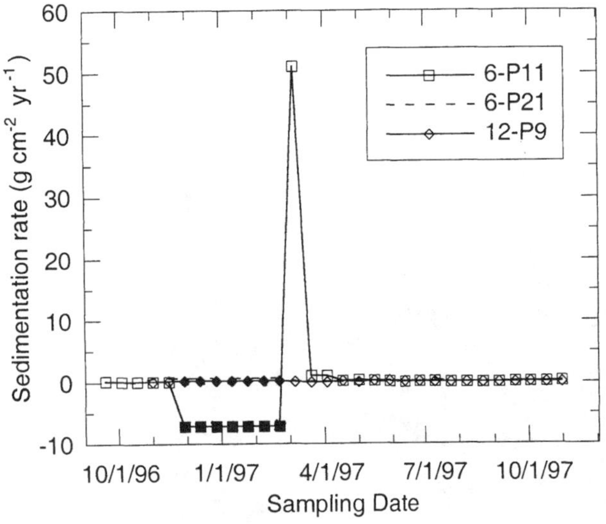

Figure 15. Time series of sedimentation and erosion on the high marsh tract that may have been created by historic animal activity. The only significant geomorphic changes occurring there today are the result of overbank deposition during flash floods.

gradation, but can alter the biogeomorphic structure of a delta. There is an important distinction between frontal progradation and vertical accretion of the delta plain that must be recognized. Animals can combat vertical accretion on the delta plain wherever there are interdistributary areas that are isolated from the primary sediment transport conduits. Such areas exist at Otter Point Creek and over the long term are two steps behind the rest of the delta plain in habitat succession and geomorphic evolution. The more isolated the site that is impacted by animal activity, the less able the system will be to restore the conditions there, as transport capacity and sediment supply are both low in these locales. Over the long term, disturbed sites provide habitat diversity, and thus animals serve an important function in the tidal freshwater river ecosystem.

Acknowledgments. This research was supported by the National Oceanic and Atmospheric Administration, the Maryland Department of Natural Resources, and Harford County. I gratefully acknowledge the assistance of numerous students and community volunteers in field monitoring and mapping efforts, including members of the OPC Alliance. This project benefited from discussions with Professor Grace Brush and Dr. Bill Hilgartner of The Johns Hopkins University as well as Heather Helm, the park naturalist at Otter Point Creek. Critical reviews by Robb Jacobson, David Montgomery, and an anonymous reviewer substantially improved this chapter.

REFERENCES

Berry, S.S., *Observations on a Montana beaver canal*, Annual Report of the Board of Regents of the Smithsonian Institution, Government Printing Office, Washington, D.C., 1922.

Brush, G.S., Patterns of recent sediment accumulation in Chesapeake Bay (Virginia-Maryland, U.S.A.) tributaries, *Chem. Geol.*, 44, pp. 227-242, 1984.

Brush, G.S. and F.W. Davis, Stratigraphic evidence of human disturbance in an estuary, *Quat. Res.*, 22, pp. 91-108, 1984.

Butler, D.R. and G.P. Malanson, Sedimentation rates and patterns in beaver ponds in a mountain environment, *Geomorphology*, 13, pp. 255-269, 1995.

Costa, J.E., Effects of agriculture on erosion and sedimentation in the Piedmont Province, Maryland, *Geol. Soc. Amer. Bull.*, 86, pp. 1281-1286, 1975.

Davis, F.W., Historical changes in submerged macrophyte communities of upper Chesapeake Bay, *Ecology*, 66(3), pp. 981-993, 1985.

Fernald, M.L., *Gray's Manual of Botany*, D. Van Nostrand Company, New York, 1970.

Folk, R.L., *Petrology of Sedimentary Rocks*, Hemphill Publishing Company, Austin, 1974.

Froomer, N.L., Sea level changes in the Chesapeake Bay during historic times, *Mar. Geol.*, 36, pp. 289-305, 1980.

Gottschalk, L.C., Effects of soil erosion on navigation in upper Chesapeake Bay, *Geogr. Rev.*, 35, pp. 219-238, 1945.

Gurnell, A.M., The hydrogeomorphological effects of beaver dam-building activity, *Prog. Phys. Geogr.*, 22(2), pp. 167-189, 1998.

Hilgartner, W.B., *Habitat development in a freshwater tidal wetland: a paleoecological study of human and natural influences*, Ph.D. dissertation, Johns Hopkins University, Baltimore, 1995.

Hooke, R.LeB., Spatial distribution of human geomorphic activity in the United States: comparison with rivers, *Earth Surf. Proc. Landf.*, 24, pp. 687-692, 1999.

Ives, R.L., The beaver-meadow complex, *J. Geomorphol.*, 5, pp. 191-203, 1942.

Jacobson, R.B. and D.J. Coleman, Stratigraphy and recent evolution of Maryland Piedmont flood plains, *Amer. J. Sci.*, 286, pp. 617-637, 1986.

Jenkins, S.H., A size-distance relation in food selection by beavers, *Ecology*, 61(4), pp. 740-746, 1980.

Johnston, C.A. and R.J. Naiman, Aquatic patch creation in relation to beaver population trends. *Ecology*, 71(4), pp. 1617-1621, 1990.

Khan, H. and G.S. Brush, Nutrient and metal accumulation in a freshwater tidal marsh. *Estuaries*, 17(2), pp. 345-360, 1994.

Mitsch, W.J. and J.G. Gosselink, *Wetlands, 2nd Edition*, Van Nostrand Reinhold, New York, 1993.

Mueller-Dombois, D. and H. Ellenberg, *Aims and Methods in Vegetation Ecology*, John Wiley and Sons, New York, 1974.

Naiman, R.J., G. Pinay, C.A. Johnston, and J. Pastor, Beaver influences on the long-term biogeochemical characteristics of boreal forest drainage networks, *Ecology*, 75, pp. 905-921, 1994.

Pasternack, G.B., *Physical dynamics of tidal freshwater delta evolution*, Ph.D. dissertation. The Johns Hopkins University, Baltimore, 1998.

Pasternack, G.B. and G.S. Brush, Sedimentation cycles in a river-mouth tidal freshwater marsh, *Estuaries*, 21(3), pp. 407-415, 1998.

Pasternack, G.B., G.S. Brush, W.B. Hilgartner, Impact of historic land-use change on sediment delivery to a Chesapeake Bay subestuarine delta, *Earth Surf. Proc. Landf.*, 26, pp. 409-427, 2001.

Pasternack, G.B., W.B. Hilgartner, and G.S. Brush, Biogeomorphology of an upper Chesapeake Bay river-mouth tidal freshwater marsh, *Wetlands*, 20(3), pp. 520-537, 2000.

Ranwell, D.S., The salt marsh to tidal woodland transition, *Hydrobiol. Bull. (Amsterdam)*, 8, pp. 139-151, 1974.

Ruedemann, R. and W.J. Schoonmaker, Beaver-dams as geologic agents, *Science*, 88, pp. 523-525, 1938.

Snodgrass, J.W., Temporal and spatial dynamics of beaver-created patches as influenced by management practices in a south-eastern North American landscape, *J. Appl. Eco.*, 34, pp. 1043-1056, 1997.

Trimble, S.W., *Man-induced soil erosion on the southern Piedmont, 1700-1970*, Soil Conservation Society of America, Ankeny, Iowa, 1974.

Whitney, G.G., *From Coastal Wilderness to Fruited Plain*, Cambridge University Press, Cambridge, 1994.

Gregory B. Pasternack, Department of Land, Air, and Water Resources, University of California, Davis, CA 95616-8628

Spatial Variations in Channel Morphology at Segment and Reach Scales, Middle Fork John Day River, Northeastern Oregon

Patricia F. McDowell

Department of Geography, University of Oregon, Eugene, Oregon

Like many montane rivers in the western U. S., the Middle Fork of the John Day River is characterized by alternating narrow- and wide-valley segments. Although the river has a low gradient (generally ≤ 0.01) alluvial channel, changes in valley width exert a clear influence on channel morphology. A spatial hierarchical model of channel morphology was applied to the Middle Fork to understand reach-level and segment-level patterns of morphologic variation. Channel unit data for a 55-km long section of the mainstem channel and detailed profiles at segment transitions where the valley width widens or narrows sharply were used in the analysis. Wide-valley segments, where not significantly modified by human activities, have higher sinuosity, 80% more pool area, and pools up to 40% deeper, on average, than narrow-valley segments. Going downstream from a wide-valley segment to a narrow-valley segment, channel morphology changes relatively abruptly from pool-riffle morphology, coarse pebble bed-material, and a gradient of less than 0.005, to plane bed morphology, cobble bed material, and a gradient of 0.005 to 0.010. Within one wide-valley segment, distinct reaches were delineated based on channel pattern and valley position. The more sinuous mid-valley reaches have several times more pool area, and pools more than 25% deeper, than the less sinuous valley-edge reaches. Segment-level differences appear to be related primarily to long term geologic controls, but these differences have been modified by human impacts. Observed reach-level differences are related to both geomorphic controls and human impacts. In this montane setting, the hierarchical spatial model is a useful way of analyzing and understanding spatial patterns of channel morphology and aquatic habitat.

1. INTRODUCTION

The Middle Fork John Day River (referred to as the Middle Fork; Fig. 1) is a medium-sized (fourth- to fifth-order) montane river system in which channel form is influenced by multiple controls, most notably variations in valley form and human impacts. The river flows through a series of alternating wide-valley and narrow-valley segments. These differences in valley width have influenced land ownership and use, with private ranch lands used for cattle pasture dominating in wide-valley segments, and national forest lands in narrow-valley segments. In wide-valley segments, the river channel is generally meandering but has been affected by placer mining, channel straightening, and bank stabilization. In narrow-valley segments, the

Geomorphic Processes and Riverine Habitat
Water Science and Application Volume 4, pages 159-172

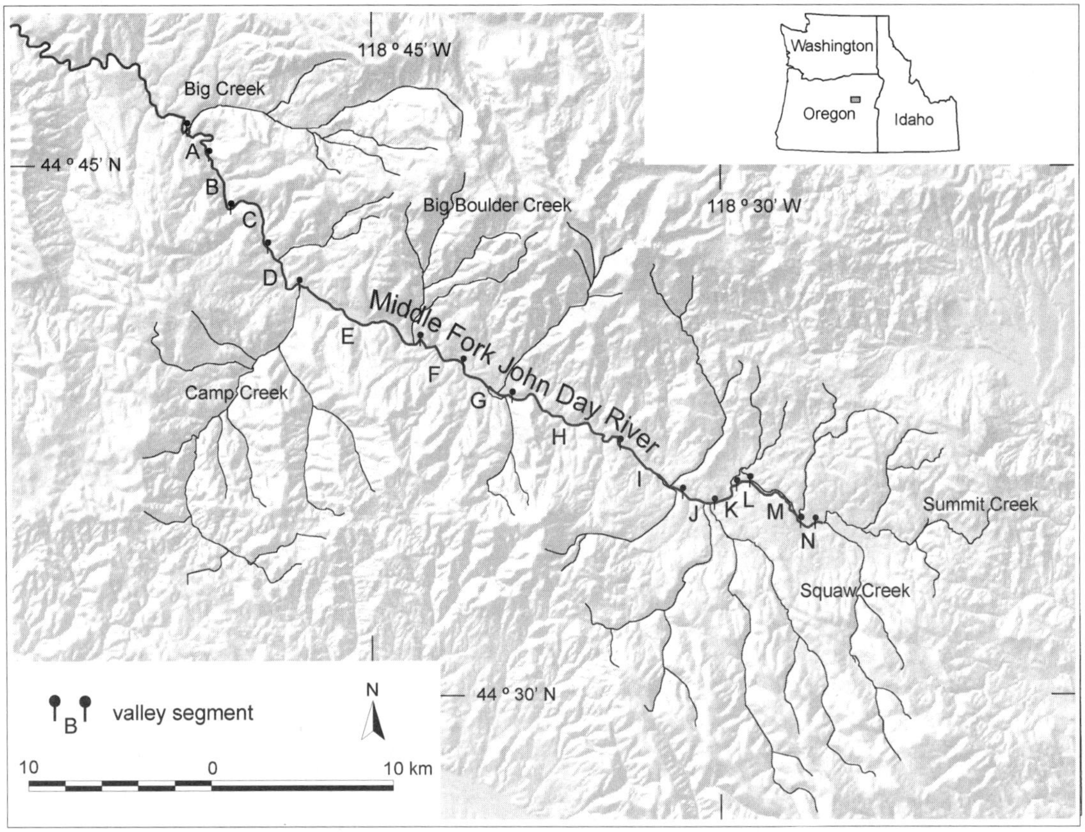

Figure 1. Map of the Middle Fork John Day River study area, with valley segments.

channel is straighter but shows less direct modification by humans [*McDowell*, 2000]. Such irregular downstream variations are probably typical of many montane rivers in the western United States. Using the Middle Fork as a case study, this paper addresses the structure and controls of the spatial variations in channel morphology. The implications of these spatial variations for fish habitat and management goals will also be discussed.

A useful approach for analyzing spatial patterns of channel morphology is the hierarchical model in which the river channel is seen as nested elements of increasing scale from the individual pool to the watershed channel network [*Frissell et al.*, 1986; *Montgomery and Buffington*, 1998]. This model also recognizes that controls of channel form are different at different scales. In this study, the hierarchical model is applied to the Middle Fork. The degree to which reaches and segments are distinct, and the controls of reach-level and segment-level differences in channel morphology will be examined.

1.1. *The Spatial Hierarchy in Channel Morphology*

Stream channel systems can be seen as a hierarchy of nested morphological units of increasing spatial scale: channel unit, reach, segment, and watershed channel network [*Frissell et al.*, 1986; *Montgomery and Buffington*, 1998]. A channel unit is a short section of the reach (typically a few channel widths in length) with distinct bed morphology, such as a pool, glide, riffle, or cascade [*Bisson and Montgomery*, 1996]. A reach is a section of channel, containing tens to hundreds of channel units, that has relatively homogenous channel pattern, slope, sediment size, and bed morphology. A segment is a significant section of the drainage network, containing one or more reaches, that has relatively uniform valley slope and valley form. Segment boundaries are typically defined by a significant change in bedrock lithology, valley form, or drainage area (i.e., at the mouth of a major tributary) [*Frissell et al.*, 1986]. The watershed channel network is composed of several to many segments.

The term reach has been used in two different ways. An *ad hoc* reach is a predetermined length of channel defined for use as a sampling unit (used by, for example, *Kaufmann and Robison* [1998] and *Fitzpatrick et al.* [1998]). An intrinsic reach is a discrete length of channel differing in morphology or size from adjacent reaches (used by, for example, *Bisson and Montgomery* [1996], *Rosgen* [1996], *U. S. Forest Service* [1996], *Overton et al.* [1997]). A finite number of non-overlapping intrinsic reaches comprise the entire length of the segment. In this paper, reach is used primarily in the sense of intrinsic reach.

Channel morphology can be analyzed at the channel unit, reach, or segment scales. For example, geomorphologists commonly characterize channel pattern at the reach scale, whereas downstream hydraulic geometry analysis is done at the watershed scale, spanning significant increases in drainage area that occur across multiple segments. Geomorphologists have long recognized the morphological distinctness of channel units, but little attention has been paid to whether and how much channel morphology varies at the reach and segment levels. Geomorphologists have not, until recently, considered variation in channel morphology across spatial scales. Using the scale-explicit framework of reaches and segments should help to identify the major controls of channel morphology at different scales. *Frissell et al.* [1986] and *Montgomery and Buffington* [1998] argued that the characteristics and processes controlling channel morphology change as scale changes. They suggested that, when moving from channel units to reaches to segments, the dominant controls of channel form and habitat are expected to shift toward phenomena that operate at larger spatial and temporal scales.

In the case of the Middle Fork, initial observation suggested that channel units are likely controlled by local factors such as flow structures and hydraulic conditions at bankfull stage, large woody debris, and individual man-made structures (bridges, bank stabilization or instream structures), whereas reaches are likely defined and controlled by differences in mass movement inputs, bounding landform, bank material, and riparian vegetation. Human impacts, such as land-use effects, channelization, and diversion, were expected to be expressed mainly at the reach level. Segment-level differences are likely associated with controls such as bedrock geology and valley form. The specific goals of this study are to (1) determine whether morphologically distinct segments and reaches are present in the Middle Fork, (2) identify the controls of channel form and habitat characteristics at the segment and reach levels, and (3) determine the scale(s) at which human impacts on channel morphology are expressed.

The hierarchical spatial model is also useful for understanding aquatic ecology and for river management. At the unit level, steep, fast-water channel units (such as rapids and riffles) are generally important for reaeration, invertebrate production, and fish spawning, whereas flatter, slower water units (various types of pools) provide resting and holding habitat, lower summertime water temperatures, and better cover from predators than adjacent units [*Bisson et al.*, 1982; *Hawkins et al.*, 1993; *Bisson and Montgomery*, 1996]. Important ecological interactions occur at the reach scale across multiple channel units. For example, over the course of a day, fish will use adjacent pools, riffles, and other units for different activities. *Frissell* et al. [1986] argued that the spatial hierarchical model provides a framework for representative sampling and monitoring in aquatic ecosystems. Methods of assessing current status and predicting sensitivity to disturbance or potential for recovery are typically applied at the reach scale [*Maddock* et al., 1995; *Rosgen*, 1996; *Montgomery and Buffington*, 1998; *Thorne*, 1998]. Land ownership and management units within federal lands typically vary at the reach or segment scale.

1.2. *Downstream Changes in Channel Morphology*

In contrast to the hierarchical model of nested reaches and segments described above, many previous researchers in geomorphology have emphasized progressive downstream changes in channel morphology, controlled by downstream changes in water and sediment inputs (for example, *Schumm* [1977], *Church* [1996]). The most familiar example of this view is downstream hydraulic geometry analysis [*Leopold and Maddock*, 1953], in which channel dimensions are described as a function of increasing discharge or drainage area. In ecology, the river continuum model [*Vannote et al.*, 1980] emphasizes progressive downstream changes in physical-habitat structure and aquatic-ecosystem structure.

Although valid for lowland rivers, the progressive downstream change model may be less appropriate in montane rivers. Many montane or bedrock-influenced alluvial rivers have relatively abrupt changes in form and pattern that are independent of drainage area (see for example, *Heritage et al.* [1999]; *Grams and Schmidt* [1999]). The most obvious control on these irregular spatial patterns is valley width. Valley form influences channel morphology through direct bedrock influence on the channel, or more indirectly through secondary processes. In many rivers, channels in narrow-valley segments are continuously or intermittently flowing on bedrock. Compared to alluvial channels, bedrock-influenced channels tend to have relatively high gradients, low sinuosity, limited capacity for morphological adjustment, and limited pool development [*Montgomery et al.*, 1996; *Montgomery and Buffington*, 1998; *Tinkler and Wohl*, 1998; *Wohl*, 1998]. Even where bedrock influences are not present, however, the effects of mass wasting processes and peak flow events are different in narrow and wide valleys. Mass wasting from valley walls and debris flows from tributaries have a stronger influence on channel form in narrow-valley segments than in wider valleys, often adding boulders and woody debris to the channel and constricting the channel [*Grant and Swanson*, 1995]. Differences in valley width may control spatial patterns of erosion during geomorphically effective floods. Wide valleys can decrease the peak discharge for a given event through decreased velocity in overbank flows that results in temporary storage of some of the runoff [*Woltemade and Potter*, 1994]. Even for a given discharge, narrow-valleys have higher stage, stream power, and shear stress than wide valleys [*Miller*, 1995]. These two effects combine to produce greater stream power and shear stress in narrow-valleys, conditions that are reflected in floodplain structure [*Nanson and Croke*, 1992] and presumably also in channel form. Alluvial channels in narrow-valleys may be expected to have steeper channel gradient, lower sinuosity, and coarser bed material than in wide valleys.

2. STUDY AREA

The John Day River is a tributary of the Columbia River, with headwaters in the Blue Mountains of Oregon. The Middle Fork John Day River is one of three main headwater branches of the John Day system. This paper focuses on the mainstem of the Middle Fork from Big Creek upstream to the junction of Summit Creek and Squaw Creek, including about 55 km of channel length (Fig. 1). The Middle Fork and other headwaters branches of the John Day system provide significant spawning and rearing habitat for both resident and anadromous salmonid fishes. The John Day system, the largest watershed in the Mid-Columbia Basin, lies above only three of the Columbia River dams and is itself one of the longest free-flowing rivers in the United States. Within the Middle Fork, anadromous steelhead (*Oncorhyncus mykiss*) and bull trout (*Salvelinus confluentus*) are listed as threatened under the Endangered Species Act, and spring chinook salmon (*O. tshawytscha*), a species of special concern, also use the study area for spawning and rearing habitat. Although spring chinook in the Middle Fork are not listed under the Endangered Species Act, present populations are estimated to be only 20 to 50% of their historic (ca. 1800 A.D.) numbers. The section of the Middle Fork studied in this paper is the zone used for spawning and holding (occupance during the period of several weeks after the fish return to the spawning site but before spawning begins) by spring chinook [*Torgersen et al.*, 1999].

The Middle Fork is typical of many montane rivers in the western United States in its relief, vegetation, and land use. Drainage area is 915 km^2 at the downstream end of the study area. The elevation is about 2300 m above sea level on the northern drainage divide. Valley floor elevations range from 1300 m at the upstream end of the study area to less than 1000 m at the downstream end. The geology of the watershed includes both Mesozoic metamorphic rocks and Tertiary volcanic rocks, all relatively resistant to weathering and erosion compared to softer lithologies in this region. Located in the Blue Mountains ecoregion [*Clarke and Boyce*, 1997], the uplands of the watershed are forested with Ponderosa pine and other conifers, and the valley bottom is a mosaic of forest and meadow. Large woody debris loading is low (less than 40 pieces per km) throughout most of the study section, and it therefore has little influence on channel morphology (except in one segment, discussed in section 4.2).

For most of its length, the mainstem channel of the Middle Fork is alluvial with a bed dominated by coarse gravels to cobbles. Pool-riffle and plane-bed channel reach types predominate. Bankfull width ranges from about 25 m at the downstream end of the study area to less than 10 m at the upstream end. Individual pools and riffles are commonly two to four bankfull widths in length, but their sizes vary widely. Riffles 10 to 50 bankfull widths in length were identified throughout the system; the longer riffles are in effect plane-bed reaches.

3. METHODS AND DATA SOURCES

Fourteen valley segments, labeled A (downstream) to M (upstream), were defined *a priori* on the basis of changes in valley width evident on topographic maps [*Frissell et al.*, 1986; *Church*, 1996] (Fig. 1). Specific placement of segment boundaries was guided by tributary confluences and cultural features (bridges, mine tailings), in addition to changes in valley width. After identification of segment boundaries, U. S. Geological Survey digital raster graphic files of 1:24,000-scale topographic maps were used to measure drainage area for each segment and valley width at 200 m intervals along the main channel. A reach-level analysis was conducted within segment E. The channel in segment E was divided into six intrinsic reaches based on pattern and channel morphology recognizable in aerial photos.

Independently of segment and reach identification, channel unit data for the entire study area were obtained in a field survey in late summer 1996, following standard stream inventory procedures [*U.S. Forest Service, 1996*; *Bisson and Montgomery*, 1996]. More than 1000 channel units were identified in the 55-km long section of mainstem channel. Data recorded for each channel unit included unit type (pool, riffle, etc.), length, wetted width, average water depth, maximum water depth for pools, presence of woody debris, and presence of channel modifications such as bank stabilization with rip-rap. These data, which represent physical habitat characteristics during summer conditions (low flow), were used to examine segment-level and reach-level variations in channel morphology and habitat characteristics. Because summer water temperatures in the Middle Fork are predominately above optimal or even limiting temperatures for anadromous salmonids, and abundant and deep pool habitat is particularly important for the survival of these fish [*Torgersen et al.*, 1999], analysis focused on pool characteristics.

Channel planform was investigated using 1939 and 1990 aerial photos [*U. S. Forest Service,* 1939, 1990]. The aerial photos were scanned and rectified to the U.S. Geological Survey digital raster graphic files. The mainstem channel and side channels were digitized from the rectified 1939 and 1990 images. The channel unit data were linked to the 1990 stream channel using a dynamic segmentation procedure with GIS software [*Environmental Systems Research Institute*, 1996; *Radko*, 1997]. Channel length for each segment was calculated from unit lengths.

After identification of valley segments and initial analysis of segment-level differences, three segment transitions (*ad hoc* reaches spanning the boundary between two segments) were selected for detailed field study. At another transition where a continuous survey was not appropriate, noncontiguous, representative *ad hoc* reaches within each segment were surveyed. Each transition included one reach in the downstream part of the upstream segment, and the adjacent reach in the downstream segment. Longitudinal profiles were surveyed using an electronic total station, and bed-material size measurements [*Wolman*, 1954] were done at three representative riffles in each reach.

Data visualization with boxplots was used in the analysis [*Cleveland*, 1993]. Formal statistical tests (analysis of variance or covariance, or the equivalent nonparametric tests) were not used because at least one of the underlying assumptions (independence, normality, homogeneity of variance) is violated in each case, even after the variables are suitably transformed. For example, Bartlett's and Levene's tests for homogeneity of variance [*Underwood*, 1997] both indicate that the variance of average depth differs significantly across segments ($p < 0.001$ in both cases), and the results are the same for log transformed values of average depth. The non-normality and heterogeneity of variance are evident in the boxplots (presented in sections 4.2 and 6).

4. SEGMENT-LEVEL ANALYSIS

4.1. *Description of Segment-level Differences in Valley Form*

Valley segments in the Middle Fork study area are 1.3 to 8.5 km long (Table 1), and average valley width by segment ranges from 100 m to 400 m (Fig. 2a). There is no pattern of increasing valley width going downstream. Within each segment, valley width is variable, but many segment boundaries show a distinct shift in valley width. The study area has a distinct pattern of alternating wide- and narrow-valley segments, shown most clearly in segments E through I; channel morphology in these segments will be analyzed in more detail below.

At most segment boundaries, there is an abrupt transition (less than 400 m long) in valley width. The specific controls of these width changes are not evident, but lithologic change and fault structures probably control most segment boundaries. About one-half of the segment boundaries are related to geologic boundaries (Table 2; *Brown and Thayer* [1966]; *Walker and MacLeod* [1991]), but similar geologic changes also occur within segments and without any expression in valley width. There are relatively few faults mapped in the study area, and none of the boundaries is related to a mapped fault. Neither is there any apparent influence of geomorphic controls, such as terraces or tributary junctions, on valley width changes at segment boundaries.

In narrow-valley segments A through D, F, H, and J, valley width is 10 bankfull channel widths or less, and the channel is intermittently in contact with the valley wall, colluvial deposits, or the embankment of a county road built down the valley floor. Thus meanders do not develop fully. In the wide-valley segments, particularly segments G, I, K and N, valley width is 10 to 30 bankfull channel widths, and the channel is generally unconfined and not in contact with the valley wall. Even in the narrow-valley segments of the Middle Fork, however, the channel is an alluvial channel without significant bedrock control. Bedrock in the channel bed was observed at only a few isolated points in segments A, E, and L.

4.2. *Differences in Channel Morphology*

The spatial pattern of valley form described above is reflected in the channel morphology of the Middle Fork. Based on field inspection, wide-valley segments tend to be dominated by pool-riffle reaches. Narrow-valley segments have more plane-bed reaches, although they also contain

Table 1. Segment characteristics

Segment	Segment length, km	Number of channel units	Drainage area at downstream end (km^2)	Valley type	Land ownership and recent land use on valley floor	Major human channel modifications
A	4.05	50	915	Narrow	Private lands; limited grazing	Road encroachment
B	3.05	46	908	Intermediate	Mixed national forest and private; grazing	Channel straightening
C	4.40	71	886	Narrow	Mixed national forest and private; grazing and residential	Placer mining, straight channel constructed
D	3.24	41	857	Narrow	Mixed national forest and private; grazing	None
E	8.49	112	662	Wide	Private lands; grazing, nature preserve	Channel straightening, rip-rap
F	3.20	53	580	Narrow	National forest; limited grazing	None
G	3.40	76	554	Wide	Private lands; grazing	Placer mining, straight channel constructed
H	8.34	191	465	Narrow	National forest	None
I	5.69	133	403	Wide	Private lands; grazing	Bank stabilization with barbs
J	1.50	32	317	Narrow	Private; former sawmill site	Channel straightening
K	1.31	45	171	Wide	Private lands; grazing	Channel straightening
L	1.52	40	150	Narrow	Private lands; grazing	None
M	4.21	143	140	Narrow	Mainly national forest; grazing	None
N	1.32	51	113	Wide	Private lands; grazing	None

pool-riffle reaches. Wide-valley segments are likely to have more fully developed meanders and broader floodplains. Sinuosity in 1939 (before substantial channel straightening and dredge mining) was about 1.5 in wide-valley segments E, G and I, and 1.2 in narrow-valley segments F and J (Fig. 2B). Narrow-valley segment H also had high total sinuosity in 1939, due to several reaches with two active channels. Marked decreases in sinuosity occurred between 1939 and 1990, due to channel straightening, loss of side channels, and road encroachment in a few places [*McDowell*, 2000]. The result of these human impacts is that the wide-valley segments today have sinuosity no higher than the narrow-valley segments (Fig. 2B).

The wide-valley segments also tend to have higher pool area (Fig. 2C), although human impacts in some segments have modified this pattern. Of particular interest is the downstream sequence of segments I, H, G, and F that shows the strongly alternating pattern of pool area. Wide-valley segments G, I, and N all have more than 50 % of their channel area in pools during summer flow conditions. Apart from segment M, the narrow-valley segments have more of their low-flow channel area occupied by riffles and, in smaller proportion, glides. Narrow-valley segment M has anomalously high pool area, probably because of abundant large wood pieces in the channel. (Large woody debris loading is 137 pieces/km in segment M. Large woody debris loading is generally less than 40 pieces per km in other segments, and has little or no influence on channel morphology.) In the wide-valley segments (excluding segment E), the percentage of the channel area in pools is 80% or more greater than in the narrow-valley segments.

Direct human modification of the channel (Table 1) has influenced pool abundance in segments C, E, G, J, and K. In wide-valley segment E, several sections of channel have been slightly straightened and rip-rapped, and several me-

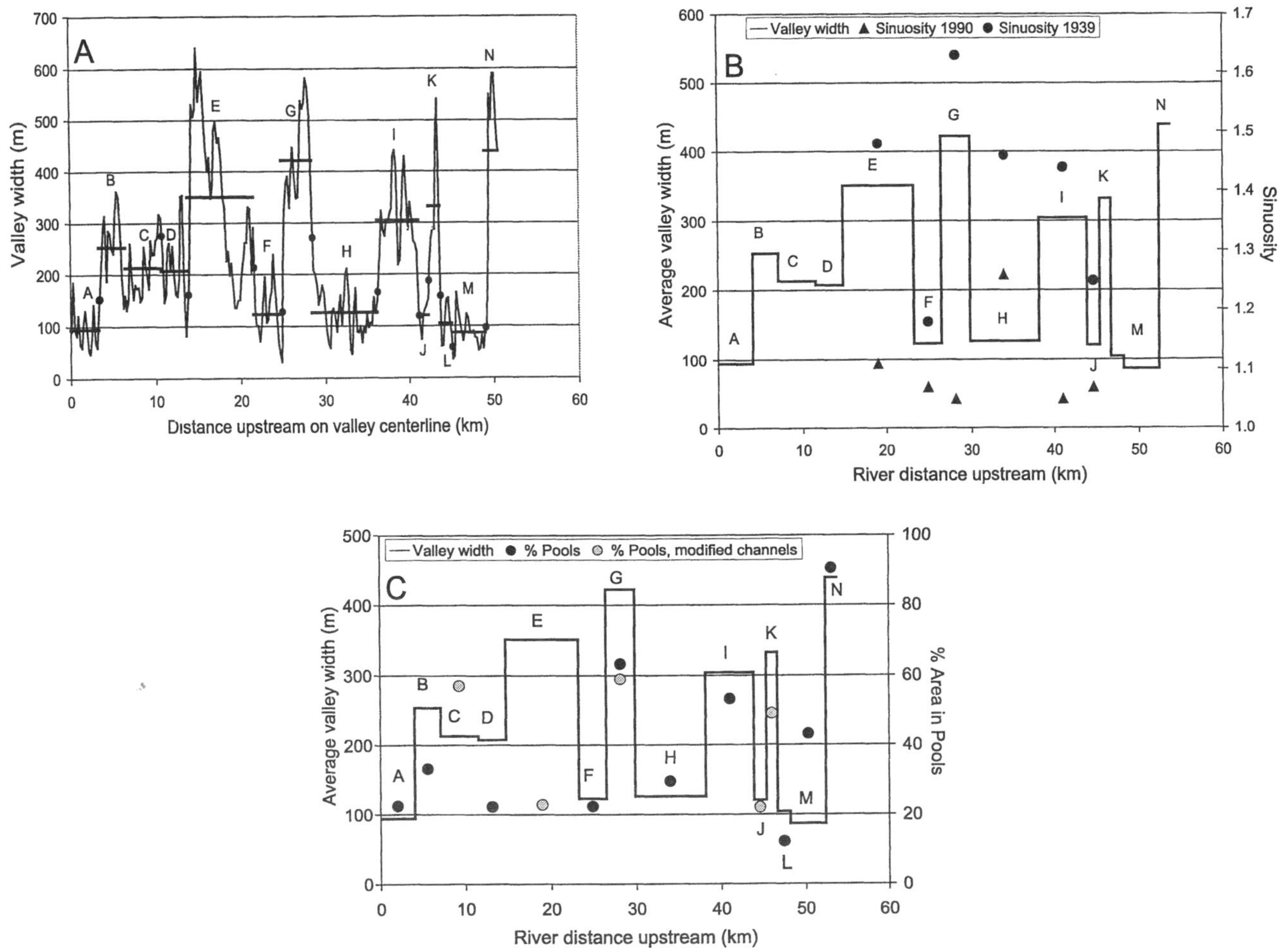

Figure 2. Valley form and channel form by segment. (A) Valley width variations. Horizontal bars represent the average valley width for each segment. Segment boundaries are shown by black dots. (B) Channel sinuosity in 1939 and 1990. (C) Percent area in pools. Channel with major human modifications (explained in Table 1) are shown by gray symbols. The two points for segment G represent the straightened channel through dredge tailings (gray symbol) and a reach of the secondary channel that was the main channel before dredging (black symbol).

anders have been cut off [*McDowell*, 2000]. Today segment E has pool area similar to narrow-valley segments, but it is probable that segment E had pool area similar to segments G and I before human alterations. Segments C and G both have channels created by dredge mining of placer gold deposits in the 1940s and 1950s; the dredged channels have moderately high pool area (gray circles on Fig. 2C). Segment G also contains a meandering channel that is now a side channel but was the main channel before dredge mining. This meandering channel has high pool area typical of wide-valley, meandering channels (black circle on Fig. 2C).

Five alternating wide- and narrow-valley segments, E through I, were analyzed in more detail. The boxplots (Fig. 3A) show that maximum pool depth is greater in wide-valley segments than in adjacent narrow-valley segments. Although there is considerable overlap in the range of maximum pool depth across these five segments, in all three wide-valley segments (E, G, and I), the median value of maximum pool depth is greater than the 75th percentile for the narrow-valley segments (F, H). The magnitudes of differences between wide-valley and narrow-valley segments were evaluated by comparing the average values. Maximum pool depth is 20 to 30% greater in wide-valley segments than in narrow-valley segments. Along with its lower sinuosity and pool area compared to other wide-valley segments, segment E also has lower maximum pool depth than segments G and I. This may be due to human-caused channel straightening, particularly cutting off of some large meander loops and loss of the large deep pools

Table 2. Geologic controls of valley width changes

Segment Boundary	Geologic control
A ⇒ B	Picture Gorge Basalt ⇒ Elkhorn Ridge Argillite
B ⇒ C	None; in Elkhorn Ridge Argillite
C ⇒ D	Elkhorn Ridge Argillite ⇒ Clarno Fm.
D	Intermixed Elkhorn Ridge Argillite, Clarno Fm.
D ⇒ E	Elkhorn Ridge Argillite ⇒ Clarno Fm.
E ⇒ F	None; within Clarno Fm.
F ⇒ G	None; within Clarno Fm.; possibly a structural control
G ⇒ H	None; within Clarno Fm.
H	Clarno Fm. ⇒ Strawberry Volcanics boundary within H
H ⇒ I	None; within Strawberry Volcanics
I ⇒ J	None; within Strawberry Volcanics
J ⇒ K	None; within Strawberry Volcanics
K ⇒ L	Strawberry Volcanics ⇒ Tertiary fluvial/lacustrine deposits
L ⇒ M	Tertiary fluvial/lacustrine deposits ⇒ Strawberry Volcanics
M ⇒ N	Strawberry Volcanics ⇒ Tertiary fluvial/lacustrine deposits

that typically form at meander bends. Average depth of pools was also compared, and it has a similar pattern to the pattern of maximum pool depth.

Average wetted (summer flow) depth of all channel unit types (Fig. 3B) shows a similar but less distinct pattern; wide-valley segments tend to have channels 16 to 48% deeper during summer flow conditions than do narrow-valley segments. The pattern for wetted (summer flow) width (Fig. 3C) shows both downstream increase and alternation between wide-valley (I, G) and narrow-valley (H, F) segments. Narrow-valley segments have channels 18 to 37% narrower than adjacent wide-valley segments (except for segment E).

5. SEGMENT TRANSITIONS

To confirm observed differences between wide- and narrow-valley segments in the segment-average data, detailed total-station field surveys of adjacent *ad hoc* reaches in wide- and narrow-valley segments were done. At three sites, continuous surveys were made through the transition from wide- to narrow-valley or vice versa. At a fourth transition (segments H to G), a continuous survey was not appropriate because there was a short intervening reach strongly influenced by bedrock constraint and a bridge, so two nearby representative *ad hoc* reaches were surveyed. Longitudinal profiles of the noncontiguous reaches, in segments G and H, are shown in Fig. 4A. Segment H has a narrow valley, and segment G has a wide valley. The surveyed reach in segment G is on the meandering channel that was not dredge mined. The surveyed reach in G displays clear pool-riffle morphology, whereas the surveyed reach in segment H is a plane-bed reach with a few pools present. The surveyed reach in G also has a slightly lower slope and generally finer bed material than that in H. These two reaches are representative of the difference between narrow- and wide-valley segments.

For two of the three transitions with continuous surveys, channel morphology changes abruptly at the segment boundary. The downstream transition from wide-valley segment N to narrow-valley segment M corresponds to a decrease in pool depth, increase in slope and increase in bed-material size (Fig. 4B). Likewise, the downstream transition from wide-valley segment E to narrow-valley segment D also corresponds to a decrease in pools and an increase in slope and bed-material size (Fig. 4C). The detailed surveys at segment boundaries, therefore, show spatially abrupt changes in channel morphology that are consistent with the segment-level differences between wide and narrow valleys. The expected progressive downstream changes in channel morphology are reversed in these two examples.

The fourth transition survey, going from narrow-valley segment F downstream to wide-valley segment E (Fig. 4D) does not show a change in channel characteristics. Bed-material size and slope are relatively constant across the transition, and pools are relatively infrequent throughout the two reaches. The absence of channel change at this transition appears to be due to local geomorphic factors not captured in the process of delineating segments. Although the valley widens at the segment E/F boundary, much of the valley floor in this area is occupied by a large alluvial fan built by a major tributary, Big Boulder Creek (Fig. 5). Big Boulder Creek enters on the right side, and the Middle Fork mainstem channel occupies a narrow floodplain strip along the left valley wall. Consequently, the Middle Fork is effectively confined in the surveyed reach in E by the alluvial fan, creating narrow-valley conditions.

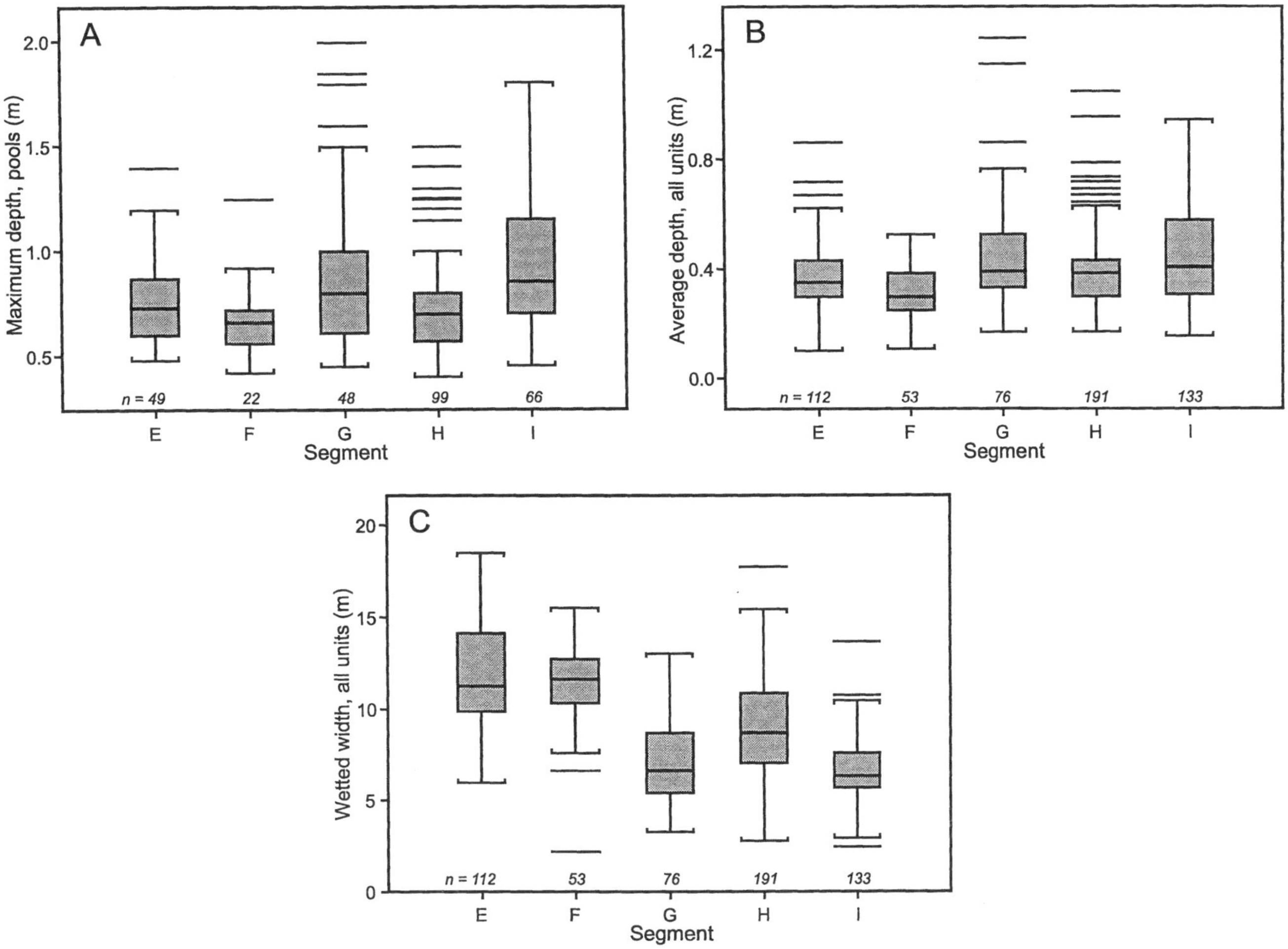

Figure 3. Box-and-whiskers plots showing segment-level differences in channel morphology. The shaded box encloses the 25th and 75th percentiles of the data. The median value is shown as a line within the box. The whiskers are drawn from the upper and lower quartiles to the nearest data point not beyond 1.5 times the inter-quartile range. Outliers, if any, are shown as a horizontal line beyond the whiskers. (A) Maximum wetted pool depth. (B) Average wetted depth, all channel units. (C) Average wetted width, all channel units.

6. REACH-LEVEL ANALYSIS

Segment E was selected for an analysis of reach-level differences in channel morphology because it is a long, wide-valley segment with space to develop unconfined meanders. It also has had substantial human alteration of the channel and is therefore a good place to evaluate the relative influence of natural controls and human influences on reach-level morphology. The goals of the reach-level analysis were to determine whether there are morphologically distinct reaches within a segment, whether reach-level morphological differences are more or less distinct than segment-level differences, and what are the controls of morphological differences among reaches. Examination of the channel in segment E on aerial photos and in the field revealed that the most obvious differences in channel form were differences in channel pattern (well developed meanders vs. low sinuosity) and channel position (against the valley wall or in mid-valley), so these characteristics were used as the basis for delineating reaches. Six intrinsic reaches (Table 3), ranging from 500 to 3400 m in length, were identified on aerial photos. Reach E1 is at the downstream end of segment E, and reach E6 is at the upstream end.

In segment E, reaches alternate between very low sinuosity reaches and moderate sinuosity reaches with alternating bars. Sinuosity is related to channel position and constraint by valley walls. The three longest reaches,

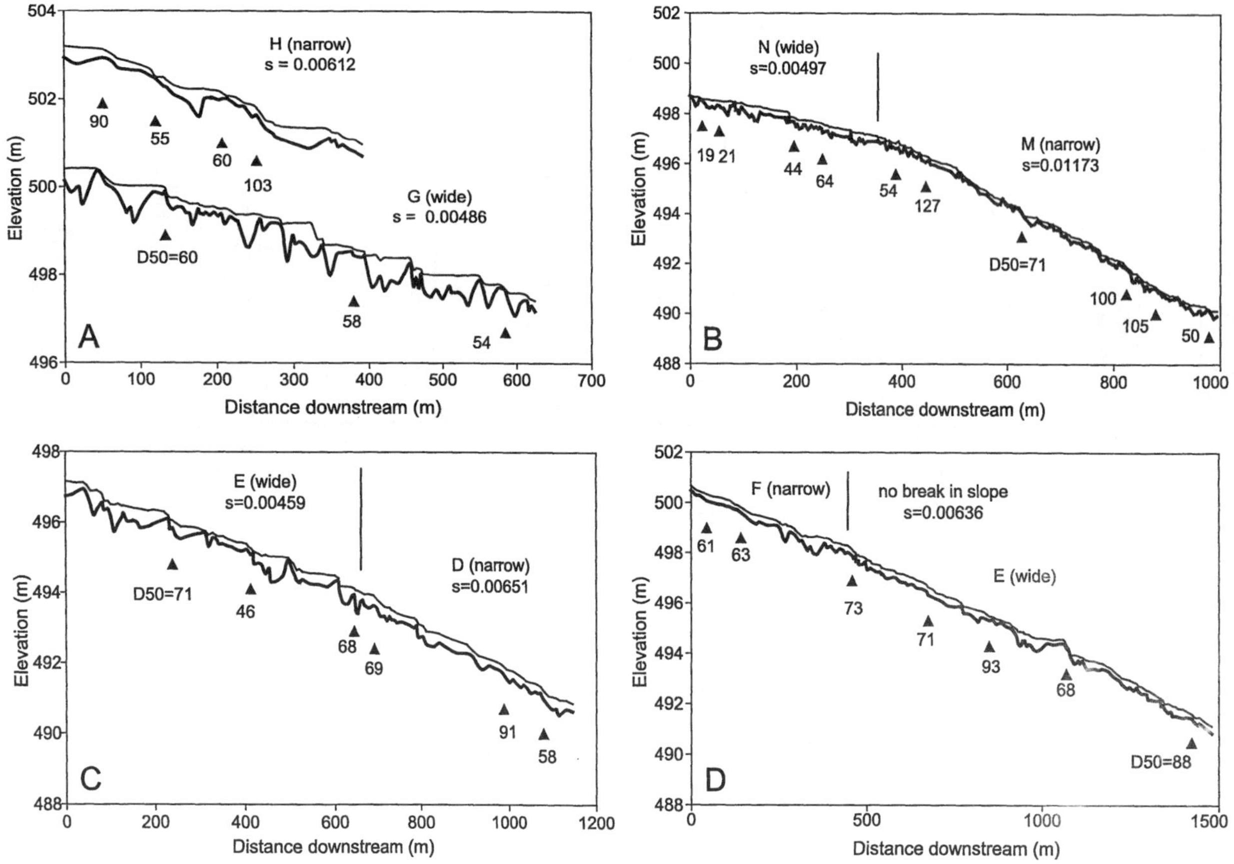

Figure 4. Channel longitudinal profiles and slopes (S) at segment transitions. The upper line is the water surface, and the lower line is the channel bed. Elevations are tied to an arbitrary site datum. Triangles show locations of bed material samples. The median size in mm (D50) is shown for each sample. (A) Noncontiguous *ad hoc* reaches near the segment H-G boundary. (B) Transition of segments M to N. (C) Transition of segments D to E. (D) Transition of segments E to F.

E2, E4, and E6, all have low sinuosity and are located at or very near to the valley edge (Table 3). E2 and E4 were significantly straightened by direct human action between 1939 and 1990 [*McDowell*, 2000]; large meander loops were cut off. These two reaches account for most of the loss in sinuosity in segment E (Fig. 2B). Reach E6 was already low sinuosity and located against the valley wall in 1939. In part of reach E6, however, old meander scars are visible on the 1939 and 1990 aerial photos, indicating that it did have higher sinuosity in the recent past. The other three reaches, E1, E3, and E5, are located in mid-valley positions, are bounded by floodplain deposits and have higher sinuosity (Table 3). All of the reaches in segment E are also significantly affected by bank structures (rip-rap and bank barbs) constructed of rock.

Reach-to-reach differences in channel morphology and habitat characteristics are significant in segment E. Pool area is 17% greater in reach E1, and several times greater in E3 and E5, than in the low sinuosity, valley-edge reaches (Table 3). In reach E5, many of the pools are associated with bank barbs (which tend to create scour and therefore force pools), but these pools are mainly located on meander bends where self-formed pools would be expected. In reaches E1 and E3, however, pools are generally self-formed, and few are associated with bank barbs.

Maximum pool depth is 26 to 39% greater, and average depth of all units is 11 to 44% greater, in reaches E1, E3, and E5 than in the low sinuosity reaches (Fig. 6A and B). Wetted width shows a decrease going downstream, probably because of irrigation diversions and some subsurface loss (Fig. 6C). The higher sinuosity reaches (E1, E3, and E5) have widths 1 to 19% narrower than the low sinuosity reaches. The greater widths in reaches E2, E4, and E6

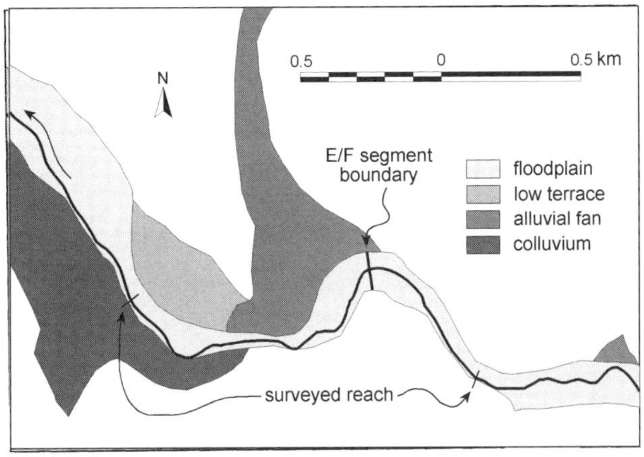

Figure 5. Geomorphic map of Segment E-F transition. The boundary between segments (based on valley-width change) is shown by the heavy line, and the ends of the surveyed reaches are shown by lighter lines. Flow is from right to left.

may be associated with a shift to plane-bed channel type and with increased riffle area, but the reasons for the width difference are not clear.

7. DISCUSSION AND CONCLUSIONS

The Middle Fork displays a distinct alternating pattern of wide- and narrow-valley segments, and morphologically distinct reaches are present within segment E. The segments contain tens to hundreds of channel units, and the reaches contain up to a few tens of channel units, consistent with the hierarchical spatial model of channel morphology [*Frissel et al.*, 1986; *Montgomery and Buffington*, 1998].

Clear differences in channel morphology and habitat characteristics are evident between wide- and narrow-valley segments. Wide-valley segments without major human modifications have higher sinuosity, more pool area, and deeper pools than narrow-valley segments. Wide-valley segments with the most intense human modifications of the channel have channel morphology similar to that of narrow-valley segments. Detailed surveys at segment transitions show that the channel morphologic changes from wide-valley to narrow-valley segments occur abruptly. Going downstream from a wide-valley to a narrow-valley segment, the alluvial channel becomes slightly steeper and coarser in bed material, and bed morphology changes from pool-riffle to plane bed. Thus, even the moderate valley width changes observed in the Middle Fork, without bedrock control on the channel, can produce reversals of typical downstream patterns of channel morphology and habitat characteristics [*Leopold and Maddock*, 1953, *Vannote et al.*, 1980].

Reach-level differences in channel morphology are also distinct in segment E. Compared to the less sinuous valley-edge reaches, the more sinuous mid-valley reaches have two to five times more pool area, and the pools are deeper by 25% or more. The valley-edge reaches have channel characteristics similar to narrow-valley segments. In the Middle Fork, segment-level differences in channel morphology are about equal in magnitude to reach-level differences within segment E.

The specific causes of the observed spatial variations in channel morphology in the Middle Fork are complex. Segment-level differences in channel morphology appear to originate primarily in natural controls. Human influences have, to a certain extent, overwritten the effects of natural controls. Land use has been more intensive, and therefore it has had more impact on channel morphology, in wide-valley segments than in narrow-valley segments. Reach-level differences in segment E are due to a combination of natural controls (channel position) and human

Table 3. Reach characteristics in segment E

Reach	Length (m)	Number of channel units	Channel position and Morphology	Sinuosity[a] 1989	Structures[b] (% length of units)	Area in pools (%)
E1	934	17	At valley contraction; mid-valley; pool-riffle	1.35	36.9	29.8
E2	1654	19	At right valley footslope; plane bed with some pools	1.00	49.8	25.5
E3	939	21	Crossing valley; pool-riffle	1.33	54.4	50.2
E4	1054	19	At left valley footslope; pool-riffle	1.06	41.8	9.7
E5	519	9	Mid-valley; pool-riffle with forced pools at bank barbs	1.14	59.5	65.7
E6	3390	27	At left valley footslope; plane bed	1.01	56.7	12.3

[a] Total sinuosity (including length of side channels with water at low flow conditions)
[b] Structures include rip-rap and bank barbs.

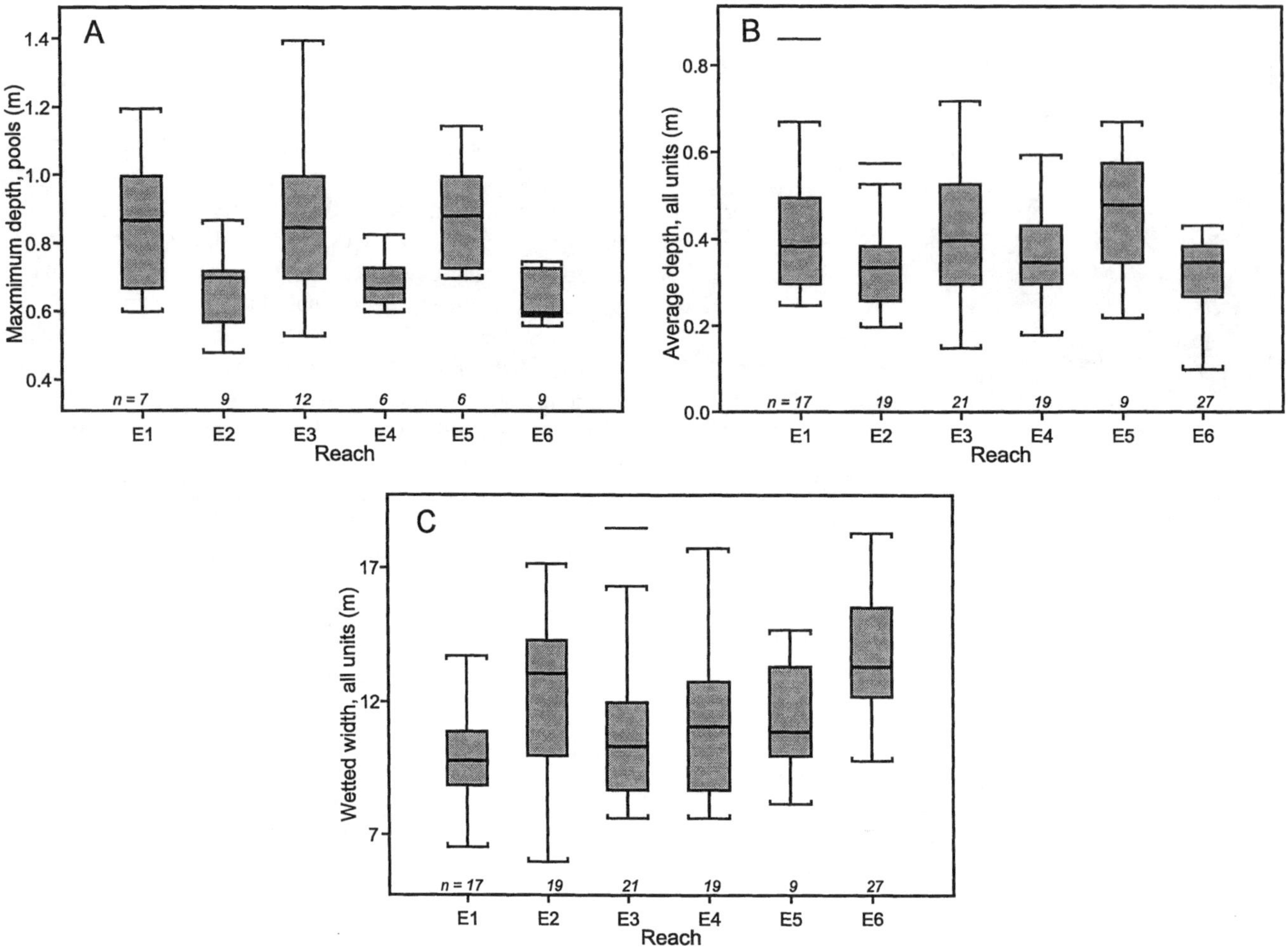

Figure 6. Box-and-whiskers plot showing reach-level differences in channel morphology within segment E. See Fig. 3 for explanation. (A) Maximum pool depth. (B) Average wetted depth, all channel units. (C) Average wetted width, all channel units.

influences (straightening, bank structures). In this study, there are not clear differences in the controls of channel morphology at reach- and segment-level.

Torgersen et al. [1999] found that habitat use by spring chinook in the Middle Fork is unevenly distributed at both the channel unit level and reach level. Pools, particularly deep pools, are intensely used by adult salmon for holding. Pools typically are relatively cool, and in much of the study area, deep pools are the only form of cover available. Analyzing *ad hoc* 1-km long reaches, *Torgersen et al.* found that chinook selected relatively cool reaches with low W:D ratios and more abundant and deeper pools. In the present analysis, these characteristics are found to be associated with wide-valley segments, and with mid-valley reaches within segment E. Segment-level and reach-level patterns of channel morphologic variation therefore are potentially important for fish habitat.

The results of this study show that channel morphologic characteristics that are important for fish habitat have uneven spatial patterns. Channel morphology varies in a non-progressive manner from segment to segment and from reach to reach. This heterogeneity in geomorphic characteristics creates large to small patches of more favorable habitat for salmonids. For example, wide-valley segments have higher habitat potential, in that they have the ability to develop more and deeper pools than narrow-valley segments. Some reaches within wide-valley segments, however, may have fewer and less deep pools, with pool characteristics similar to narrow-valley segments. In segment E of the Middle Fork these reaches with reduced pool area and depth were associated with human modifications and valley edge positions. The best of the wide-valley segments and reaches may provide models of "restoration potential" [*Torgersen et al.*, 1999]. Recognizing the spatial scale of these morphological segments and reaches is also important for aquatic habitat management and for planning projects to improve aquatic habitat. Reach to reach differences in channel morphology are important, and reaches are potentially useful as small scale management units. Management prescriptions and restoration projects should be designed at the reach level.

Acknowledgments. This research was supported by The Environmental Protection Agency Water and Watersheds Program, grant R-824774-01 to University of Oregon. Bruce McIntosh and Christian Torgersen of Oregon State University provided the stream inventory data. Steven Jett, Hudson Henry, Jennifer Pierce and Andrew Mowry of University of Oregon assisted in GIS analysis and field work.

REFERENCES

Bisson, P. A., J. L. Nielson, R. A. Palmason, and L. E. Gore, A system of naming habitat types in small streams, with examples of habitat utilization by salmonids during low stream trout, in *Acquisition and Utilization of Aquatic Habitat Inventory Information*, edited by N. B. Armantrout, pp. 62-73, Proceedings Western Division of the American Fisheries Society, Portland, Ore., 1982.

Bisson, P. A., and D. R. Montgomery, Valley segments, stream reaches, and channel units, in *Methods in Stream Ecology*, edited by F. R. Hauer and G. A. Lamberti, pp. 23-52, Academic Press, San Diego, Calif., 1996.

Brown, C. E., and T. P. Thayer, Canyon City Quadrangle, northeastern Oregon, U.S. Geol. Surv. Map I-447, scale 1:250,000, 1966.

Church, M., Channel morphology and typology, in *Rivers Flows and Channel Forms*, edited by G. Petts and P. Calow, pp. 185-202, Blackwell Scientific, Oxford, U. K., 1996.

Clarke, S. E., and S. Boyce, editors, Hierarchical subdivisions of the Columbia Plateau and Blue Mountains Ecoregions, Oregon and Washington, 114 pp., Pacific Northwest Research Station General Technical Report PNW-GTR-395, U.S.D.A. Forest Service, 1997.

Cleveland, W. S., *Visualizing Data*, 360 pp., Hobart Press, Summit, N. J., 1993.

Environmental Systems Research Institute, ARC/INFO GIS version 7.0.4, Environmental Systems Research Institute, Inc., Redlands, Calif., 1996.

Fitzpatrick, F. A., I. R. Waite, P. J. D'Arconte, M. R. Meador, M. A. Maupin, and M. E. Gurtz, Revised Methods for Characterizing Stream Habitat in the National Water-Quality Assessment Program, 67 pp., *U.S. Geological Survey Water-Resources Investigations Report* 98-4052, 1998.

Frissell, C. A., C. J. Liss, C. E. Warren, and M. D. Hurley, A hierarchial framework for stream habitat classification: viewing streams in a watershed context, *Environmental Management* 10, 199-214, 1986.

Grams, P. E., and J. C. Schmidt, Geomorphology of the Green River in the Eastern Uinta Mountains, Dinosaur National Monument, Colorado and Utah, in *Varieties of Fluvial Form*, edited by A. J. Miller and A. Gupta, pp. 81-111, John Wiley & Sons, Chichester, U. K., 1999.

Grant, G. E. and F. J. Swanson, Morphology and processes of valley floors in mountain streams, Western Cascades, Oregon, *Natural and Anthropogenic Influences in Fluvial Geomorphology*, edited by J. E. Costa *et al.*, pp. 83-101, Amer. Geophys. Union Spec. Mono. 89, 1995.

Hawkins, C. P., J. L. Kershner, P. A. Bisson, M. D. Bryant, L. M. Decker, S. V. Gregory, D. A. McCullough, C. K. Overton, G. H. Reeves, R. J. Steedman, and M. K. Young, A hierarchial approach to classifying stream habitat features, *Fisheries* 18, 3-12, 1993.

Heritage, G. L., A. W. van Niekerk, and B. P. Moon, Geomorphology of the Sabie River, South Africa: an incised bedrock-influenced channel, in *Varieties of Fluvial Form*, edited by A. J. Miller and A. Gupta, pp. 53-79, Wiley, Chichester, U. K., 1999.

Kaufmann, P. R., and E. G. Robison, Physical Habitat Characterization, in *Environmental Monitoring and Assessment Program – Surface Waters: Field Operations and Methods for Measuring the Ecological Condition of Wadeable Streams*, pp. 77-118, edited by J. M. Lazorchak, D. J. Klemm, and D. V. Peck, EPA/620/R-94/004F. U.S. Environmental Protection Agency, Washington, D. C., 1998.

Leopold, L. B., and T. J. Maddock, Jr., The hydraulic geometry of stream channels and some physiographic implications, 56 pp., U. S. Geol. Surv. *Prof. Pap.* 252, 1953.

Maddock, I., G. Petts, and M. Bickerton, River channel assessment – a method for defining channel sectors on the River Glen, Lincolnshire, U. K., in *Man's Influence on Freshwater Ecosystems and Water Use*, edited by G. Petts, pp. 219-226, Inter. Assoc. Sci. Hydr. Pub. 230, 1995.

McDowell, P. F., Human impacts and river channel adjustment, northeastern Oregon: Implications for restoration, in *Riparian Ecology and Management in Multi-Land Use Watersheds*, edited by P. J. Wigington, Jr., and R. L. Beschta, pp. 257-261, Amer. Water Res. Assoc., Middleburg, Vir., TPS-00-2, 2000.

Miller, A. J., Valley morphology and boundary conditions influencing spatial patterns of flood flow, *Natural and Anthropogenic Influences in Fluvial Geomorphology*, edited by Costa, J.E., *et al.*, pp. 57-81, Amer. Geophys. Union Spec. Mono. 89, 1995.

Montgomery, D. R., T. B. Abbe, J. M. Buffington, N. P. Peterson, K. M. Schmidt, and J. D. Stock, Distribution of bedrock and alluvial channels in forested mountain drainage basins, *Nature* 381, 587-589, 1996.

Montgomery, D. R., and J. M. Buffington, Channel processes, classification, and response, in *River Ecology and Management: Lessons from the Pacific Coastal Region*, edited by R. J. Naiman, R. E. Bilby, and S. Kantor, pp. 13-42, Springer, N. Y., 1998.

Nanson, G. C., and J. C. Croke, A genetic classification of floodplains, *Geomorphology* 4, 459-486, 1992.

Overton, C. K., S. P. Wollrab, B. C. Roberts, and M. A. Radko,

R1/R4 (Northern/Intermountain Regions) Fish and Fish Habitat Standard Inventory Procedures Handbook, 73 pp., U.S. Forest Service Intermountain Research Station *General Technical Report* INT-GTR-346, 1997.

Radko, M. A., Spatially linking basinwide stream inventories to arcs representing streams in a geographic information system, 22 pp., U.S. Forest Service Intermountain Research Station *General Technical Report* INT-GTR-345, 1997.

Rosgen, D., *Applied River Morphology,* 1 vol., Wildland Hydrology, Inc., Pagosa Springs, Colo., 1996.

Schumm, S. A., *The Fluvial System*, 338 pp., John Wiley & Sons, New York, 1977.

Tinkler, K. J., and E. E. Wohl, A primer on bedrock channels, in *Rivers over Rock: Fluvial Processes in Bedrock Channels*, edited by K. J. Tinkler, and E. E. Wohl, pp. 1-18, Am. Geophys. Union, Geophysical Monograph 107, Washington DC, 1998.

Thorne, C. R., *Stream Reconnaissance Handbook*, 133 pp., John Wiley & Son, Chichester, U. K., 1998.

Torgersen, C. E., D. M. Price, H. W. Li, and B. A. McIntosh, Multiscale thermal refugia and stream habitat associations of chinook salmon in northeastern Oregon, *Ecolog. Applications* 9, 301-319, 1999.

U.S. Forest Service, Umatilla National Forest Project no. F. S. 6-21, symbol BTU [aerial photos], scale 1:20,000, 1939.

U.S. Forest Service, Malheur National Forest Project no. 616040A [aerial photos], scale 1:12,000, 1990.

U.S. Forest Service, *Stream Inventory Handbook, Level I & II, Pacific Northwest Region*, version 9.5, 76 pp., U.S.D.A. Forest Service, Portland, Ore., 1996.

Underwood, A. J., *Experiments in Ecology : their logical design and interpretation using analysis of variance*, 504 pp., Cambridge Univ. Press, Cambridge, U. K., 1997.

Vannote, R. L., G. W. Minshall, K. W. Cummins, J. R. Sedell, and C. E. Cushing, The river continuum concept, *Canadian J. of Fisheries and Aquatic Sci.* 37, 130-137, 1980.

Walker, G. W., and N. S. MacLeod, *Geologic Map of Oregon*, U.S. Geol. Surv., scale: 1:500,000, 1991.

Wohl, E. E., Bedrock channel morphology in relation to erosional processes, in *Rivers over Rock: Fluvial Processes in Bedrock Channels*, edited by K. J. Tinkler, and E. E. Wohl, pp. 133-151, Am. Geophys. Union, Geophys. Mono. 107, Washington D. C., 1998.

Wolman, M. G., A method of sampling coarse river-bed material, *Am. Geophys. Union Trans.* 35, 951-956, 1954.

Woltemade, C. J., and K. W. Potter, A watershed modeling analysis of fluvial geomorphic influences on flood peak attenuation, *Water Resour. Res.* 30, 1933-1942, 1994.

Patricia F. McDowell, Dept. of Geography, University of Oregon, Eugene, OR, 97403-1251

The Downstream Effects of Dams on the Regeneration of Riparian Tree Species in Northern Japan

Futoshi Nakamura

Department of Forest Science, Graduate School of Agriculture, Hokkaido University, Sapporo, Japan

Nozomi Shin

Sapporo Salmon Museum, Sapporo, Japan

Dams for water regulation substantially alter the regeneration habitats of riparian species, thereby affecting their recruitment, composition and growth rate. This paper examines the effects of dams on the dynamics of riparian forests. First, the distribution of dominant riparian tree species was investigated in the Rekifune River, where no dams have been built, and their regeneration habitats were classified into five geomorphic surfaces based upon flooding frequency. *Chosenia arbutifolia* and *Toisusu urubaniana* appear on the most flood prone areas such as gravel bars and lower floodplains. *Poplus maximowiczii* also prefers coarse sediment, but establishes on upper floodplains. The late successional species, such as *Ulmus davidiana* var. *japonica* and *Fraxinus mandshurica* var. *japonica*, occur on stable surfaces referred to as terraces where sand and fine gravel make up the soil. *Salix* spp. dominates secondary channels. Second, the effects of dams on the regeneration process were examined in the Satsunai River where a flood control dam was completed in 1997. Spring floods in snowy regions mobilize riverbed sediment and create the moist, alluvial substrate ideal for seed germination. These habitats are essential for the recruitment of salicaceous species. Dams, in general, attenuate spring flooding and/or stabilize summer flows. This flow regime limits the development of gravel bars and reduces flood frequency. A decrease in flood frequency and magnitude may reduce the densities of pioneer species and shift species distribution from pioneer to late successional species.

1. INTRODUCTION

With the recent elucidation of the effects of streamside, riverside and wetland forests on river and stream ecosystems [*Swanson et al.*, 1982; *Gregory et al.*, 1991; *Nakamura*, 1995], discussions of river management now focus on not only river channels themselves, but also surrounding riparian zones. "Conservation and improvement of river environments" was included in the purpose statement of the Japanese River Law, which was revised in 1997, and it is clearly stated in the Law that riparian areas should be protected. Most streams and rivers in Japan have been altered for land development purposes or through the construction of maintenance facilities, so natural riparian forests are disappearing in all parts of the nation [*Sakio* and *Suzuki*, 1997].

Reservoir dams, erosion control dams and other dam structures are thought to have a considerable effect on the structure, species diversity and the regeneration mechanisms of riparian forests by regulating river flows and trapping sediment. How-

Geomorphic Processes and Riverine Habitat
Water Science and Application Volume 4, pages 173-181

ever, very few studies in Japan have clearly shown such effects. Many studies are conjectural, estimating the effects from the dynamics of natural riparian forests. As there are very few streams and rivers that have not been altered by man-made structures in Japan, there is a strong possibility that many of the target areas treated as natural riparian forests have actually been affected by upstream dam structures. Differentiation of areas around channels with frequent gravel movement and high terraced floodplains may be the signs of the impact of such structures. Such effects are already being observed throughout catchment areas [*Nakamura* and *Kikuchi*, 1996; *Nakamura et al.*, 1997; *Lee et al.*, 1998].

Unlike in Japan, where the progress of research has been relatively slow, interesting papers on the downstream effects of dams have been presented in other countries since the 1980s [e.g., *Reily* and *Johnson*, 1982; *Harris et al.*, 1987; *Rood & Heinze-Milne*, 1989]. It is especially difficult to obtain immediate results concerning the effects of river management on forests because long-term observation is required. The effects of flood disturbances and dam structures may also vary for trees in different developmental stages, such as seed germination, seedlings, saplings and canopy trees [*Ito* and *Nakamura*, 1994]. Thus, the habitat condition of tree species should be examined in relation to life history, and dam effects should be discussed in the same regard.

The purpose of this paper is to clarify the habitat characteristics and life histories of riparian tree species in northern Japan and to elucidate the impact of dams on the regeneration process of riparian trees. Specific dam effects examined were alterations in frequency of flood disturbance, seasonal discharge pattern with respect to seed dispersal, and resultant habitat condition.

2. STUDY SITES AND METHODS

2.1. Study Sites

This study was conducted in the Tokachi District at the Rekifune River (catchment area: 540 km^2) which flows south to the Pacific Ocean and the Satsunai River (catchment area: 560 km^2) which runs north into the Tokachi River (Figure 1). Both rivers originate in the Hidaka Mountain Range, and have similar annual and seasonal water discharge and riparian tree species. The Satsunai River Dam was completed in 1997 for flood control, water utilization and power generation. In contrast, the natural river environment has been preserved in the Rekifune River which has incised-meandering channels in the middle stretches and is braided in its lower reaches. The channel floor in the braided section is approximately 1 km wide (Figure 2).

Mature riparian forests are widely distributed along the banks of both rivers, and their species composition is relatively constant throughout the investigated areas. They consist of pioneer species such as *Chosenia arbutifolia, Toisusu urubaniana, Poplus maximowiczii, Salix sachliensis, Salix pet-susu* and *Alnus hirsta*, and late successional species, including *Ulmus davidiana* var. *japonica, Fraxinus mandshurica* var. *japonica* and *Quercus mongolica* var. *grosseserrata*. *C. arbutifolia,* a rare species found in Nagano Prefecture (a mountainous region in central Japan) and the eastern part of Hokkaido (northern Japan), is considered to be a surviving species of the last glacial period. This species dominates braided gravel-bed rivers, but is disappearing because of embankment, channelization and dam construction projects.

First, the natural distribution of the above tree species was investigated in the Rekifune River. The total length of the research sections stretches approximately 3.5 km. In the upper incised meander section, the channel floor width was approximately 200 m, the gradient was approximately 0.0051, and the mean diameter of bed materials, obtained through the linear-lattice sampling method, was 8.8 cm. In the lower braided section, the mean floor width was approximately 1 km and the gradient was approximately 0.0046. The mean diameter of bed materials was 11.5 cm. Second, to ascertain the downstream effects of dams on the regeneration process of tree species, a large gravel bar, extending 350 m in length, and about 2.0 km downstream from the dam site in the Satsunai River, was studied.

2.2. Methods

The geomorphic surfaces of the riverbed were classified based on inundation frequency by estimating water levels under various probabilities. A total of 11 cross-sectional profiles of the riverbed were surveyed in Rekifune River, and the area was measured by setting the water level at intervals of 0.5 m from the bottom of the channel floor on each transect line. The water discharge was calculated by multiplying cross-sectional

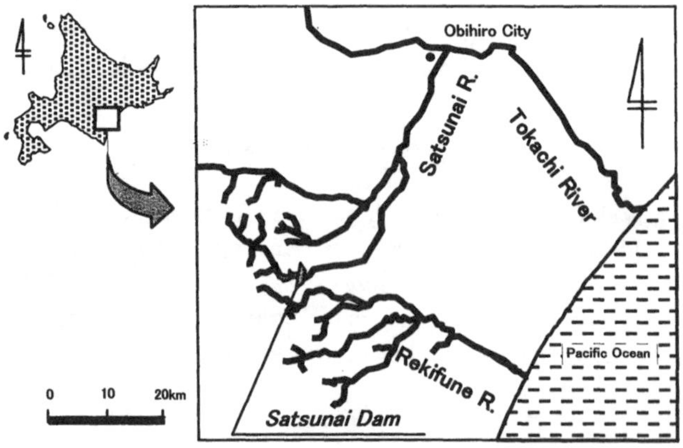

Figure 1. Map of Rekifune River and Satsunai River investigated. Flood control dam was built in the Satsunai River in 1997.

area and velocity. The velocity at each cross-section was estimated by the Manning formula with the Manning coefficient being set at 0.04 in this study.

Based on the water level and the calculated discharge, a curve representing the relationship between water level and flow volume was produced. Using this curve, the water level at which the largest flow of the year (highest yearly water level) runs through the channel was estimated. Subsequently, a straight line expressing the relationship of highest yearly water level and probability at which these levels are exceeded was produced to obtain the highest water level under various probabilities. Two water levels, corresponding to discharges with recurrence intervals of 2 and 20 years, were used as criteria to classify geomorphic surfaces.

The river floor was divided into five geomorphic surfaces: gravel bar, lower and upper floodplains, secondary channel and terrace. The gravel bar is an area submerged every year. The lower floodplain is along the mainstream channel and will be submerged at the flood recurrence interval of once every two years (expressed by 1/2). The secondary channel runs through riparian forests and will be submerged at the same probability as the lower floodplain. The upper floodplain is the area that will be submerged at a probability of 1/20. The terrace will not be submerged even at this probability. A map showing the distribution of these five surfaces was produced based on the classification shown on transect lines and by filling gaps between these lines through reconnaissance, aerial photos and analysis of topographic maps.

Ten quadrats were randomly set on each geomorphic surface, for a total of 50 quadrats. The field surveys were conducted from 1997 to 1999. Three types of quadrat, 5m x 5m or 10m x 10m or 20m x 20m, were used depending on the crown height of the species present. In each quadrat, the diameters of trees at breast height were measured and species were identified for trees taller than 1.3 m. Also, five individual trees with larger diameters at breast height were selected from each quadrat and their ages were determined using an increment borer. We identified the original root mat by removing buried sediment and tree ages were adjusted by adding years calculated from their initial growth rate.

Soil moisture was measured using a portable instrument for soil moisture percentage on dried soil basis (TRIME-FM) at all quadrats on sunny days in summer. This soil moisture percentage was then converted into soil moisture tension (pF) to compare the amount of available water content for plants under different soil textures, by building a relationship between pF and soil moisture percentage for each geomorphic surface.

The relationships between environmental variables and the dominant tree species were analyzed using direct environmental gradient analysis. As environmental gradients, two variables were considered. One was flooding probability representing disturbance frequency and the other one was pF representing water stress for plants. Flooding probability for terrace areas, which is greater than once every 20 years, was defined as once every 40 years by taking a younger age class of sampled trees. These two variables were combined using Principal Component Analysis (PCA). The relative dominance of a dominant species was calculated as percentage of total cross-sectional area at breast height.

The age of each geomorphic surface was determined from the average age of sampled trees. The area decay curve was expressed as an exponential function as follows [*Everitt*, 1968]:

$$F(t) = a \exp[-b.t] \tag{1}$$

where t is the age of sediment in a year, F(t) is the area of floodplain sediment aged t, and a and b are the constants determined by regression analysis. The average lifetime of geomorphic surfaces can be estimated by 1/b by assuming the steady state condition of the age distribution [*Nakamura*, 1986].

In the Satsunai River, return periods of water discharge before and after the dam construction were determined by Satsunai River Dam Construction Office, Obihiro Development and Construction Department in 1998. We used this information to understand a change in flooding frequency before and after regulation. A decrease in water level after Satsunai River Dam at a large gravel bar, located 2.0 km downstream from the dam site, was calculated by the same methods applied to the Rekifune River for the classification of the geomorphic surfaces. No vegetation quadrats were set in the Satsunai River.

3. RESULTS AND DISCUSSION

3.1. Distribution of Riparian Tree Species Along an Environmental Gradient

Among tree species that appeared in the Rekifune River, eight species were frequently observed, and were examined using gradient analysis (Figure 3). A gradient was made by the 1st axis of principal component explaining 53% of variance. The disturbance frequency decreases and available water content increases along the gradient. The elevated geomorphic surfaces such as terraces rarely suffer severe flood disturbance. Tree trunks on these surfaces act as large roughness elements, reducing stream power of the water during a flood. Under these conditions, suspended fine sediment can settle down on geomorphic surfaces and keep available water content at a higher level [*Nakamura et al.*, 1997].

C. arbutifolia and *T. urubaniana* appeared on the most flood prone areas such as gravel bars and lower floodplains where substrate was composed of cobble-sized sediment. *P. maximowiczii* also occurred on coarse sediment, but on higher floodplains. The late successional species, such as *U. davidiana*

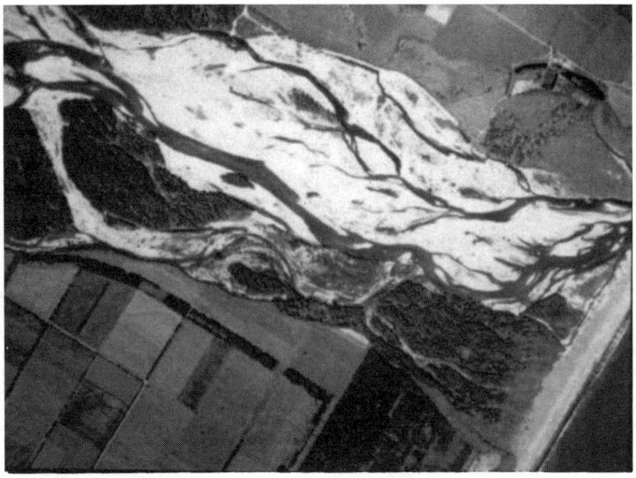

Figure 2. Braided river channel observed in the Rekifune River.

var. *japonica* and *F. mandshurica* var. *japonica*, were established on terraces with sand and fine gravel substrates. *S. sachalinensis, S. pet-susu* and *A. hirsta* were situated at mid-elevations compared to the above pioneer and late-successional species, dominating the secondary channels.

Several studies have been conducted regarding the effects of the frequency and magnitude of disturbances on riparian vegetation [*Lindsey et al.*, 1961; *Sigafoos*, 1961; *Teversham* and *Slaymaker*, 1976; *Bell* and *del Moral*, 1977; *Bell*, 1980; *Hupp* and *Osterkamp*, 1996]. Most of these studies used elevation above the thalweg as a primary gradient, because the frequency and magnitude of flood disturbance are related to relative elevation. *Bell* [1980] and *Nakamura et al.* [1997] indicated that the distribution of species composing riparian forests can be explained by the relative elevation gradient corresponding to the frequency of floods.

The results of previous and present gradient analysis shown in Figure 3 suggest that, if flooding frequency and water level are altered by dam operation, the entire composition of riparian tree species will shift along the gradient.

3.2. Seed Dispersal and Germination Periods of Riparian Tree Species

The necessity of spring flooding for seedling establishment of *Poplus* species has been pointed out in many research papers [*Fenner et al.*, 1985; *Rood* and *Mahoney*, 1990; *Nilsson et al.*, 1991]. The relationship between the timing of spring flooding

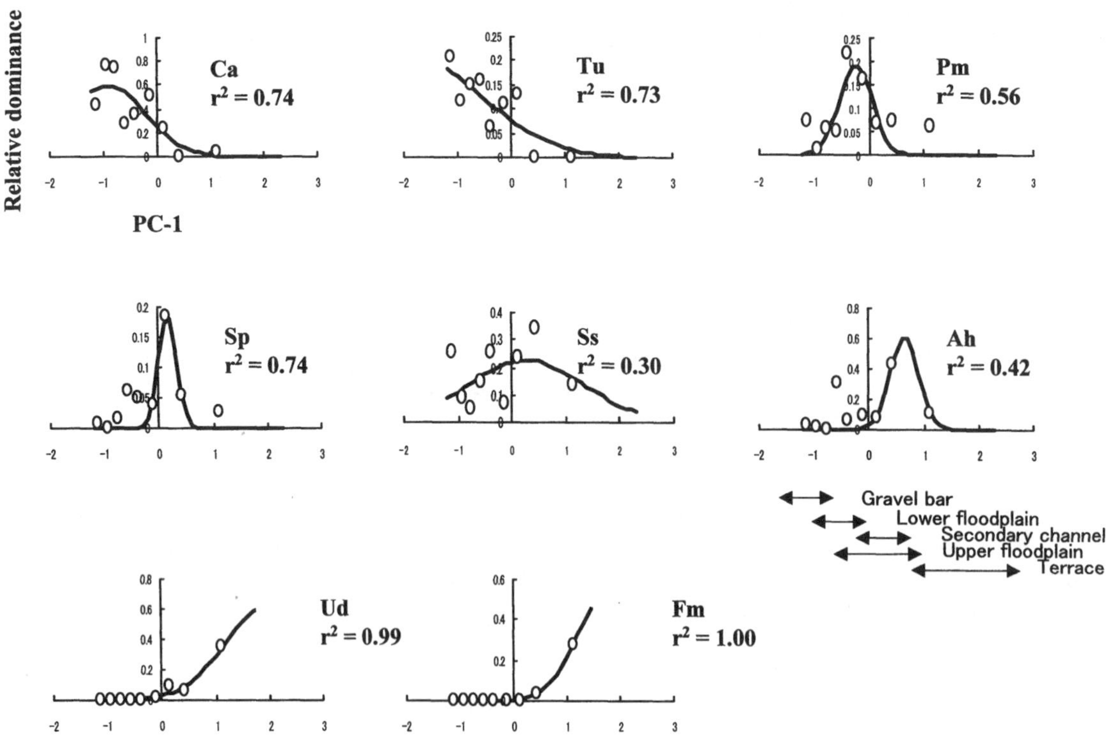

Figure 3 Relative dominance of riparian tree species along the environmental gradient expressed by PC-1. Ca: *Chosenia arbutifolia*, Tu: *Toisusu urubaniana*, Pm: *Poplus maximowiczii*, Sp: *Salix pet-susu*, Ss: *Salix sachliensis*, Ah: *Alnus hirsta*, Ud: *Ulmus davidiana* var. *japonica*, and Fm: *Fraxinus mandshurica* var. *japonica*.

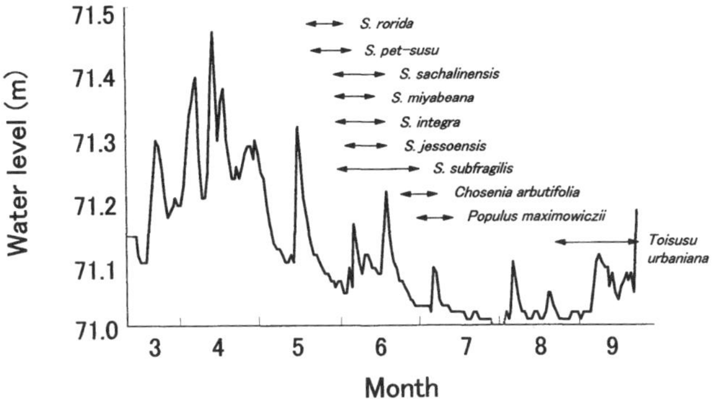

Figure 4. Variation of water level and seed dispersal periods of Salicaceous species in Hokkaido. The water level was measured at the Bannosawa gauging station in Bibai River, one of the tributary of Ishikari River in 1993 (Civil Engineering Division of Hokkaido Prefectural Government (1993)). Original data on seed dispersal periods are from Niiyama (1990) and Nagasaka (1996). They monitored seed dispersal periods at tributaries of Ishikari River.

and the seed dispersal period is thus important. *Niiyama* [1990] and *Nagasaka* [1996] discussed the timing of spring flooding and seed dispersal periods of Salicaceous plants in Japan, especially in Hokkaido (Figure 4). According to *Niiyama* [1990] and *Nagasaka* [1996], seeds of various *Salix* spp. are dispersed in May and June when the spring flood waters subside, and the dispersal period varies slightly by species. *Salix rorida* begins to disperse its seeds the earliest, in mid-May. The last species is *Salix subfragilis*, which disperses its seeds from June to early July. The difference in dispersal times by genus is rather obvious. The dispersal of *C. arbunfolia* and *P. maximowiczii* seeds is from late June to early July, which is later than that of *Salix* spp. *T. urbaniana* is even later, from August to September.

Although seeds of *Salix* spp. with earlier dispersal periods are able to invade bare lands left by flooding before other species invade, they are often submerged or washed away by small flood peaks as shown in Figure 4. In contrast, seeds of *C. arbutlfolia, P. maximowiczii* and *T. urbaniana* and other species with later dispersal periods are deposited in these areas in July or later when the water is calm and at a low level. Thus, they are dispersed in disadvantageous habitats that may have already been colonized by other *Salix* spp., or on bare land near stream channels. If spring flooding is controlled by the flow regulation of a reservoir dam, gravel bars do not form and sufficient water is not supplied in May and June, so regeneration habitats, on which seeds can germinate and take root, decrease dramatically. It is thus easy to imagine that salicaceous plants will disappear from the riparian zones.

Populus spp. are phoreatophytes that have roots deep in the ground to absorb water from the top of the water table or the area directly above it. The volume of water that can be used by *Populus* is determined by the river water level. According to *Fenner et al.* [1985], the roots of *Populus fremontii* in Arizona grow 6 mm a month on average following the lowering of water levels. It is also reported that their roots can grow to reach groundwater at depths of 70 to 160 cm by summer [*Johnson*, 1994]. Therefore, if the river water level drops faster than root growth by dam operation, saplings cannot reach the water table and will die [*Rood* and *Mahoney*, 1990].

3.3. Lifetime of Geomorphic Surfaces

The age distribution of geomorphic surfaces in a braided section of the Rekifune River is shown in Figure 5. The average lifetime of geomorphic surfaces in the Rekifune River estimated by 1/b in equation (1) was calculated to be about 28 years. This relatively short lifetime of geomorphic surfaces provides good habitat conditions for seedlings of pioneer tree species such as *C. arbutifolia* and *T. urubaniana*. These species grow and disperse their seeds 10-15 years after establishment (unpublished data), and take another 10 years to mature. The short life history of pioneer species seems to coincide with the frequency of flood disturbances estimated from the age distribution of geomorphic surfaces.

This result implies that the average disturbance interval provided by braided channel dynamics and the time span for species to complete their life histories are two important factors that can be used to explain species composition in a riparian zone. Thus, *C. arbutifolia* and *T. urubaniana* are common in the flood prone habitats provided by braided channel dynamics but late successional species such as *U. davidiana* var. *japonica*, and *F. mandshurica* var. *japonica* are limited.

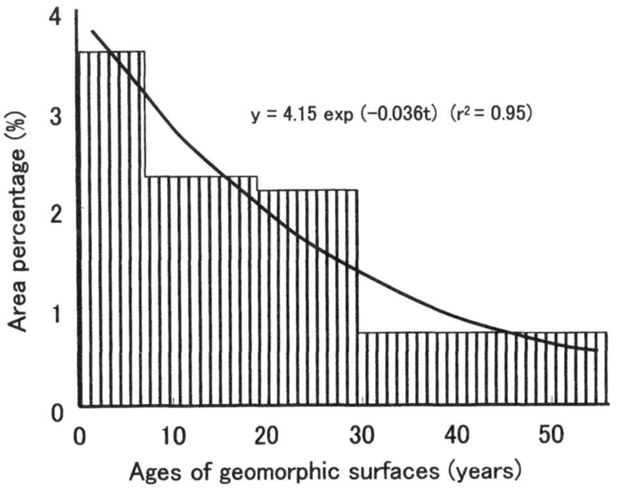

Figure 5. Age distribution of geomorphic surfaces. A decreasing trend with age was regressed by exponential curve.

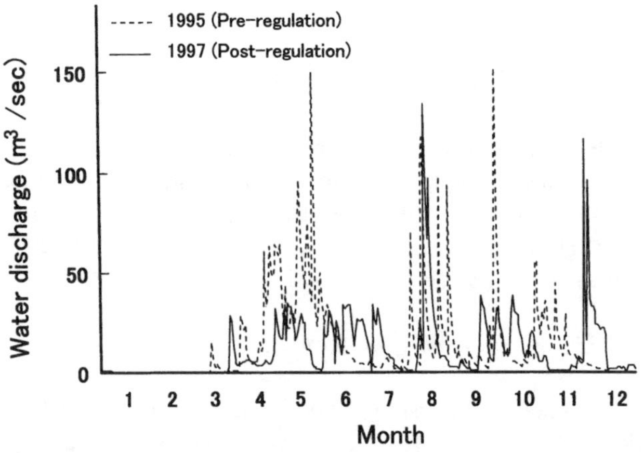

Figure 6. Annual variations of water discharge before and after Satsunai River Dam construction.

3.4. Dam Effects on the Regeneration of Riparian Tree Species

3.4.1. Seasonal changes in water discharge and sediment supply. Many of the dam structures used for flood control or irrigation serve to regulate flow rates. Thus the daily and monthly flow rates differ considerably before and after dam construction, and the peak flow rate at the time of flooding is greatly reduced by flood control. With the completion of the Satsunai River Dam in 1997, flow conditions in the river dramatically changed (Figure 6). Hokkaido, a northern region of Japan, is characterized by a peak of spring snowmelt flooding around May, in addition to floods caused by heavy rains in summer. In terms of dam operation, the thawing season is included in the non-flooding season (November - June) when water for irrigation is stored, and the peak flow rate of spring flooding decreases considerably.

Johnson et al. [1976] pointed out that the decrease in spring flooding caused by a dam has an effect on salicaceous plants, especially *Populus*. They conducted detailed research on the grain composition, available water content, nutrient content and other surface soil factors on a floodplain of the Missouri River in the U.S. They discussed the changes in habitats and dominant species by conducting a succession analysis using time as a gradient. At the same time, the effect of the Garrison Dam (a reservoir dam) located at the uppermost part of the research area was estimated by analyzing the density and annual rings of saplings. According to this research, spring flooding is extremely important for the development of *Populus deltoides* roots. Flooding supplies their seeds with sufficient moisture for germination and root growth. The elimination of spring flooding due to the construction of the dam is thought to contribute to the higher death rate of the saplings.

Not only water but sediment from upstream catchment areas accumulates behind a reservoir dam. This accumulated sediment decreases dam capacity. It also lowers the riverbed level and causes armoring of riverbed material in the downstream reaches, because it dramatically reduces the sediment supply. A decrease in sediment causes changes in the formation of gravel bars in the downstream riverbed. Gravel and silt carried by spring flooding provide germination beds that are necessary for pioneer trees [*Johnson et al.*, 1976]. If the supply of gravel is halted by a dam, land with sufficient water and nutrient conditions cannot be formed and pioneer plants will disappear [*Fenner et al.*, 1985; *Rood* and *Heinze-Milne*, 1989].

The species composition of riparian forests, especially riverside forests, has been explained in connection with substrates in many studies [*Johnson et al.*, 1976; *Hupp* and *Osterkamp*, 1985; *Niiyama*, 1987, 1989; *Ishikawa*, 1988; *Ariga et al.*, 1996; *Yajima et al.*, 1998]. *Nilsson et al.* [1991] found a positive correlation between the number of species composing riparian vegetation and the type of substrate. It can be seen from these studies that each tree species in riparian forests prefers a specific substrate condition and secures its own distribution area. The simplification of substrates caused by dams degrades the habitat heterogeneity of the substrate and thus decreases the diversity of species composing riparian forests [*Nilsson et al.*, 1991].

3.4.2. Altered disturbance regime. In general, lateral changes in channels are suppressed and the development and formation of gravel bars are limited after water regulation. Because the degradation of the riverbed due to downcutting is more prominent, the river flows down as a single channel. Such changes in hydrology and geomorphology not only affect the size distribution and moisture content of substrate, but also alter the frequency and magnitude of flood disturbances dramatically.

It is easy to imagine that the existing disturbance regime is altered and that the distribution of species in riparian forests along the flooding probability gradient is considerably affected as a result of the regulation of flow rates and sediment discharge by dams. Even if disturbances are controlled and the present cross-sectional landforms are left unchanged, the frequency and magnitude of disturbances, as well as the soil texture of the habitat surface, change greatly due to the regulation of flow rates by dams. As shown in Figure 7, the frequency of flood disturbances, of a magnitude which occurred every year before the constructon of the Satsunai River Dam, can be reduced to once every five years or so after the construction [*Satsunai River Dam Construction Office, Obihiro Development and Construction Department*, 1998]. Because the maximum discharge rate of the Satsunai River Dam is only 150 m^3/sec, it is expected that flood disturbances of a magnitude that occurred every two years before dam construction will not occur in the future.

3.4.3. Results expected after dam construction. A large gravel bar located 2.0 km downstream from the dam site was used as an example of a considerable drop of water level after

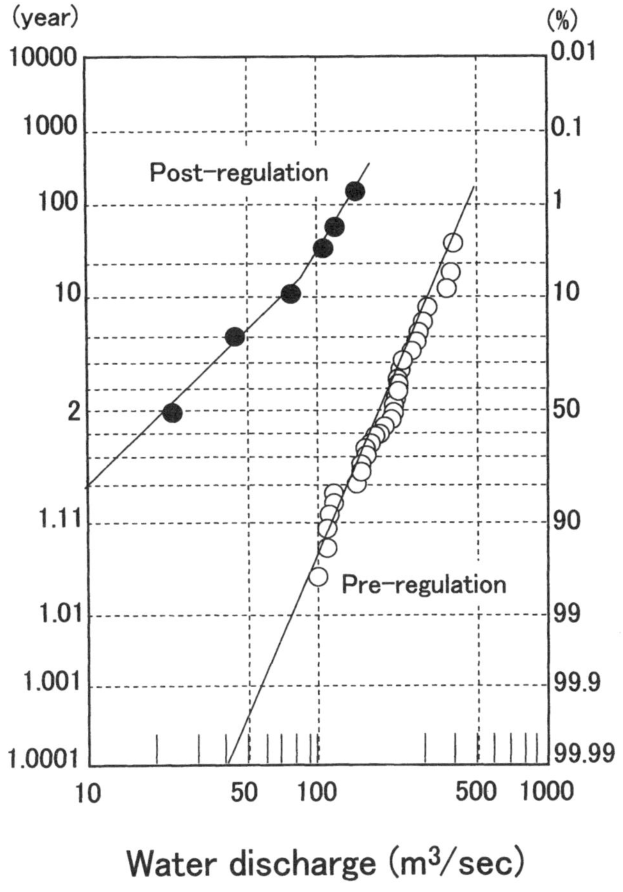

Figure 7. Return periods of peak discharge before and after Satsunai River Dam construction.

dam construction (Figure 8). Before dam construction, a flood with a recurrance interval of 5 years would submerge the entire river floor except for 8% of the higher elevations on the gravel bar. In contrast, the dam greatly reduced water levels, with only 55% of the river floor being submerged by a flood with an occurrence interval of 20 years.

Based on the results of the gradient analysis shown in Figure 3, a decrease in the frequency of flood disturbance causes a change in species distributions. Pioneer species represented by *C. arbutlfolia, T. urbaniana*, and *P. maximowiczii*, which can grow in unstable riparian areas, decline without regeneration and late successional species such as *U. davidiana* var. *japonica* and *F. mandshurica* var. *japonica* gradually dominate these areas.

The age distribution of geomorphic surfaces after dam construction was predicted by simulation. The original age distribution was the exponential curve calculated for the geomorphic surfaces in the Rekifune River. As shown in Figures 7 and 8, maximum flooding probability after water regulation is almost equivalent to once every two years under natural conditions. The age of geomorphic surfaces having 1/2 flooding probabilities was about 20 years. Under the steady state condition, the erosion rate is equal to b in equation (1) and was about 4% for all surfaces in the Rekifune River. Assuming that the duration of flooding is increased by dam operation, the erosion rate was set at 20% until 20 years (1/2 probability), and dropped to 0% for the surfaces older than 20 years because 1/2 flooding discharge before regulation became the maximum water release of the Satsunai River Dam. The bipolarized age distribution of geomorphic surfaces was created after repeating the calculation for 25 years (Figure 9). Under these circumstances, younger surfaces would be continuously maintained by erosion and deposition of sediment, but old surfaces would be segregated from the dynamics of fluvial processes.

Although the present cross-sectional landforms are left unchanged, the frequency of flooding and characteristics of geomorphic surfaces are greatly altered. The original age distribution containing five geomorphic surfaces is bipolarized into frequently flooded areas, classified as gravel bar and lower floodplain, and very stable terrace surfaces after regulation. This segregation of geomorphic surfaces will lead to the ex-

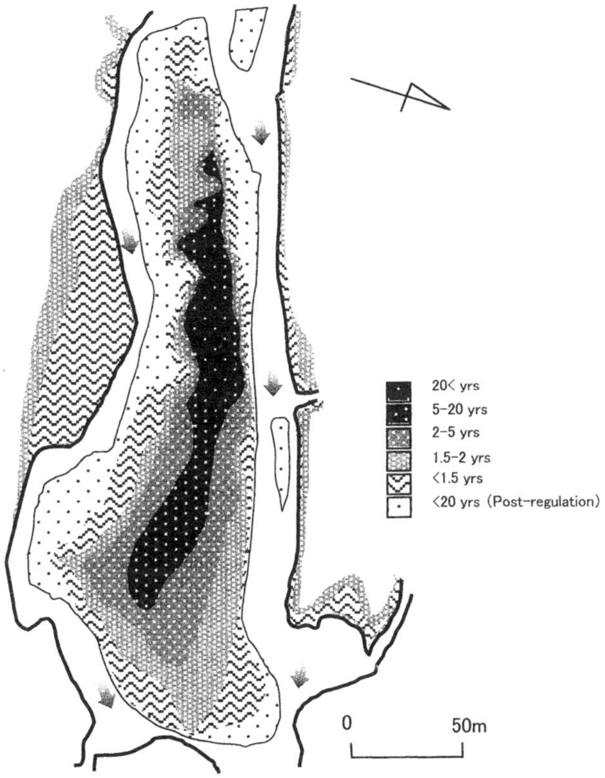

Figure 8. A decrease in water level after the regulation by Satsunai River Dam. The water level occurred once every 20 years after the regulation was far lower than the water level which had been occurred once every 1.5 years before the regulation.

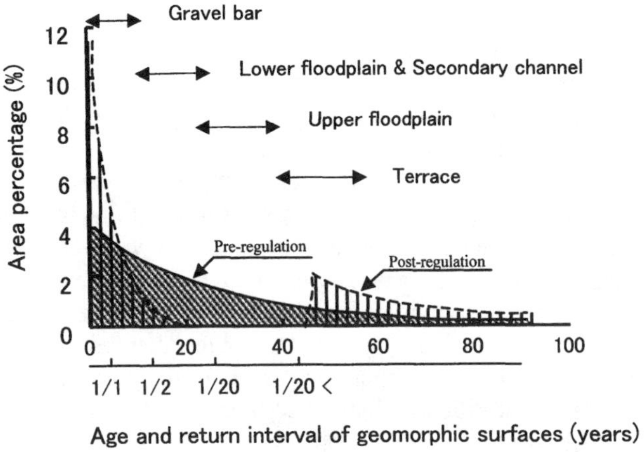

Figure 9. A shift of age distribution of geomorphic surfaces after the dam construction. A exponential curve observed in the Rekifune River will be bipolarized into two age classes after water regulation by the dam.

tinction of pioneer species. The mature pioneer trees that can produce and disperse seeds over fresh gravel bars mainly occur on upper floodplains. No age class between 20 and 40 years means a lack of upper floodplains. Although mature trees of *P. maximowiczii* can survive on stable surfaces (> 40 years old) for another decade, *C. arbutlfolia* and *T. urbaniana* will disappear from the surfaces where there used to be lower and upper floodplains. Under these conditions, pioneer tree species such as *C. arbutlfolia* and *T. urbaniana* will decrease, not being able to complete their life histories, and eventually may become extinct.

While a decrease in pioneer trees after dam construction has been reported, there are also many reports that bare land is decreasing and the expansion of woodlands is progressing in floodplains. *Harris et al.* [1987] pointed out the existence of rivers in which the crown coverage of riparian forests expanded after dam construction, and determined the cause to be the decrease in scoured areas following a decrease in the peak flow rate of flood waters. *Johnson* [1994] conducted research on the expansion of woodlands, which began after 1900, and reported that, in areas where the expansion of woodlands was most advanced, sandbars were decreasing at an annual rate of 10% and forests were expanding toward the lower reaches. One of the reasons for this expansion is the decrease in flow rates in June due to water storage in dams and the use of water for irrigation. It is thought that, due to the drop in the water level caused by these decreases, more trees take root in sandbars that would usually be submerged if the water levels were normal. Although salicaceous plants are currently growing on such sandbars, late successional trees are expected to become dominant with the stabilization of the gravel bar. The tendency of woodlands to expand has also been reported in Japan in recent years [*Lee et al.*, 1998].

The degradation of riverbeds and the re-accumulation of sand and silt are thought to accelerate the stabilization of geomorphic surfaces and increase the available water and nutrient content [*Johnson et al.*, 1976; *Nakamura*, 1996]. Therefore, the gravel bar dominated river landscape shown in Figure 2 will be greatly altered in association with forest establishment after dam construction, and tree species will shift from pioneer to late successional species. In the end, the types of forests that can be seen on hill slopes will be established on stable surfaces segregated from river dynamics, and this is in effect a dramatic reduction in the riparian zone.

4. MANAGEMENT IMPLICATION

In order to minimize the downstream effects of dams on the regeneration process of riparian tree species, the ideas of adaptive management and the possibility of an artificial flooding experiment should be discussed. The artificial flood may mobilize the stable geomorphic surfaces and may create bare sites for pioneer species to germinate. The Grand Canyon of the Colorado River has faced various environmental problems since Glen Canyon Dam was built in 1963. Prior to the dam construction, large volumes of sediment and numerous boulders were carried by spring flooding in June, and many plants were stripped from the river's banks forming sandbars. After the construction, however, significant natural seasonal changes in flow conditions were completely suppressed and replaced by a number of small daily changes caused by electric power consumption [*Long*, 1997]. Under such circumstances, the ideas of adaptive management were introduced and an artificial flooding experiment was conducted in March 1996. As a result of the artificial flooding, most of the thick sediment, which had accumulated on the riverbed due to discharge restrictions, was redistributed on high ground surfaces, contributing to the formation of a new sandbar.

Such artificial flooding experiments have also been conducted in Japan as a part of a adaptive management of dams. Although the magnitude of the experimental floods in Japan was fairly small in comparison with the above case, there is a possibility that such experiments will be conducted for riparian plant communities in the future. The question is when and how much water should be discharged to allow plant species, constituting riparian areas, to secure regeneration habitats and establish their growing conditions. We believe that direct gradient analysis in relation to the life history of plant species conducted by the present study will provide an idea about the necessary magnitude and frequency of artificial flooding for regeneration.

Acknowledgments. We are grateful to the staff of Satsunai River Dam Construction Office for their kind assistance. This research was supported in part by Grant-in Aid for Scientific Research (10460059) from the Ministry of Education, Science and Culture, and the grant from Water Resources Environment Technology Center.

REFERENCES

Aruga, M., Nakamura, F., Kikuchi, S. & Yajima, T., Characteristics of floodplain forests and their site conditions in comparison to toe slope forests in the Tokachi River. *Journal of Japanese Forestry Society*, pp. 354-362, 1996 (in Japanese with English abstract).

Bell, D. T., Gradient trends in the streamside forest of central Illinois. *Bulletin of the Torrey Botanical Club* 107, pp. 172-180, 1980.

Bell, D. T. & del Moral, R., Vegetation gradients in the streamside forest of Hickory Creek, Will County, Illinois. *Bulletin of the Torrey Botanical Club* 104, pp. 127-135, 1977.

Everitt, B. L., Use of the cottonwood in an investigation of the recent history of a floodplain. American Journal of Science 266, pp 417-439, 266.

Fenner, P., Brady, W. W. . Patton D. R., Effects of regulated water flows on regeneration of Fremont cottonwood. *Journal of Range Management* 38(2), pp. 135-138, 1985.

Gregory, S.V., Swanson, F. J., McKee, W. A. & Cummins, K. W., An ecosystem perspective of riparian zone - Focus on links between land and water -. *BioScience* 41, pp. 540-551, 1991.

Harris, R. R., Fox, C. A. . Risser, R., Impacts of hydroelectric development on riparian vegetation in the Sierra Nevada region, California, USA. *Environmental Management* 11(4), pp. 519-527, 1987.

Hupp, C. R. & Osterkamp W. R., Bottomland vegetation distribution along Passage Creek, Virginia, in relation to fluvial landforms. *Ecology* 66, pp. 670-681, 1985.

Hupp, C. R. & Osterkamp, W. R., Riparian vegetation and fluvial geomorphic processes. *Geomorphology* 14, pp. 277-295, 1996.

Ishikawa, S., Floodplain vegetation of the Ibi River in central Japan, I. Distribution behavior and habitat conditions of the main species of the river bed vegetation developing on the alluvial fan. *Japanese Journal of Ecology* 38, pp. 73-84, 1988 (in Japanese with English abstract).

Ito, S. & Nakamura, F., Forest disturbance and regeneration in relation to earth surface movement. *Japanese Journal of Forest Environment* 36(2), pp. 31-40, 1994 (in Japanese with English abstract).

Lee, S., Fujita, K., Tsukahara, T., Watanabe, S., Yamamoto, K. and Mochizuki, T., Roles of floods and fine sediment transport to woodland expansion on a gravel river-bed. *Suikougaku Ronbunshu* 42, pp. 433-438, 1998 (in Japanese with English abstract).

Long, M. E., The grand managed canyon. *National Geographic* 192(1), pp. 114-135, 1997.

Lindsey, A. A., Petty, R. O., Sterling, D. K. & VanAsdall, W., Vegetation and environment along the Wabash and Tippecanoe rivers. *Ecological Monographs* 31(2), pp. 105-156, 1961.

Nagasaka, Y., Salix species in riparian zone. *Koshunai-kiho* 101, pp. 12-17, 1996 (in Japanese).

Nakamura, F., Analysis of storage and transport processes based on age distribution of sediment. *Transactions. Japanese Geomorphological Union* 7(3), pp. 165-184, 1986.

Nakamura, F., Forest and stream interaction in riparian zone. *Japanese Journal of Ecology* 45, 295-300, 1995 (in Japanese).

Nakamura, F. & Kikuchi S., Some methodological developments in the analysis of sediment transport processes using age distribution of floodplain deposits. *Geomorphology* 16, pp. 139-145, 1996.

Nakamura F., Yajima, T.. Kikuchi, S., Structure and composition of riparian forests with special reference to geomorphic site conditions along the Tokachi River, northern Japan. *Plant Ecology* 133, pp. 209-219, 1997.

Niiyama, K., Distribution of Salicaceous species and soil texture of habitats along the Ishikari River. *Japanese Journal of Ecology* 37, pp. 163-174, 1987 (in Japanese with English abstract).

Niiyama, K., Distribution of Chosenia Arbutifolia and soil texture of habitats along the Satsunai River. *Japanese Journal of Ecology* 39, pp. 173-182, 1989 (in Japanese with English abstract).

Niiyama, K., The role of seed dispersal and seedling traits in colonization and coexistence of Salix species in a seasonally flooded habitat. *Ecol. Res.* 5, pp. 317-331, 1990.

Nilsson, C., Ekblad, A., Gardfjell, M. . Carlberg, B., Long-term effects of river regulation on river margin vegetation. *Journal of Applied Ecology* 28, pp. 963-987, 1991.

Reily, P. W. . Johnson, W. C., The effects of altered hydrologic regime on tree growth along the Missouri River in North Dakota. *Canadian Journal of Botany* 60, pp. 2410-2423, 1982.

Rood, S. B. . Heinze-Milne, S., Abrupt downstream forest decline following river damming in southern Alberta. *Canadian Journal of Botany* 67, pp. 1744-1749, 1989.

Rood, S. B. . Mahoney, J. M., Collapse of riparian poplar forests downstream from dams in western prairies: probable causes and prospects for mitigation. *Environmental Management* 14(4), pp. 451-464, 1990.

Satsunai River Dam Construction Office, Obihiro Development and Construction Department, Reports on monitoring results of Satsunai River Dam. No. 4, 4-8, 1998.

Sakio, H. & Suzuki, W., Overview of riparian vegetation: structure, ecological function and effect of erosion control works. *Journal Japan Society of Erosion Control Engineering.* 49(6), pp. 40-48, 1997 (in Japanese).

Sigafoos, R. S., Vegetation in relation to flood frequency near Washington, D. C. *Geological Survey Research 1961*, pp. C248-C249, 1961.

Swanson, F. J., Gregory, S. V., Sedell, J. R. and Campbell, A. G., Land-water interactions: The riparian zone. In: Analysis of coniferous forest ecosystems in the western United States (ed. Edmonds R.L.) *Hatchinson Ross Publishing Company*, pp. 267-292, 1982.

Teversham, J. M. & Slaymaker, O., Vegetation composition in relation to flood frequency in Lillooet River Valley, British Columbia. *Catena* 3, pp. 191-201, 1976.

Yajima, T., Nakamura, F., Shimizu, O. & Shibuya, M., Forest recovery after disturbance by the 1926 Mudflow at Mount Tokachi, Hokkaido, Japan. *Research Bulletins of the College Experiment Forests, Faculty of Agriculture, Hokkaido University* 55(1), pp. 216-228, 1998.

Futoshi Nakamura
Nozomi Shin

Section 4

Restoration Strategies

Relations Between Streamflow, Sediment Transport, and Aquatic Habitat in Regulated Rivers

John Pitlick

Department of Geography, University of Colorado, Boulder, Colorado

Peter Wilcock

Department of Geography and Environmental Engineering, Johns Hopkins University, Baltimore, Maryland

This paper reviews approaches for assessing the impacts of river regulation on stream ecosystems. These impacts can vary greatly depending on the hydrogeomorphic setting of the river, and on the change in water and sediment supply to the reach downstream. Results drawn from field studies of gravel-bed reaches of the Colorado River in Colorado and Trinity River in California illustrate the range of issues to consider in assessing the effects of flow regime and sediment transport on micro- and meso-scale habitats. Important aspects of bed-load transport at the local scale are discussed, along with ecological implications of specific transport phases and processes. For a wide range of rivers, including those that carry predominantly fine sediment (silt and sand), the loss of channel complexity and habitat heterogeneity represents a broader issue. We suggest that unless the broader-scale importance of sediment input and output is recognized and quantified, restoration efforts that focus on site-specific issues or single-species enhancement are likely to fall short of their objectives.

1. INTRODUCTION

Rivers serve a basic role in the hydrologic cycle of carrying water, sediment, and nutrients from distributed sources within a drainage basin to outlets or sinks downstream. In most large river systems, this natural flux is now interrupted by dams and diversions, and it is becoming increasingly apparent that the placement and operation of these and other engineering structures have diminished the ecological integrity of rivers throughout the world [*Calow and Petts*, 1992; *Sparks*, 1995; *Stanford et al.*, 1996]. The eco-hydrological effects of dams and water diversions can vary widely in scale, from the local detail of substrate modification or the obstruction of fish migration, to widespread impacts on suites of inter-related processes involving nutrient cycles [*Newbold*, 1992], channel-floodplain interactions [*Stanford and Ward*, 1992], riparian vegetation [*Friedman et al.*, 1998], food webs [*Wooton et al.*, 1996], sediment loads [*Andrews*, 1986], and channel geomorphology [*Ligon et al.*, 1995; *Van Steeter and Pitlick*, 1998].

Scientific interest in the environmental effects of river regulation has expanded in the last decade from focused studies (e.g. the effects of fine sediment on spawning habitats) to broader ecosystem questions concerning long reaches of rivers (e.g. the Recovery Implementation Pro-

Geomorphic Processes and Riverine Habitat
Water Science and Application Volume 4, pages 185-198

gram for Endangered Fish Species in the upper Colorado River Basin; *USFWS*, 1987]. These broader efforts require the geomorphologist to integrate information over a range of scales. Work at local scales may be heavily dependent on direct observation and can take advantage of the fact that the underlying physical processes are well known. Regional-scale investigations of river regulation can often rely on historical records of flow and channel position, and on the depositional histories left in floodplain or channel sediments. To be effective in supporting ecosystem analysis and recovery, it is necessary to link the physical basis of the local scale to the integrated record of the regional scale. A natural link between these scales is made through the principle of mass conservation for water and sediment, i.e. the supply of water and sediment must be balanced across all scales, otherwise geomorphic changes will occur. This principle provides a strong physical basis for extending results from local (at-a-site) studies to much longer reaches of a river, where the integrated effects of flow regulation and changing sediment transport regime are expressed through differences in channel morphology.

This paper considers linkages of water and sediment transport processes from local to regional scales and their role in understanding ecosystem change. The paper is organized into three sections: The first section considers some broader aspects of the interruption of the river continuum by dams and reservoirs; the second section addresses links between sediment transport and micro-scale habitat in gravel-bed rivers; and the third section illustrates a local-to-regional linkage made in evaluating habitat heterogeneity on the Colorado River. Although much of our experience (and perhaps some of our biases) comes from working on gravel-bed rivers in the western U.S., we believe that many of the points and issues discussed here are relevant to other types of rivers in other parts of the World.

2. INTERRUPTING THE WATER/SEDIMENT CONTINUUM

Rivers adjust their morphology (width, depth, slope, and channel pattern) and composition in response to the water and sediment supplied from the drainage basin. In a typical dendritic river basin, water and sediment are supplied more or less evenly along the length of the trunk stream, causing the channel to become systematically wider and deeper downstream. At the same time, the sediment composing the bed of the river (substrate) becomes smaller and the channel gradient decreases. Although large variation exists within and between basins, the physical characteristics of streams vary consistently from headwaters to mouth. The continuum of fluvial forms and processes provides the basis of essential concepts advanced in both fluvial geomorphology [*Leopold and Maddock*, 1953] and aquatic ecology [*Vannote et al.*, 1980]. (However, it is worth noting here that the utility of these concepts has not gone unchallenged or unchanged; see *Statzner and Higler* [1985]; *Ferguson* [1986]; and *Montgomery* [1999]).

Dams clearly disrupt the orderly downstream transfer of water and sediment. The effect of this interruption can vary widely depending on the size of the reservoir, the nature of the dam operations, the position of the dam within the watershed, and on the hydrology and geology of the watershed [*Williams and Wolman*, 1984; *Collier, et al.*, 1996]. As a result, approaches to river management problems may require very different strategies. In the following two sections, we discuss how regional variations in climate may buffer or amplify the hydrologic effects of a dam, and then consider how the location of the dam relative to sediment sources determines potential downstream impacts.

2.1. Hydrology

Reservoirs operated for flood control, water storage, and power generation typically alter the natural hydrograph by reducing peak flows and increasing base flows. The net effect is a flow regime with less variability. In their investigation of the downstream effects of dams, *Williams and Wolman* [1984] found that reservoir operations reduced peak flows by an average of 39% for the 29 reservoirs studied, and in the most extreme case, peaks flows were reduced by 97%. Changes in streamflow variability can impact the aquatic ecology of river systems by lowering the frequency of disturbance and reducing the diversity of riverine habitats [*Stanford et al.*, 1996]. The ecological significance of changes in streamflow may be accentuated or mitigated by other factors, including presence/absence of additional dams, contributions from unregulated tributaries, or by regional differences in the natural hydrologic regime.

Streamflow is produced by runoff originating from rainfall, snowmelt, groundwater flow, or combinations of these processes (e.g. rain-on-snow events). The intensity and variability of runoff generated by these different processes can vary significantly. Figure 1 shows a time series of annual peak discharges for two unregulated rivers in the western U.S. The Yampa River at Steamboat Springs, CO, represents a river fed largely by snowmelt, whereas the Eel River at Scotia, CA, represents a river fed almost entirely by rainfall. The contrast in peak flow variability of these two rivers is striking, especially because the river with more variable flows (Eel) is much larger and would, therefore, be expected to have a more stable flow regime. Differences in peak flow characteristics and annual runoff of unregulated rivers are strongly influenced by the type and

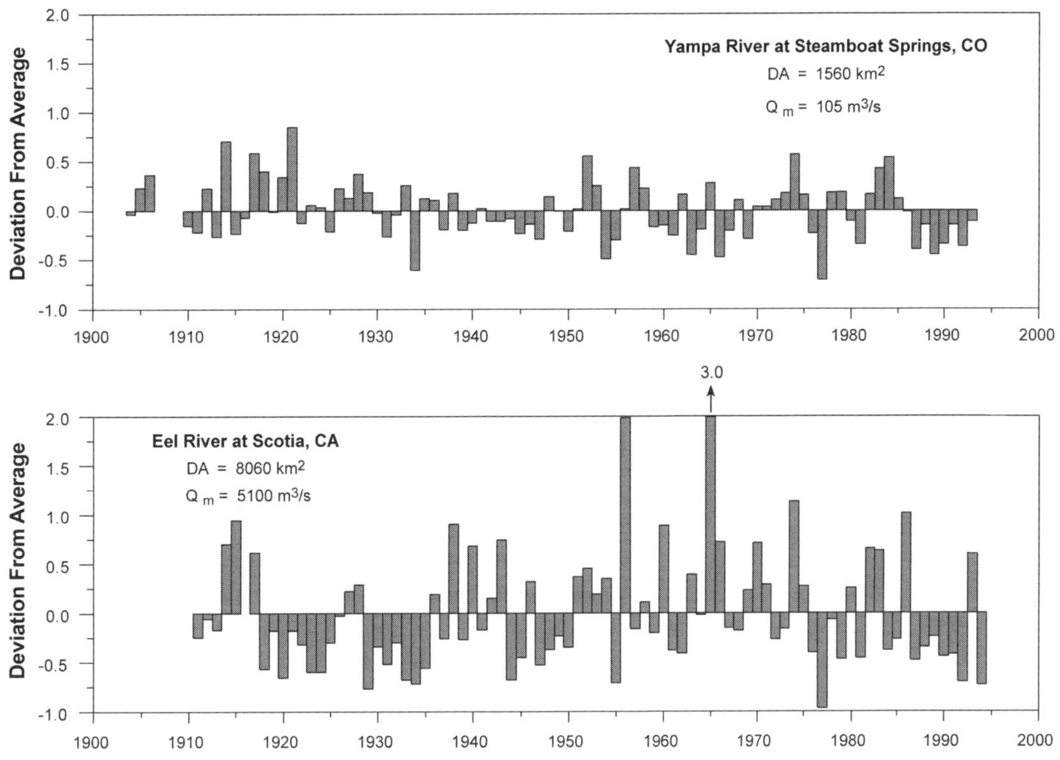

Figure 1. Annual peak discharges of the Yampa River at Steamboat Springs, Colorado, and the Eel River at Scotia, California.

intensity of precipitation, driven primarily by regional climatic processes [*Lins et al.*, 1990; *Pitlick*, 1994]. The interannual variability of streamflow in rivers fed by snowmelt or low intensity rainfall is typically small compared to runoff regimes dominated by intense seasonal rainfall (convective storms) or rain-on-snow events [*Riggs and Harvey*, 1990].

Ecologists have suggested that differences in streamflow characteristics (flow variability, flood frequency, intermittency, etc.) are likely to have a strong influence on the structure of aquatic communities [*Poff and Ward*, 1989; *Bailey and Li*, 1992; *Puckridge et al.*, 1998]. *Poff and Ward* [1989] used data from 78 streams across the continental U.S. to develop a conceptual model of stream hydrology that could be used as the basis for predicting ecological attributes in particular regions. They proposed that biotic interactions such as competition and predation may dominate over abiotic (physical and chemical) processes as the flow regime becomes more predictable and less variable. A regulated flow regime with less variability may thus give an exotic fish a competitive advantage over a native species that is better adapted to flooding. *Bayley and Li* [1992] extended Poff and Ward's model to predict ecological attributes of fish assemblages across different hydrologic regimes. Their inferred trends, shown in Figure 2, suggest that variations in streamflow will be reflected by differences in fish migration, body size, and species richness. Thus, as the flow regime becomes more predictable, species richness increases; this may not be a desirable condition if the increased numbers of species are not native to the system. Subsequent studies of fish community structure conducted by *Poff and Allan* [1995] indicated that fish with generalized feeding strategies were associated with hydrologically variable sites, whereas "specialists" (fish that require specific microhabitats or food resources) were associated with more stable sites. In studying the effects of hydropower releases on fish community structure, *Bain et al.* [1988] found that a group of small specialist fish, common to a nearby free-flowing river, were essentially absent from a reach subject to high flow fluctuations. Benthic organisms are likewise affected by the intensity and variability of streamflow, although the responses of particular species in different locations appear to vary widely. In streams in New Zealand, *Clausen and Biggs* [1997] found that invertebrate biomass and species diversity were positively correlated with indices of flow variability, whereas periphyton biomass and species richness decreased with increasing flood frequency. In streams in Manitoba, Can-

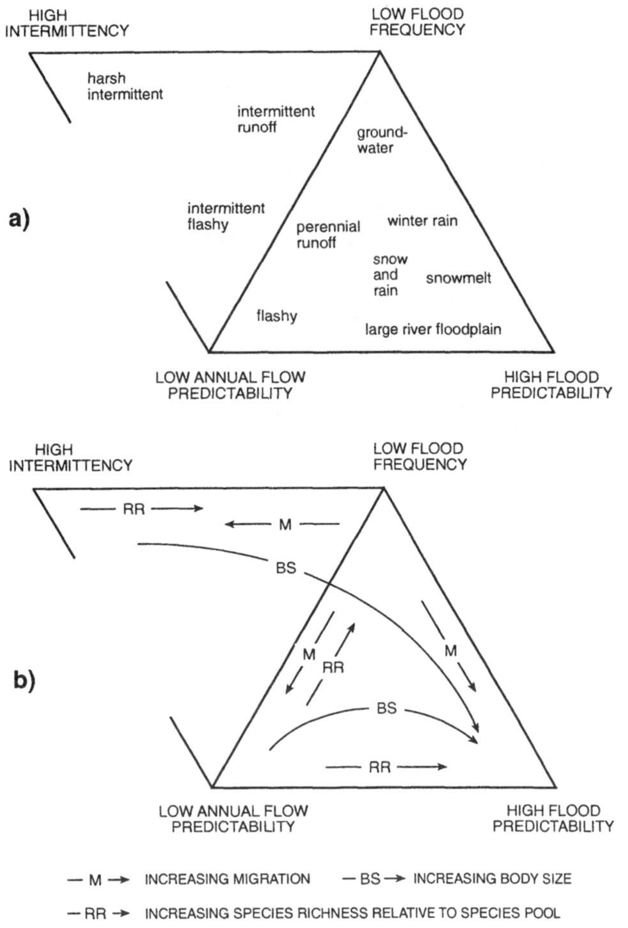

Figure 2. Proposed conceptual relations between streamflow characteristics and fish assemblage attributes (from *Bailey and Li*, 1992).

ada, *Cobb et al.* [1992] found that the density of benthic invertebrates decreased as flood discharge increased. It appears that, in some cases, the benefits of high flows and substrate disturbance (e.g. removal of interstitial fine sediment) appear to outweigh the losses due to mortality. It is likely that some of these observations reflect variation in other factors such as substrate mobility (discussed later), but at regional scales, indices of hydrologic variability can be useful predictors of aquatic community structure.

Results from the ecologically based studies cited above indicate that changes in flow variability caused by dams may shift the ecosystem toward species better adapted to more stable flows. The magnitude of this ecosystem shift is likely to be greatest in arid and semi-arid regions such as the American West, where the natural flow variability is high and the amount of water impounded by reservoirs is large in proportion to the annual flow [*Graf*, 1999].

2.2. Sediment Sources

Reservoirs are typically very effective at trapping sediment, leaving downstream reaches starved of a sediment supply. The primary effect of reduced sediment supply is to decrease the concentration of sediment carried by the river. Reduction in sediment concentration, coupled with a reduction in high flows, can reduce the total sediment load to a fraction of pre-dam values. *Williams and Wolman* [1984] provide striking examples of reduced annual sediment loads caused by individual reservoirs, and note that these effects can persist for several 100s of kilometers downstream.

The impact of sediment reductions on downstream channels and aquatic habitat can vary widely, depending on the amount of reservoir storage, the dam operations, and the location of the dam relative to sediment sources. Where there is no sediment supply immediately downstream of a dam, river bed degradation may be as much as 1-3 m [*Williams and Wolman*, 1984]. Degradation is typically accompanied by a coarsening of the river bed with associated loss of useable habitat by fish and benthic invertebrates. Regulated rivers may also decrease in width [*Williams*, 1978; *Andrews*, 1986; *Van Steeter and Pitlick*, 1998], although data presented by *Williams and Wolman* [1984] indicate that, more often than not, degrading channels tend to widen as a result of bank erosion caused by bed degradation and stage fluctuations related to hydropower releases at the dam. Much of the impact of Glen Canyon dam on ecosystem resources in Grand Canyon is attributed to the fact that approximately 90% of the sediment formerly delivered to the canyon by the Colorado River is now trapped behind the dam in Lake Powell [*Webb et al.*, 1999].

Where abundant sources of sediment exist below dams, the river channel may show net aggradation, rather than degradation, particularly when the reduction in sediment-transport capacity is large. The Trinity River below Trinity-Lewiston dams in California is a good example of a sediment-rich system. The dams on the Trinity River were constructed in the early 1960s as part of a storage/diversion project, effectively eliminating major flooding in the reaches immediately downstream [*Wilcock et al.*, 1996]. In subsequent years, the population of native chinook salmon and steelhead trout has been significantly reduced. Abundant sediment supply from downstream tributaries, together with a severe reduction in sediment-transport capacity, has produced pervasive deposition along with narrowing and simplification of the channel. Large parts of the watershed below the dam are underlain by a deeply weathered granite that decomposes into granules, which accumulate in the bed and banks of the Trinity River. The granules clog the interstices of the spawning gravels, preventing salmonid fry from emerging at the proper time. Sedimentation and colonization of vegetation on formerly active gravel bars have also narrowed the channel by an average of about 40% [*Pitlick*, 1992] and removed much of the topographic variability useful for rearing habitat [*USFWS*, 1999]. Dam releases are capable of flushing some of this sediment from the bed [*Wilcock et al.*, 1996], but are not able to remove mature vegetation and alter bars. The imbalance between sediment supply and transport capacity are sufficiently large that habitat restoration is likely to require mechanical removal of sediment by pool dredging and regrading and vegetation removal on bars and banks.

Segments of the Green River in Utah and the Colorado River in Colorado are likewise influenced by sediment sources downstream of dams, only in these cases the sources are more diffuse. The Green River and Colorado River are regulated by a series of mainstem and tributary dams that store snowmelt at high elevations in the central and southern Rocky Mountains. Both rivers then flow long distances across the Colorado Plateau, through sparsely vegetated terrain that yields little additional runoff, but large amounts of sediment [*Iorns et al.*, 1965; *Andrews*, 1986; *Van Steeter and Pitlick*, 1998]. The available data indicate that suspended-sediment concentrations at mainstem gaging stations have not changed appreciably in the last 50 yr [*Iorns et al.*, 1964; *Van Steeter and Pitlick*, 1998], suggesting that unregulated tributaries are supplying sediment more or less as they have in the past. However, because flows are now regulated, both rivers have lost some of their capacity to carry sediment, resulting in 30-50% decreases in annual loads [*Andrews*, 1986; *Van Steeter and Pitlick*, 1998] and deposition of tributary derived sediment in the mainstem [*Andrews, 1986*; *Van Steeter and Pitlick*, 1998; *Allred and Schmidt*, 1999].

The end result of this process is illustrated in Figure 3, which shows a sequence of channel change in an alluvial reach of the Colorado River inhabited by the endangered Colorado pikeminnow (*Ptychocheilus lucius*). Historical changes in channel pattern were delineated along 150 km of the river by comparing two sets of aerial photographs taken in 1937 and 1993. The outlines of features such as channel banks, islands, and secondary channels were digitized and adjusted to a common scale using Geographic Information System (GIS) techniques and software. The analysis indicates that, since the late 1930s, the average width of the main channel of the Colorado River has decreased by about 10%, while the area of secondary channels has decreased by 25-30% [*Van Steeter and Pitlick*, 1998; *Pitlick and Cress*, 2000]. The changes in channel morphology coincide with a period from the late 1950s through the late 1970s when sediment loads decreased by 40-60% because of reservoir operations [*Pitlick et al.*, 1999]. The geomorphic response to the altered flow regime is significant in an ecological sense because, as shown in Figure 3, deposition and narrowing have simplified the channel, thus reducing the amount of side-channel and back-water habitat available to the Colorado pikeminnow. The loss of these important habitats, combined with changes in water quality and the introduction of non-native fishes, has potentially had long-lasting impacts on the native fish community of the Colorado River.

3. LOCAL SCALE: SEDIMENT TRANSPORT PROCESSES AND AQUATIC HABITAT

The movement of sediment on a streambed affects habitat at various scales, ranging from the small interstices between particles, to individual morphologic units such as pools and riffles. At much larger scales, erosion and deposition control macro-habitats (e.g. secondary channels and backwaters) by altering channel form and associated flow fields. The linkages among sediment transport, channel form, and ecosystem function are many [*ASCE*, 1992], as

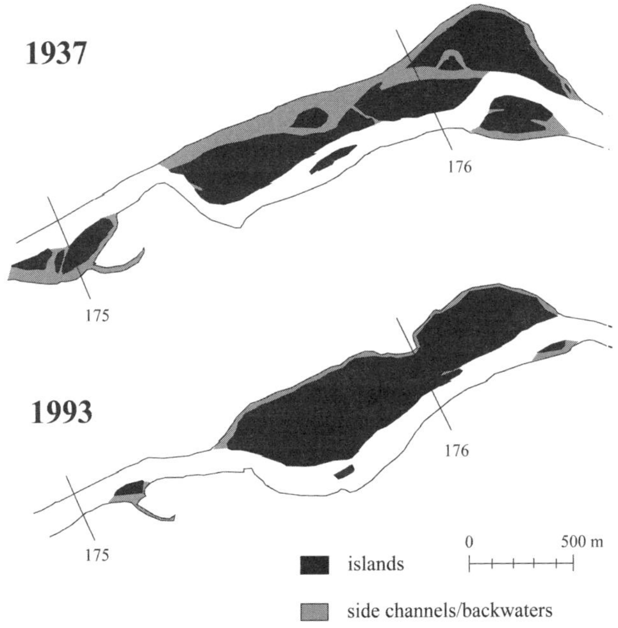

Figure 3. Digitized channel maps of a section of the Colorado River showing historical changes in channel complexity. Markers indicate mileage upstream from the Green River confluence (from *Van Steeter and Pitlick*, 1998).

are the time scales of interest [*Sear*, 1994]. Freshwater ecosystems include species that live anywhere from days to tens of years, and the habitat requirements of individual aquatic species vary widely. The movement of coarse bed material during floods may reduce invertebrate populations and species abundance, but over the long run changes in bed surface texture caused by sediment transport are often necessary to maintain habitat quality. Short-term evacuation of fine sediment from pools or deposition of coarse sediment in bars may increase the quality of habitat used for spawning, whereas long-term scour or aggradation may simplify the channel and reduce habitat heterogeneity. The processes of sediment transport are thus essential to the formation and maintenance of aquatic habitats.

The topics reviewed in this section emphasize sediment transport at the local scale, while the following section emphasizes reach-scale processes. At the local scale, the processes of entrainment, transport, and sorting of bed material are strongly influenced by local flow conditions and sediment properties. Among the hydrodynamic parameters used to evaluate these processes, the dimensionless shear stress, or Shields parameter, τ^*, is perhaps the most useful because it accounts for the influence of both flow and sediment properties. Measuring these properties accurately in the field is not often easy, and as discussed below, the physical meaning of τ^* is not the same at small and large spatial scales. Although the underlying physics remain the same at both scales, the types of data available and theoretical foundation vary with spatial scale. Bridging this range of scales is one of the primary challenges facing successful incorporation of geomorphology into ecosystem analysis.

3.1. Shields Parameter

The Shields parameter is defined as $\tau^* = \tau[(\rho_S-\rho)gD]^{-1}$, where τ is the boundary shear stress, ρ_S and ρ are the densities of sediment and water, respectively, g is gravitational acceleration, and D is grain size. The physical meaning of τ^* is clear over a small area of the bed containing perhaps several hundred of the larger grains in the substrate. At this scale, τ^* represents a ratio of the mean fluid force acting on the bed to the weight of the individual grains. Locally, the near-bed flow is highly variable in time and the forces acting on individual grains are more usefully related to the local instantaneous velocity than to the mean shear stress [*Nelson et al.*, 1995]. Nonetheless, in the absence of sharp discontinuities in the flow field, the correlation between near-bed velocity and mean shear stress is sufficiently strong that τ^* remains a suitable parameter for estimating sediment transport thresholds and sediment transport rates.

As scale increases beyond a small, relatively uniform area of the streambed, the simple physical interpretation of τ^* as a force balance loses some of its validity. In natural channels, both τ and D can vary widely [*Whiting and Dietrich*, 1991; *Johnston et al.*, 1998], resulting in potentially poor correlation between these two variables [*Lisle et al.*, 2000]. Although average values of τ and D can be defined for any case, predictions of sediment transport rates based on mean quantities of τ and D can be biased because sediment transport is a nonlinear process. For steady uniform flow, the mean shear stress is calculated from $\tau = \rho gRS$, where R is the hydraulic radius and S is the water-surface slope. However, at the scale of typical habitat units (e.g. riffles and bars) the flow can be strongly nonuniform, in which case, the total force acting on the streambed includes components related to both down-stream and cross-stream changes in depth and velocity. Estimates of τ at this scale are better derived using a one- or two-dimensional hydrodynamic model, although in many cases the data required to run such models exceed the information available. Measurements in large rivers are further constrained by flow depth, which limits the ability to sample the substrate. Thus, it is common for substrate grain size to be estimated from a single point count of 100 stones on an exposed gravel bar or in shallow water. Uncertainty in the estimates of τ and D can therefore be quite large (easily exceeding a factor of 2), a fact that is often not taken into account when reporting results from field studies or when developing habitat maintenance plans.

Although there are theoretical and practical limitations to characterizing the details of transport at field scales, τ^* does incorporate a measure of flow strength and substrate size that can provide a useful index of sediment transport intensity. Figure 4 shows the variation of τ^* with discharge, Q, derived from a series of field measurements in a gravel-bed reach of the Colorado River near Grand Junction, Colorado. These measurements were made as part of on-going studies of the effects of reservoir regulation on habitats used by endangered fish. The open circles indicate modeled values of τ^* at each cross section, while the closed circles indicate reach-average values. The individual values were derived from a one-dimensional nonuniform flow model, calibrated to flows ranging from 76 to 1100 m^3s^{-1}. Note that the shear stress at any particular cross section can differ significantly from the reach average, particularly at low discharges; these differences reflect the lengthwise variations in channel geometry, slope, and grain size discussed above. The relation between Q and τ^* is nonlinear, with the shear stress increasing rapidly in the range from low to intermediate discharges, and more slowly thereafter. The overall shape of the curve is similar

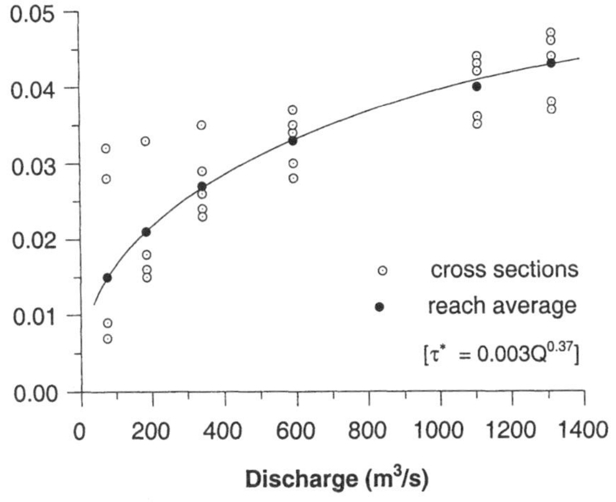

Figure 4. Relation between discharge and modeled values of dimensionless shear stress for a reach of the Colorado River near Grand Junction, Colorado. Open circles indicate modeled values of τ^* at individual cross sections; closed circles represent averages for the reach.

to a stage-discharge relation, although in this case, the slope (0.37) is considerably less than that of a typical rating curve. As discussed below, the variation of τ^* with Q can be used to provide some basis for evaluating habitat maintenance alternatives.

3.2. Incipient Motion

Incipient motion of river bed material has several ecologically significant functions. Entrainment of surface clasts is needed to remove fine sediment from the bed in order to maintain spawning habitat or interstitial void space for invertebrates and juvenile fish [*Diplas*, 1994; *Kondolf and Wilcock*, 1996; *Milhous*, 1998]. Entrainment also helps to maintain gravels in a loose state such that salmonids can successfully build redds. Incipient motion also defines the minimum flow needed to initiate transport and maintain an active channel with topographic variability [*Pitlick and Van Steeter*, 1998].

The value of τ^* associated with incipient motion is termed the critical dimensionless shear stress, τ^*_c. Data compiled by *Buffington and Montgomery* [1997] suggest that the range in τ^*_c for the median grain size, D_{50}, varies by at least a factor of two. This variability may be apportioned between real differences in conditions needed to initiate transport of different sediments, and apparent differences due to problems in defining and observing incipient motion. For typical fluvial gravels in a loose state, sufficient data exist to show that sediment characteristics (size distribution, sorting and shape) strongly influence the value of τ^*_c [*Wilcock*, 1993; *Gomez*, 1994]. Additional factors, such as sand content, can affect incipient motion. *Wilcock* [1998] has demonstrated that τ^*_c for the gravel fraction can vary from 0.045 to 0.010, depending on the amount of sand on the bed. Values of τ^*_c can also vary over the course of an event because of changes in bed compaction and structure, and additions of organic material. *Church et al.* [1998] have demonstrated that the development of bed surface structure alone can increase τ^*_c by at least a factor of two.

3.3. Partial Transport

It has been shown that large portions of the bed surface can remain entirely immobile throughout a transport event, even when the shear stress is well above that needed for incipient motion [*Wilcock and McArdell*, 1997]. This condition is defined as partial transport, meaning that only part of the grain population of the bed surface is mobilized at any flow [*Wilcock and McArdell*, 1993]. Partial transport occurs for all coarse sizes and demonstrates that the presence of any size in transport does not indicate that all grains of that size are in motion. Results from flume experiments by *Wilcock and McArdell* [1997] showed that individual size fractions went from near complete immobility to complete mobilization over a range in τ^* of a factor of two, and that the range of sizes in a state of partial transport at any flow was also approximately a factor of two. In natural channels, the range of flows and grain sizes in a state of partial transport would likely be larger because of the spatial variations in flow conditions and sediment properties.

As discharge and shear stress increase more and more coarse grains are entrained from the bed, until eventually most all particles on the surface are mobilized. Complete mobilization does not occur in gravel-bed rivers until somewhere near bankfull flow [*Pitlick and Van Steeter*, 1998]; however, at this stage much of the surface layer is re-worked to depths equaling or exceeding the coarsest particles. Field measurements of entrainment and scour in gravel-bed rivers show that the depth of bed material exchange is spatially variable and sensitive to the excess shear stress [*Carling*, 1987; *Lisle*, 1989; *Wilcock et al.*, 1996; *Haschenburger*, 1999]. *Wilcock et al.* [1996] observed that a modest dam release on the Trinity River produced partial transport whose intensity varied with local flow and topography along the channel. Results from an extensive study of scour by *Haschenburger* [1999] show that frequency distributions of scour depths are modeled well by the exponential function, with the mean and variance of scour depth increasing as the Shields stress in-

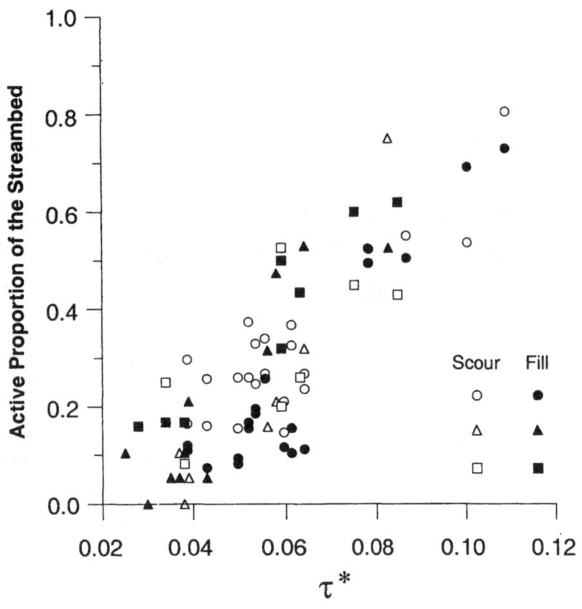

Figure 5. Change in the active proportion of the bed surface with increasing dimensionless shear stress. Different symbols represent data from three different sites, two in Canada and one in England (from *Haschenburger*, 1999).

creases. Data from this study (Fig. 5) indicate that at Shields stresses near incipient motion ($\tau^* = 0.03\text{-}0.04$) typically no more than about 20% of the bed surface is active. As the stress increases, proportionally more of the bed becomes active, but even in the range well above critical ($\tau^* > 0.09$), much less than 100% of the bed may be reworked.

Some of the uncertainty in estimating τ^* or τ^*_c can be addressed with field measurements. Verification of incipient motion may be accomplished using tracer grains placed on the bed surface, or through simple observations of surface particle mobility. Tracer deployment and recovery is labor intensive, but may be appropriate in locations of particular interest, such as spawning sites. *Wilcock et al.* [1996] used tracer gravels to document entrainment at potential spawning sites, and to evaluate discharges needed for flushing fine sediment. In the more general case, simple field observations may verify entrainment. For example, where the upper side of clasts develop an algal stain, the distribution of stained vs. unstained clasts can provide a map of entrainment. Patches of rocks, painted in place, provide similar information when coupled with flow measurements. The sensitivity of bed load transport to local flow processes and sediment properties provides strong motivation for verifying the extent of entrainment and scour in the field.

3.4. Ecological Implications

Among the results discussed above, the observation that only a portion of the bed surface is mobilized during typical high-flow events has perhaps the most important implications for aquatic ecology. Field research conducted by freshwater ecologists suggests that the importance of substrate heterogeneity and streambed disturbance varies widely within and between streams. *Erman and Erman* [1984] found that substrate heterogeneity had no effect on the abundance or species richness of macroinvertebrates in small gravel-bed streams; likewise, *Bain et al.* [1988] found that substrate heterogeneity was not important in distinguishing habitats used by fish in a small river in Vermont. In contrast, research in New Zealand suggests that the diversity of invertebrate species is higher in streams with coarser particles and greater substrate heterogeneity [*Scarsbrook and Townsend*, 1993; *Quinn and Hickey*, 1994; *Clausen and Biggs*, 1997]. Some of the conclusions concerning the effects of substrate heterogeneity are most certainly related to other factors, such as substrate stability. *Minshall and Minshall* [1977] repeatedly sampled benthic invertebrates in a small creek in Idaho and found that species abundance was much higher in a riffle (which was characterized as 'erosional') than in a nearby pool (which was characterized as 'depositional'). Results from this and other studies indicate that periodic, low-level disturbance is a desired attribute of many aquatic ecosystems. However, with increasing discharge and sediment-transport intensity, refuge opportunities for mobile organisms decrease, thus the total density of invertebrates and number of species commonly decreases as substrate mobility increases [*Cobb et al.*, 1992; *Scarsbrook and Townsend*, 1993; *Death and Winterbourn*, 1995]. Field data from 11 sites of varying stability in New Zealand (Fig. 6) indicate that the number of species collected in seasonal samples of invertebrates declined linearly as substrate instability increased [*Death and Winterbourn*, 1995]. Data reported by *Cobb et al.* [1992] show that the density of aquatic insects in gravel-bed reaches of a stream in Manitoba, Canada, declined by several orders of magnitude as discharge and substrate mobility increased. In the extreme case of flash floods in desert streams, *Fisher et al.* [1982] observed that flows capable of moving the bed material reduced total invertebrate density by as much 98%. In some settings, the influence of seasonal changes in nutrient loading and water temperature can exceed that due to floods and sediment transport [*Boulton et al.*, 1992], however, the results noted above suggest that bed-load transport events of low to intermediate intensity and frequency are very important for maintaining aquatic habitats and species diversity.

4. REGIONAL SCALE: SEDIMENT TRANSPORT AND CHANNEL COMPLEXITY

Ecologists describe the piedmont reaches of unregulated alluvial rivers as hot spots of biodiversity [*Stanford et al.*, 1996] where native plant and animal communities are adapted to periodic overbank flooding and reworking of the channel. These processes maintain a mosaic of features, including bars, islands, secondary channels, and semi-permanent embayments such as backwaters. The habitat heterogeneity found in complex, laterally active reaches is thought to be essential for maintaining diversity in aquatic communities [*Power et al.*, 1995].

The traditional engineering approach of managing rivers to reduce flooding and improve navigation has the undesired ecological effects of decreasing channel complexity and habitat heterogeneity. Structural modifications (levees, rip-rap and dikes), combined with reservoir operations provide an effective one-two punch that significantly decreases the spatial and temporal variability of fluvial systems (as shown previously in Figure 3). The process of channel simplification typically occurs at regional scales, thus proposed restoration and habitat-maintenance programs should emphasize the long-term, reach-wide behavior of a river. In this section, we present results from a case study of the Colorado River demonstrating how data collected at regional scales can be used to guide management decisions regarding habitat maintenance and restoration.

4.1. Mass Balance Considerations

The key to maintaining channel complexity, or preventing further simplification, is to maintain the mass balance of sediment carried by the river. Sediment that is not transported through a reach will be deposited somewhere, resulting in further channel simplification and loss of habitat. If this basic principle is not acknowledged, then flow recommendations based on site-specific study and analysis will be valid only in so far as the processes at that site are representative of the river as a whole. A broad-based analysis of sediment transport is particularly important in the case of migratory species (e.g. anadromous or potanadromous fish), or generalist species that utilize a wide range of habitats.

The long-term load or mass balance of sediment carried by a river can be estimated on the basis of flow and sediment transport relations (sediment rating curves) or on the basis of observed changes in channel morphology. Assuming sufficient hydrologic data are available, rating-curve relations can be used to estimate sediment loads over seasonal, annual or decadal time scales, an advantage that may highlight important periods when sediment loads were higher or lower than the long-term average. Figure 7 illustrates this point by comparing serial trends in the annual discharge and suspended-sediment load of the Colorado River near Cameo, Colorado. Subdividing these records into separate time periods highlights the interval from 1950 through 1982 when annual loads were about 30% lower than the period before or after. The drop in sediment loads coincides with the mid-century period of reservoir construction in the upper Colorado River basin, which resulted in systematic reductions in the magnitude and frequency of high flows [*Van Steeter and Pitlick*, 1998]. A substantial volume of sediment would have been deposited in the Colorado River during this time, and it is likely that this process had adverse effects on the native fish population. Further analyses indicated that, under present conditions, 65 to 78% of the annual suspended sediment load is carried by flows that exceed one half the bankfull discharge. To maintain the mass balance of sediment, therefore, and prevent further loss of habitat, *Pitlick et al.* [1999] recommended that flows exceeding this level should occur with the same average frequency (30 days per year) as in the past 20 years.

Mass balances derived from morphologic measurements of scour and fill, or the translation of sediment waves and bars, provide important information on spatial patterns of sediment storage and channel change [*Pitlick*, 1993; *Madej and Ozaki*, 1996; *Mclean and Church*, 1999]. These estimates may require several decades of observation, coupled with extensive field measurements, but the approach is well suited to large rivers because the features of interest are of a size and scale that are measurable on maps and aerial photographs. Results from the studies cited above

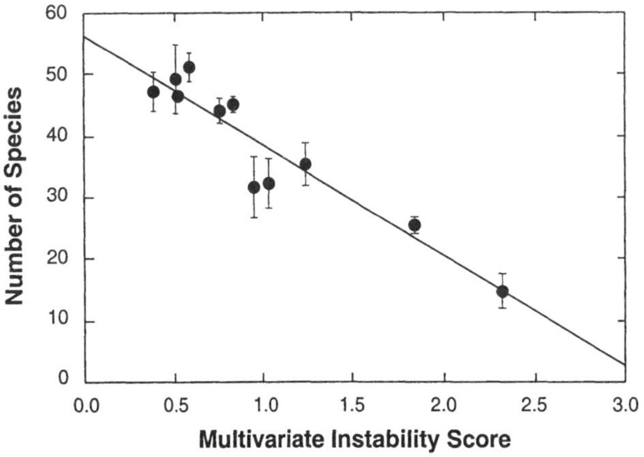

Figure 6. Decrease in number of invertebrate species with increasing habitat instability. The instability score is derived from six variables characterizing substrate properties, water temperature, depth, velocity and shear stress (from *Death and Winterbourn*, 1995)

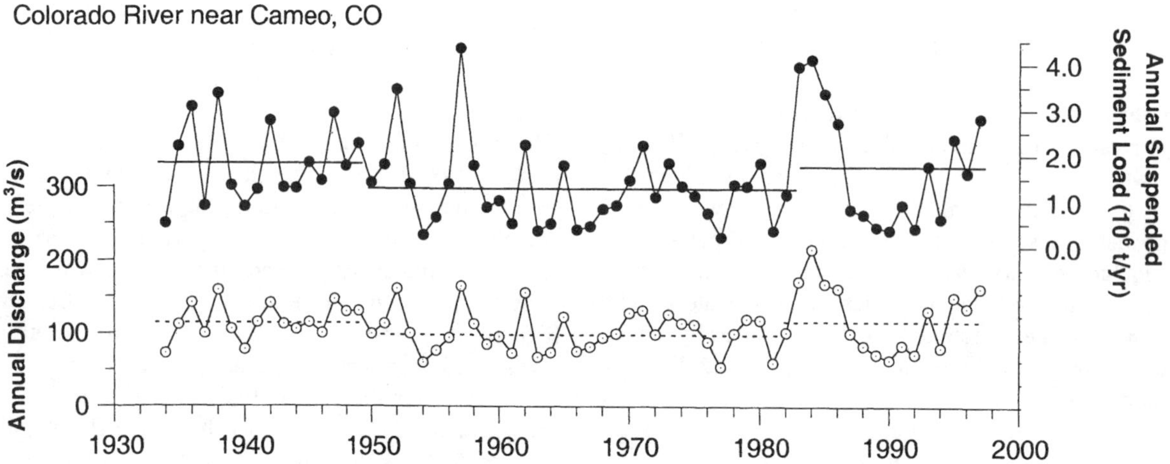

Figure 7. Trends in annual suspended sediment load (closed circles) and average annual discharge (open circles) of the Colorado River near Cameo, Colorado. Horizontal lines indicate mean values for respective time periods (from *Van Steeter and Pitlick*, 1998).

show that the spatial patterns of sediment storage are often nonuniform, with distinct depositional zones developing in discrete places along a river. The juxtaposition of laterally active depositional zones with more stable transfer zones appears to be a common (and potentially desirable) attribute of large alluvial rivers. The benefit of using morphologic measurements to derive a sediment mass balance, therefore, is to gain additional information on channel complexity and geomorphology that is not provided by a rating-curve analysis of sediment loads at spatially distinct points.

4.2. Connection to Channel Geomorphology

To illustrate the connection between sediment transport and channel morphology, we draw again on results from studies of fish habitats in alluvial reaches of the Colorado River, and its major tributary, the Gunnison River. *Pitlick et al.* [1999] measured evenly spaced cross sections along separate 100-km segments of these rivers, and used the resulting data to develop quasi-continuous hydraulic geometry relations for bankfull width and bankfull depth. Corresponding measurements of water-surface slope were made using a global positioning system with sub-meter accuracy. Point counts of the surface bed material (armor) and bulk samples of the subsurface bed material were taken at dozens of locations at low flow. The cross section data were then used with the slope and bed-material data to formulate downstream relations for the bankfull dimensionless shear stress, τ^*_b, and the bankfull discharge, Q_b. The results of this analysis indicate that, on both rivers, the bankfull width changes little downstream, whereas the bankfull depth increases by almost a factor of two [*Pitlick et al.*, 1999]. On both rivers, average slopes decrease by roughly a factor of two over the respective study reaches, whereas the D_{50} of the bed material decreases by only about 10%. Combining these results, Figure 8 shows downstream trends in τ^*_b, formulated from individual values of bankfull depth and reach-average values of slope and grain size. These data reveal no obvious trends, indicating that both rivers have adjusted their hydraulic geometry to give a consistent τ^*_b, with nearly identical reach-average values of 0.046 for the Colorado River and 0.045 for the Gunnison River. This result has several important implications for managing flows to improve fish habitat. First, the range in τ^*_b indicates that the bankfull channel dimensions are set by discharges that produce moderately intense bed load transport; on average, τ^*_b is well above typical values for incipient motion (0.030-0.040). Second, the consistency in τ^*_b within and between reaches suggests that prescribed discharges are likely to achieve the desired results (e.g. incipient motion or flushing fines) in many places. As noted earlier, there are limitations in formulating τ^* at reach scales, but as long as this is understood, the parameter provides a useful index of sediment transport intensity. Taking this a step further, the consistency in τ^*_b suggests that these reaches are in approximate equilibrium with the present-day sediment-transport regime, altered as it is. Sediment transport functions typically incorporate the variable τ^* (or the related term, excess shear stress, $\tau^* - \tau^*_c$), and since this is roughly constant, we infer that the channel has developed a morphology (width, depth and slope) that maintains the mass balance of sediment.

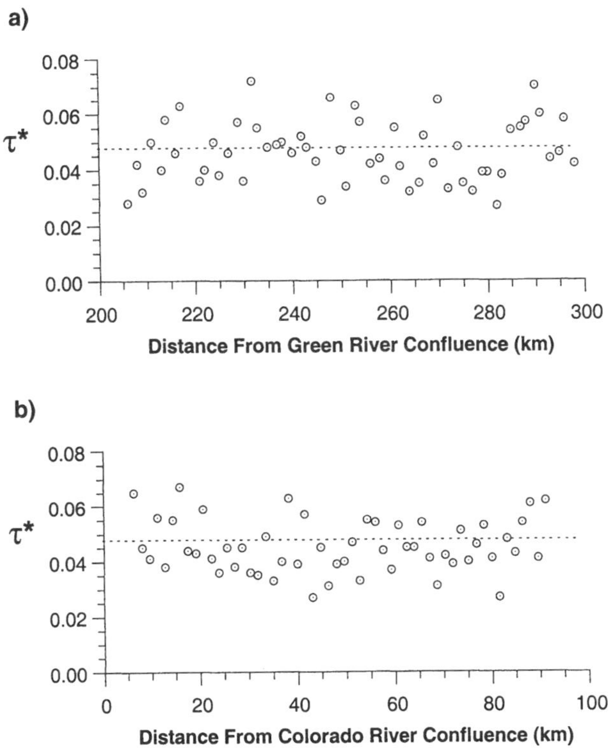

Figure 8. Trends in the bankfull dimensionless shear stress, τ^*_b, in alluvial reaches of the Colorado River (a), and nearby reaches of the Gunnison River (b). The individual values of τ^*_b are formulated from cross section measurements of bankfull depth, and sub-reach measurements of slope and median grain size (from *Pitlick et al.*, 1999).

The results discussed above can be extended to provide a potential method for restoring habitat heterogeneity and channel complexity through non-mechanical means. Historical analysis of aerial photographs of the Colorado River indicates that bars and secondary channels have been lost, thus another potential long-term restoration goal is to provide additional channel complexity. To do this, *Pitlick et al.* [1999] have suggested that the river must first become wider to create the space for new bars and secondary channels to form. A straightforward theory for predicting conditions required to initiate channel widening in gravel-bed rivers is discussed in a series of papers by *Parker* [1978a; 1978b; 1979]. Parker showed that as discharge and depth increase, there comes a point when the stress in the near-bank region exceeds that required for incipient motion ($\tau^* > \tau^*_c$), causing particles on the banks to move, thus widening the channel. Widening will continue, and the depth and shear stress will decrease until particles on the banks are no longer moving. By this reasoning, the threshold for widening and instability should exceed the threshold for incipient motion by some small amount. *Parker*'s [1979] analysis indicates that straight gravel-bed channels of uniform depth can maintain stable bank profiles for Shields stresses up to ~$1.2\tau^*_c$. Subsequent laboratory flume experiments by *Ikeda et al.* [1988] and *Diplas* [1990] support this result.

The field data presented above show that alluvial reaches of the Colorado River and Gunnison River are similarly adjusted to a gravel transport threshold, although in this case the threshold appears to be somewhat higher (~$1.5\tau^*_c$) than that observed in laboratory channels. Streamflow records for the last 20 years indicate that this threshold has been exceeded for about 5 days per year, on average [*Pitlick et al.*, 1999]. The final recommendation for improving fish habitats in these reaches is, therefore, to increase the frequency of bankfull discharges to 6-10 days per year. This could be accomplished in moderately wet years by allowing some of the runoff from the upper basin to bypass reservoirs. It is not expected that this opportunity will arise every year, nor that these flows will cause widespread and extensive channel change. However, the consistency in hydraulic geometry suggests that flows above this threshold will cause local widening, followed by the formation of new bars and secondary channels.

5. CONCLUDING REMARKS

Results from hydro-ecological studies described here and elsewhere in this volume underscore the importance of conducting interdisciplinary research in large river systems. Along with the need for interdisciplinary science is the need to recognize that there are important differences in fluvial/aquatic processes in different environments. Conceptual models of river system behavior, as outlined in the "River Continuum Concept" of Vannote et al. [1980] or the theory of hydraulic geometry presented by Leopold and Maddock [1953], provide useful starting points, but the differences among rivers, or even within rivers, are often sufficiently large to prevent us from making such generalizations. Considering some of the differences in river systems we offer the following comments and conclusions:

1. Sediment source areas vary greatly from river to river. Rivers in mountainous areas are often fed by coarse sediment from discrete sources, such as landslides, whereas rivers in lowland areas typically carry fine sediment supplied from diffuse sources, including the surrounding watershed and the reaches upstream. Some rivers, particularly those in the western U.S., carry mixtures of both fine and coarse sediment, and the substrate can grade quickly from one type of material to another. The placement and opera-

tion of reservoirs within these systems can have very different eco-hydrological effects, depending on whether the sediment source areas lie upstream of or downstream from the reservoir.

2. Sediment transport is essential for many aquatic organisms. Benthic invertebrates, and fish that rely on coarse substrates for spawning, benefit from periodic flushing flows or floods. For a flushing flow to be effective, however, the fine material must be fully suspended and removed from the reach of interest. At the same time, the coarse material cannot become so mobile that it causes high mortality among aquatic organisms or becomes depleted. It is therefore important to distinguish accurately whether the various sizes of sediment will move as bed load or suspended load, and whether the transport rates will significantly alter the mass balance of sediment. Physically based flow and sediment transport relations can provide reliable solutions to these problems, but not without a considerable amount of effort and field data.

3. Results derived from focused, site-specific studies should be placed in the broader context. Field studies of the ecological effects of flow and sediment transport often focus on a single species, or a particular stage in the life cycle of a single species (e.g. salmonid spawning). While it is true that some organisms have evolved to utilize rather specific habitats, the conditions needed to support other organisms, or to benefit the same species at a different stage in life, may differ significantly from those in the target habitat. For example, what is the benefit of improving spawning habitat at a particular location if the recommended flow negatively affects habitats used by another species at another location? Focused field studies will, no doubt, continue to be an important part of future collaborative research, however, there should be a parallel effort aimed at putting site-specific results into a meaningful historical and spatial context.

Acknowledgments. Portions of the research described in this paper were supported by the USFWS Trinity River Flow Study, USFWS Colorado River Recovery Implementation Program, the U.S. Forest Service Stream Systems Technology Center, and the National Science Foundation (SBR-9305191). We are grateful for insights gained from discussions with colleagues Matt Kondolf, Tom Lisle, Jonathan Nelson, and Doug Osmundson, and the energetic efforts of our graduate students Brent Barkett, Alan Barta, Robert Cress, Margaret Franseen, Chris Johnston, Steve Kenworthy, Brian McArdell, Conner Shea, and Mark Van Steeter. We thank Judy Haschenburger for supplying the data for Figure 5, and Peter Birkeland, John Hoopes, David Montgomery, and Kevin Scott for providing thoughtful comments on an earlier version of this paper.

REFERENCES

ASCE Task Committee on Sediment Transport and Aquatic Habitats, Sediment and aquatic habitat in river systems, *J. Hydraul. Eng.*, 118, 669-687, 1992.

Allred, T.M. and J.C. Schmidt, Narrowing by vertical accretion along the Green River near Green River, Utah, *Geol. Soc. Amer. Bull.*, 111, 1757-1772, 1999.

Andrews, E.D., Downstream effects of Flaming Gorge Reservoir on the Green River, Colorado and Utah, *Geol. Soc. Amer. Bull.*, 97, 1012-1023, 1986.

Bailey, P.B. and H.W. Li, Riverine fishes, *in The Rivers Handbook: Hydrological and Ecological Principles*, edited by P. Calow and G.E. Petts, pp. 251-281, Blackwell Scientific Publications, London, 1992.

Bain, M.B., J.T. Finn, and H.E. Booke, Streamflow regulation and fish community structure, *Ecology*, 69, 382-392, 1988.

Boulton, A.J., C.G. Peterson, N.B. Grimm, S.G. Fisher, Stability of an aquatic macroinvertebrate community in a multiyear hydrologic disturbance regime, *Ecology*, 73, 2192-2207, 1992.

Buffington, J.M. and D.R. Montgomery, A systematic analysis of eight decades of incipient motion studies, with special reference to gravel-bedded rivers, *Water Resour. Res.*, 33, 1993-2029, 1997.

Calow, P. and G.E. Petts (eds.), *The Rivers Handbook: Hydrological and Ecological Principles*, Blackwell Scientific Publications, Boston, 1992.

Carling, P.A., Bed stability in gravel streams, with reference to stream regulation and ecology, *in River Channels: Environment and Process*, edited by K.S. Richards, pp. 321-347, Inst. of Brit. Geogr., Oxford, England, 1987.

Church, M., M.A. Hassan, and J.F. Wolcott, Stabilizing self-organized structures in gravel-bed streams, *Water Resour. Res.*, 34, 3169-3179, 1998

Clausen, B. and B.J.F. Biggs, Relationships between benthic biota and hydrological indices in New Zealand streams, *Freshwater Biol.*, 38, 327-342, 1997.

Cobb, D.G., T.D. Galloway, and J.F. Flannagan, Effects of discharge and substrate stability on density and species composition of stream insects, *Can. J. Fish. Aquat. Sci.*, 49, 1788-1795.

Collier, M., R.H. Webb, and J.C. Schmidt, Dams and rivers: Primer on the downstream effects of dams, *U.S. Geol. Surv. Circ. 1126*, 94 pp., 1996.

Death, R.G. and M.J. Winterbourn, Diversity patterns in stream benthic invertebrate communities: The influence of habitat stability, *Ecology*, 76, 1446-1460, 1995.

Diplas, P., Characteristics of self-formed straight channels, *J. Hydraul. Eng.*, 116, 707-728, 1990.

Diplas, P., Modeling fine and coarse sediment interaction over alternate bars, *J. Hydrol.*, 159, 335-351, 1994.

Erman, D.C. and N.A. Erman, The response of stream macroinvertebrates to substrate size and heterogeneity, *Hydrobiologia*, 108, 75-82, 1984.

Ferguson, R., Hydraulics and hydraulic geometry, *Prog. Phys. Geog.*, 10, 1-31, 1986.

Fisher, S.G., L.J. Gray, N.B. Grimm, and D.E. Busch, Temporal succession in a desert stream ecosystem following flash flooding, *Ecol. Mono.*, 52, 93-110, 1982.

Friedman, J.M., W.R. Osterkamp, M.L. Scott, and G.T. Auble, Downstream effects of dams on channel geometry and bottomland vegetation: Regional patterns in the great Plains, *Wetlands*, 18, 619-633, 1998.

Gomez, B., Effects of particle shape and mobility on stable armor development, *Water Resour. Res.*, 30, 2229-2239, 1994.

Graf, W.L., Dam nation: A geographic census of American dams and their hydrologic impacts, *Water Resour. Res.*, 35, 1305-1311, 1999.

Haschenburger, J.K., A probability model of scour and fill depths in gravel-bed channels, *Water Resour. Res.*, 35, 2857-2869, 1999.

Ikeda, S., G. Parker and Y. Kimura, Stable width and depth of straight gravel rivers with heterogeneous bed material, *Water Resour. Res.*, 24, 713-722, 1988.

Iorns, W.V., C.H. Hembree, D.A. Phoenix, and G.L. Oakland, Water resources of the upper Colorado River basin- Basic Data, *U.S. Geol. Surv. Prof. Pap. 442*, 1036 pp., 1964.

Iorns, W.V., C.H. Hembree, and G.L. Oakland, Water resources of the upper Colorado River basin- Technical Report, *U.S. Geol. Surv. Prof. Pap. 441*, 1965.

Johnston, C.E., E.D. Andrews, and J. Pitlick, *In situ* determination of particle friction angles of fluvial gravels, *Water Resour. Res.*, 34, 2017-2030, 1998.

Kondolf, G.M. and P.R. Wilcock, The flushing flow problem: Defining and evaluating objectives, *Water Resour. Res.*, 32, 2589-2599, 1996.

Leopold, L.B. and T. Maddock, The hydraulic geometry of stream channels and some physiographic implications, *U.S. Geol. Surv. Prof. Paper 252*, 57 pp., 1953.

Ligon, F.K., Dietrich, W.E., and Trush, W.J., Downstream ecological effects of dams, *Bioscience*, 45, 183-192, 1995.

Lins, H.F., F.K. Hare, and K.P. Singh, Influence of the atmosphere, *in Surface Water Hydrology; v. O-1, Geology of North America*, edited by M.G. Wolman and H.C. Riggs, pp. 11-53, Geological Society of America, Boulder, 1990.

Lisle, T.E., Sediment transport and resulting deposition in spawning gravels, north Coastal California, *Water Resour. Res.*, 25, 1303-1319, 1989.

Lisle, T.E., J.M. Nelson, J. Pitlick, M.A. Madej, and B.L. Barkett, Variability of bed mobility in natural gravel-bed channels and adjustments to sediment load at local an reach scales, *Water Resour. Res.*, 36, 3743-3755, 2000.

Madej, M.A. and V. Ozaki, Channel response to sediment wave propagation and movement, Redwood Creek, California, USA, *Earth Surf. Processes Landforms*, 21, 911-927, 1996.

Mclean, D. G. and M.A. Church, Sediment transport along lower Fraser River 2. Estimates based on the long-term gravel budget, *Water Resour. Res.*, 35, 2549-2559, 1999.

Milhous, R., Modelling of instream flow needs; the link between sediment and aquatic habitat, *Regulated Rivers*, 14, 79-94, 1998.

Minshall, G.W. and J.N. Minshall, Microdistribution of benthic invertebrates in a Rocky Mountain (USA) stream, *Hydrobiologia*, 55, 231-249, 1977.

Montgomery, D.R., Process domains and the river continuum, *J. Amer. Water Resour. Assoc.*, 35, 397-410, 1999.

Nelson, J. M., R. L. Shreve, S. R. McLean, and T. G. Drake, Role of near-bed turbulence in bed load transport and bed form mechanics, *Water Resour. Res.*, 231, 2071-2086, 1995.

Newbold, J.D., Cycles and spirals of nutrients, *in The Rivers Handbook: Hydrological and Ecological Principles*, edited by P. Calow and G.E. Petts, pp. 379-408, Blackwell Scientific Publications, London, 1992.

Parker, G., Self-formed straight rivers with equilibrium banks and mobile bed. Part 1. The sand-silt river, *J. Fluid Mech.*, 89, 109-125, 1978a.

Parker, G., Self-formed straight rivers with equilibrium banks and mobile bed. Part 2. The gravel river, *J. Fluid Mech.*, 89,127-146, 1978b.

Parker, G., Hydraulic geometry of active gravel rivers, *J. Hydraul. Eng.*, 105, 1185-1201, 1979.

Pitlick, J., Stabilizing effects of riparian vegetation during an overbank flow, Trinity River, California, *EOS Transactions American Geophysical Union*, 73(43), 231, 1992.

Pitlick, J., Response and recovery of a subalpine stream following a catastrophic flood, *Geol. Soc. Amer. Bull.*, 105, 657-670, 1993

Pitlick, J., Relation between peak flows, precipitation, and physiography for five mountainous regions in the western USA, *J. Hydrol.*, 158, 219-240, 1994

Pitlick, J. and M.M. Van Steeter, Geomorphology and Endangered Fish Habitats of the Upper Colorado River 2: Linking Sediment Transport to Habitat Maintenance, *Water Resour. Res.*, 34, 303-316, 1998.

Pitlick, J., M.M. Van Steeter, B. Barkett, R. Cress, and M. Franseen, Geomorphology and Hydrology of the Colorado and Gunnison Rivers and Implications for Habitats Used by Endangered Fishes, *U.S. Dept. Interior Fish and Wildlife Service Final Report*, Grand Junction, Colo., 57 pp., 1999.

Pitlick, J. and R. Cress, Longitudinal Trends in Channel Characteristics of the Colorado River and Implications for Food-Web Dynamics, *U.S. Dept. Interior Fish and Wildlife Service Final Report*, Grand Junction, Colo., 45 pp., 2000.

Poff, N.L. and J.D. Allan, Functional organization of stream fish assemblages in relation to hydrological variability, *Ecology*, 76, 606-627, 1995.

Poff, N.L. and J.V. Ward, Implications of streamflow variability and predictability for lotic community structure: a regional analysis of streamflow patterns, *Can. J. Fish. Aquat. Sci.*, 46, 1805-1818, 1989.

Power, M.E., S. Sun, G. Parker, W.E. Dietrich, and J.T. Wooten, Hydraulic Food-Chain Models, *Bioscience*, 45, 159-167, 1995

Puckridge, J.T., F. Sheldon, K.F. Walker, A.J. Boulton, Flow variability and the ecology of large rivers, *Mar. Freshwater Res.*, 49, 55-72, 1998.

Quinn, J. M. and C.W. Hickey, Hydraulic parameters and benthic invertebrate distributions in two gravel-bed New Zealand rivers, *Freshwater Biol.*, 32, 489-500, 1994.

Riggs, H.C. and K.D. Harvey, Temporal and spatial variability of streamflow, *in Surface Water Hydrology; v. O-1, Geology of North America*, edited by M.G. Wolman and H.C. Riggs, pp. 81-96, Geological Society of America, Boulder, 1990.

Sear, D.A., River restoration and geomorphology, *Aquat. Cons.*, 4, 169-177, 1994.

Scarsbrook, M.R. and C.R. Townsend, Stream community structure in relation to spatial and temporal variation: a habitat template study of two contrasting New Zealand streams, *Freshwater Biol.*, 29, 395-410, 1993.

Sparks, R.E., Need for ecosystem management of large rivers and their floodplains, *Bioscience*, 45, 168-182, 1995.

Stanford, J.A. and J.V. Ward, Management of aquatic resources in large catchments: Recognizing interactions between ecosystem connectivity and environmental disturbance, *in Watershed Management, Balancing Sustainability and Environmental Change*, edited by R.J Naiman, pp. 91-124, Springer-Verlag, New York, 1992.

Stanford, J.A., J.V. Ward, W.J. Liss, C.A. Frissell, R.N. Williams, J.A. Lichatowich, and C.C. Coutant, A General Protocol for Restoration of Regulated Rivers, *Reg. Rivers Research and Management*, 12, 391-413, 1996.

Statzner, B. and B. Higler, Questions and comments on the river continuum concept, *Can. J. Fish. Aquat. Sci.*, 42, 1038-1044, 1985.

U.S. Fish and Wildlife Service, Recovery implementation program for endangered fish species in the upper Colorado River basin, *U.S. Dept. Interior Fish and Wildlife Service Final Report*, Denver, Colo., 82 pp., 1987.

U.S. Fish and Wildlife Service and Hoopa Valley Tribe, Trinity River Flow Evaluation, *U.S. Dept. Interior Fish and Wildlife Service Final Report*, Arcata, Calif., 511 pp., 1999.

Van Steeter, M.M. and J. Pitlick, Geomorphology and Endangered Fish Habitats of the Upper Colorado River 1. Historic Changes in Streamflow, Sediment Load and Channel Morphology, *Water Resour. Res.*, 34, 287-302, 1998.

Vannote, R.L., G.W. Minshall, K.W. Cummins, J.R. Sedell, and C.E. Cushing, The river continuum concept, *Can. J. Fish Aquat. Sci.*, 37, 130-137, 1980.

Webb, R.H., J.C. Schmidt, G.R. Marzolf, and R.A. Valdez (eds.), *The controlled flood in Grand Canyon*, Amer. Geophys. Union Monogr. 110, Washington, D.C., 1999.

Whiting, P.J. and W.E. Dietrich, Convective accelerations and boundary shear stress over a channel bar, *Water Resour. Res.*, 27, 783-796, 1991.

Wilcock, P. R., The critical shear stress of natural sediments, *J. Hydraul. Eng.*, 119, 491-505, 1993.

Wilcock, P.R., Two-fraction model of initial sediment motion in gravel-bed rivers, *Science*, 280, 410-412, 1998.

Wilcock, P.R. and B.W. McArdell, Surface-based fractional transport rates: Mobilization thresholds and partial transport of a sand-gravel sediment, *Water Resour. Res.*, 29, 1297-1312, 1993.

Wilcock, P.R. and B.W. McArdell, Partial transport of a sand/gravel sediment, *Water Resour. Res.*, 33, 235-245, 1997.

Wilcock, P.R., Barta, A.F., Shea, C.C., Kondolf, G.M., Matthews, W.V.G., and Pitlick, J., Observations of flow and sediment entrainment on a large gravel-bed river, *Water Resour. Res.*, 32, 2897-2909, 1996.

Williams, G.P., The case of the shrinking channels, *U.S. Geol. Surv. Geol. Circ. 781*, 48 pp., 1978.

Williams, G.P. and M.G. Wolman, Downstream Effects of Dams on Alluvial Rivers, *U.S. Geol. Surv. Prof. Paper 1286*, 83 pp., 1984.

Wooten, J.T, M.S. Parker, M.E. Power, Effects of disturbance on river food webs, *Science*, 273, 1558-1561, 1996.

John Pitlick, Department of Geography, Box 260, University of Colorado, Boulder, CO 80309-0260
email: pitlick@spot.colorado.edu

Peter Wilcock, Department of Geography and Environmental Engineering, Johns Hopkins University, Baltimore, MD 21218
email: wilcock@jhuvms.hcf.jhu.edu

Fluvial Processes and Passive Rehabilitation of the Lisbon Bottom Side-Channel Chute, Lower Missouri River

Robert B. Jacobson and Mark S. Laustrup

U.S. Geological Survey, Columbia, Missouri

Michael D. Chapman

U.S. Army Corps of Engineers, Kansas City, Missouri

Multiple large floods in 1993-1997 on the Lower Missouri River carved a side-channel chute through the river bottom at Lisbon, Missouri. Although similar in some respects to engineered side-channel chutes designed for habitat rehabilitation projects, the Lisbon Bottom chute has been unique in that it was allowed to evolve for more than four years with minimal stabilization. During the wet years, 1996-1999, the chute was subjected to abnormally high discharges and passed as much as 20% of the total discharge of the Missouri River. Relatively unrestrained fluvial processes during this time created a wide channel with highly diverse habitats. The upper one-half of the chute established a shallow, braided channel morphology similar to the pre-managed Missouri River. The lower half established a dynamically migrating, single-thread channel, and an incipient flood plain. Compared to the adjacent navigation channel, the chute established substantial areas of shallow, slow-velocity aquatic habitat that is considered to be in short supply in the present-day Lower Missouri River. The short-term biological benefits have been mixed: the chute has fewer waterbird and benthic macroinvertebrate taxa than adjacent riverine habitats, but greater numbers of fish species compared to the navigation channel.

1. INTRODUCTION

1.1. Large-River Rehabilitation

Stream restoration is defined as the re-establishment of the structure and function of the riverine ecosystem, with the goal of achieving a condition as close as possible to pre-disturbance conditions [*Federal Interagency Stream Restoration Working Group*, 1998]. In large, multi-purpose rivers the pre-disturbance goal is difficult to meet; such rivers are so altered that restoration, in a strict sense, is practically unachievable. In these cases, the concept of *rehabilitation* – recovery of some ecosystem functions by manipulating selected components of the riverine ecosystem [after *Federal Interagency Stream Restoration Working Group*, 1998]-- is a more applicable concept [*Gore and Shields*, 1995].

The lower Missouri River is a large, multiple-use system (Figure 1). The lower reaches of the Missouri drain 1,300,000 km^2 (525,000 mi^2), amounting to nearly 1/6 of the United States land area. A system of reservoirs in S. Dakota, N. Dakota, and Montana provide nearly 92,500 km^3 (75 million acre feet) of water storage and annually

Geomorphic Processes and Riverine Habitat
Water Science and Application Volume 4, pages 199-216

Published in 2001 by the American Geophysical Union

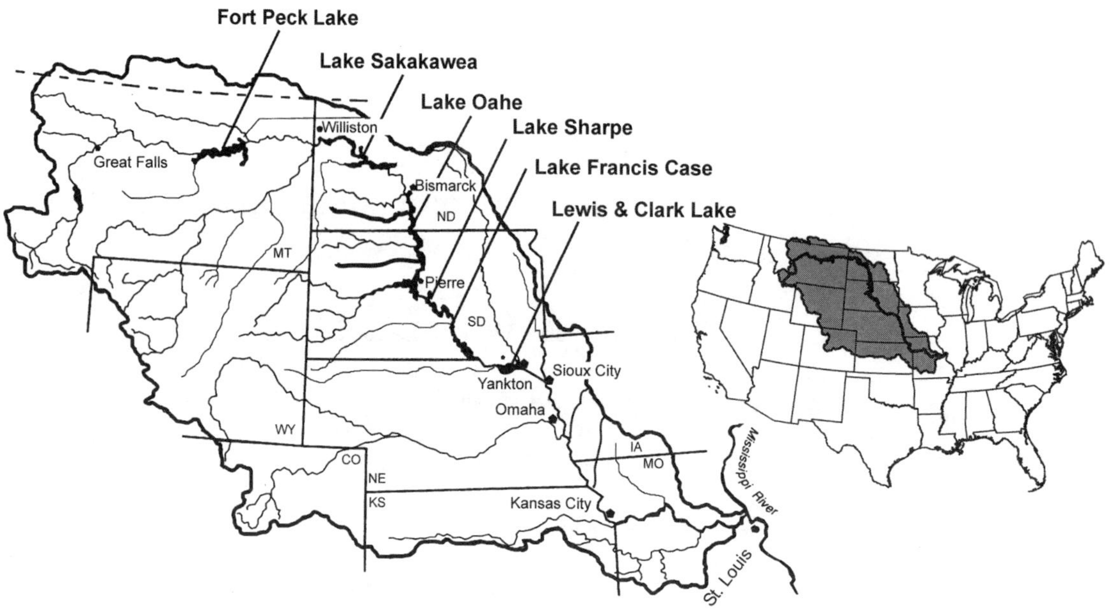

Figure 1. Missouri River basin and the Lower Missouri River. The Lower Missouri River extends downstream of Gavins Point Dam near Yankton, South Dakota, to the junction with the Mississippi River near St. Louis.

Table 1. Estimated major economic benefits of Missouri River system regulation

[Data from *U.S. Army Corps of Engineers*, 1998]

Management component	Estimated average annual benefits, $millions
Hydropower	$ 676
Water supply	541
Flood control	415
Recreation	85
Navigation	7
Total	$ 1,724

produce nearly 10 million kilowatt hours of hydroelectric power. The managed river system provides annual benefits estimated to be greater than $1.7 billion dollars (table 1).

However, management of the system for economic benefits also has been associated with substantial loss of habitats and native riverine biota [*Funk and Robinson*, 1974; *Hesse and Sheets*, 1993]. Clearing, snagging, and stabilization of the Missouri River began in the early 1800s, and have continued to the present. Most of the engineering works along the river are the direct result of the Missouri River Bank Stabilization and Navigation project, part of the Pick-Sloan Act of 1944 [*Ferrell*, 1993]. These structures have stabilized the riverbanks, and narrowed and focused the current to maintain a self-scouring channel with navigable depths of at least 2.7 m (9 feet) from St. Louis, Missouri, 1,170 km (730 mi) upstream to Sioux City, Iowa. The result has been to create a narrow, swift, and deep channel from what was historically a shallow, shifting, braided river (Figure 2). Furthermore, reservoir regulation has decreased flow variability and seasonality since the mid 1950's (Figure 3).

Recognition of the scope of habitat loss has increased interest on rehabilitating parts of the Missouri River [*Latka et al.*., 1993; *U.S. Government*, 1986]. Approaches and designs differ widely, but they can be described generally as resulting from three sets of questions:

- What are the rehabilitation objectives? Designs will vary greatly depending on whether the choice is – for example -- to (a) recover some semblance of naturally dynamic ecosystem functions, (b) create specific habitats for recreational species, or (c) create specific habitats for threatened and endangered species.
- Should rehabilitation focus on altering system hydrology, through reservoir release policies, or on altering riverine geomorphology? Hydrology determines how much water comes down the river, when, and for how long. Geomorphology, however, determines how that water is allowed to be distributed in space and create aquatic habitats. On intensively engineered rivers, hydrologic alterations alone may not be sufficient to produce more available habitat.
- Should rehabilitation employ passive or intensive approaches? Passive approaches allow the river to create

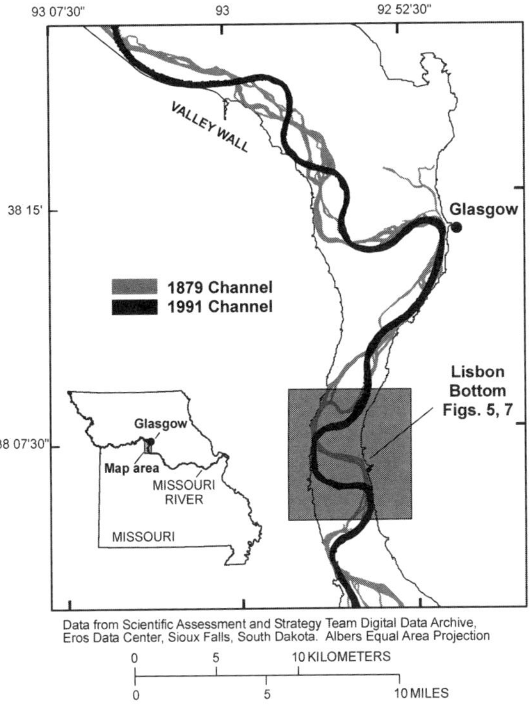

Figure 2. Channel positions of the Missouri River near Glasgow, Missouri, 1879 and 1991. Flow regulation and navigation structures have changed the river from a braided system with transient sand islands and multiple shifting, shallow channels to a meandering stream with stable banks, greater average depths, higher average velocities, and many fewer connections between the channel and flood-plain habitats.

dynamic habitats, presumably at least cost, but result in less control over the characteristics and timing of habitats. Intensive approaches – for example, diking wetlands and pumping water to create optimum waterfowl habitat – result in stable, controlled habitats, generally at greater cost.

Answers to these three questions require an informed, quantitative understanding of the ecological benefits of the alternatives. This paper addresses ecological benefits that can be achieved through alteration of riverine geomorphology using passive approaches.

It is generally accepted that physical habitat determines a template for aquatic ecosystem functions, but realization of the potential is highly dependent on other ecological processes [*Plafkin et al.*., 1989; *Schlosser*, 1987]. Unlike many biotic characteristics, however, physical habitat is directly amenable to management through reservoir regulation or channel engineering. Therefore, current management efforts on many large rivers emphasize physical habitat creation and maintenance.

Typically, aquatic habitat rehabilitation activities on the Lower Missouri River have been designed for two purposes: to provide specific flood-plain wetland habitats for recreational hunting, and to provide off-channel aquatic habitat for fishes and shorebirds. Design criteria for off-channel aquatic habitats generally have been based on the premise that rehabilitation should work to reverse the engineered simplification of the channel, to provide greater channel complexity (Figure 2). Hence, efforts have been focused on recreating side-channel chutes and increasing channel topwidth [*Harberg et al.,* 1993; *Latka et al.,* 1993; *Rothe*, 1995].

Considerable uncertainty exists about the ecological benefits of river rehabilitation projects and their long-term performance [*Federal Interagency Stream Restoration Working Group*, 1998]. The uncertainty is greater for large rivers than for small rivers because of inherent spatial and temporal complexities, the relative lack of empirical data, and shortcomings of predictive computational models or theoretical framework [*Burke* and *Robinson*, 1979; *Shields*, 1984; *Forrest* and *Ettema*, 1993; *Lubinski* and *Gutreuter*, 1993]. While there is a considerable foundation of understanding on how to build secondary channel structures [for example, *Harberg et al.,* 1993; *Klein et al.,* 1994; *Schropp*, 1995], there is very little understanding of their integrated ecological benefits or their long-term performance under highly variable hydrologic and sediment transport regimes [*Cals et al.,* 1998]. The study described here is intended to develop a stronger empirical basis for evaluating the benefits, costs, and performance of passive side-channel chute

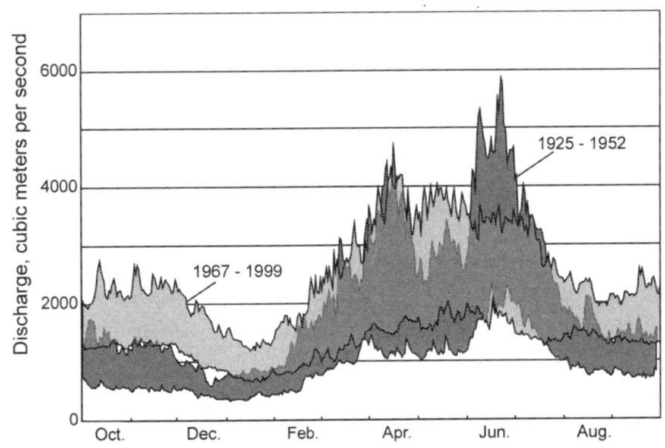

Figure 3. Duration hydrographs for the Missouri River at Boonville, Missouri. The shaded band of discharge values for preregulation (1925-52) and postregulation (1967-99) periods is the range of values between the 25 and 75 percentile of flows for each day of the year. Reservoir regulation has decreased variability of flows and shifted seasonality to maintain navigation in the lower river, April - November.

rehabilitation projects on large, multi-purpose river systems.

1.2. Purpose and Scope

This report documents evolution of physical habitat in a passively managed side-channel chute rehabilitation project on the Lower Missouri River. The chute was cut through Lisbon Bottom, the area of alluvial valley floor delineated by the valley wall and the Missouri River near the town of Lisbon, Missouri (Figures 4,5). This case study is intended to provide insights into fluvial processes and rates of habitat alteration in these heretofore poorly documented features. The relatively natural geomorphic evolution of this chute provides an opportunity to assess the quality and quantity of physical habitat that can be created on a large, multi-purpose river using least cost, passive approaches. Limited biological data cited in this report provide some insight into short-term biological responses in the chute.

Lisbon Bottom is in a high-sinuosity segment of the Lower Missouri River (Figures 4B, C). Results of this study should be applicable to geomorphically similar segments of the Lower Missouri River, and general conclusions may help inform rehabilitation projects on other large, highly managed rivers. The temporal scope of this report includes formation of a secondary side-channel chute and documentation of changes through a 4-year time interval during which human intervention was minimal. This must be considered a preliminary analysis because of the short time frame compared to adjustment rates of large rivers. Moreover, in June 2000 a hydraulic control structure was completed at the head of the Lisbon Bottom chute in order to ensure an acceptable level of risk to river navigation, thereby defining new boundary conditions for the next stage of this experiment.

2. METHODS

Multiple physical datasets have been combined to monitor and evaluate evolution of the Lisbon Bottom chute. Remotely sensed multispectral data (Landsat Thematic Mapper - TM, SPOT imagery) and aerial photography provide a broad-scale understanding of changes in the chute and surrounding river corridor. Detailed maps of the chute have been compiled from field mapping using post-processed differential global positioning system methods (DGPS); typically, these maps achieve sub-meter x-y accuracy, a level that is sufficient for monitoring morphologic changes given the large (multi-meter) annual changes in the chute boundaries.

Physical habitat in the chute was assessed using an integrated assessment system contained on a 5.8 m, shallow-draft boat. The system and methods are explained in detail in *Jacobson and Laustrup* [2000]. This system includes real-time, sub-meter DGPS for georeferencing, a high-resolution, 208-kHz echosounder, a bed-material classification system, and an acoustic Doppler current profiler (ADCP). The bed-material classification system uses the waveforms of the echosounder returns to calculate hardness and roughness parameters for the bed [for example, *Rukavina*, 1997]. In turn, these parameter values are classified into bed material based on calibration datasets obtained from independent identification of bottom materials. ADCP data were collected using a 600 kHz, broadband instrument; identical configurations were used in the chute and navigation channel. ADCP methods generally followed those of *Morlock* [1996]. The data were used to map habitats within the chute and adjacent navigation channel, and to compare habitats quantitatively.

Assessment of rehabilitation projects typically involves comparison to a reference condition to determine if the project has significantly improved habitat or biological characteristics. Historical reference conditions for Lower Missouri River physical habitats can be inferred from maps from 1879 [*Missouri River Commission*, 1879], and 1920 [*War Department*, 1920]. The historical reference condition documented in these maps present a well-defined goal for rehabilitation projects, although the goal is limited in that it displays only selected measures of habitat (land cover, planform, bathymetric transects) and only a few snapshots in time. Moreover, historical channel-morphology data do not resolve the effects of hydrologic characteristics – in particular, seasonality -- that have been altered by upstream reservoir regulation. Hence, historical channel geometry provides only one dimension of a reference condition. Notably, pre-management biological data are insufficient to provide an historical biological reference condition.

Reference conditions also can be defined by the present-day conditions in adjacent, non-rehabilitated parts of the river. Comparison to this spatial reference condition defines whether the project has improved ecological characteristics measurably compared to the managed river. The status quo reference condition is especially useful when historical data are insufficient to define a reference, or too different from achievable conditions to provide a meaningful goal. For this study, physical and biological spatial reference conditions are defined in the adjacent navigation channel of the Missouri River.

3. HISTORICAL CONTEXT OF THE LISBON BOTTOM CHUTE

The historical record of channel changes at Lisbon Bottom illustrates the natural processes of erosion and deposi-

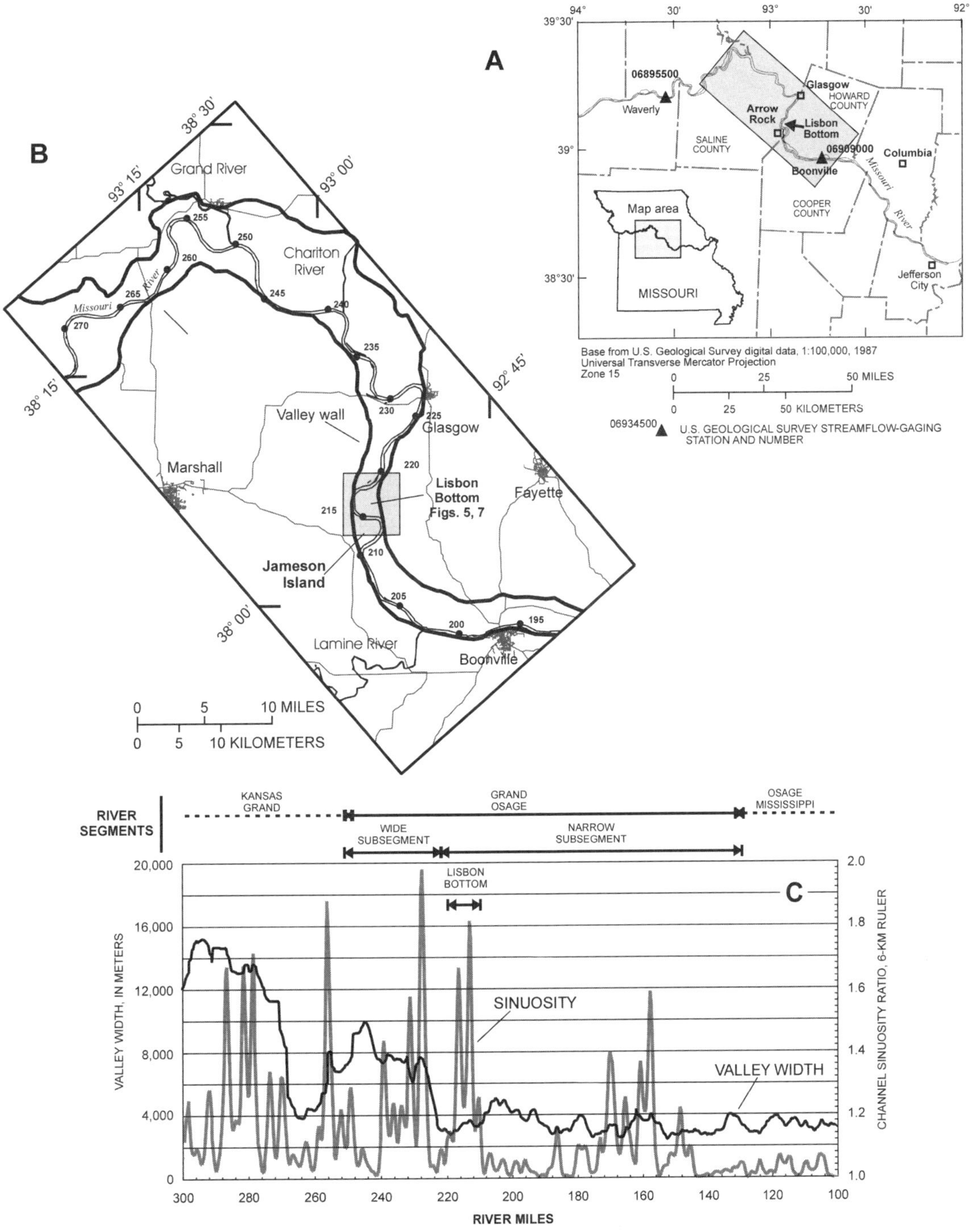

Figure 4. (A) General location map, Lower Missouri River (LMR). (B) Detailed location map showing part of the LMR valley and Lisbon Bottom location. (C) Graph of channel sinuosity (channel length per 6 km straight-line distance), valley width, and river segmentation, part of the LMR valley.

tion, and indicates the magnitude of change imposed by river management. Historical maps of channel positions in 1879 show that the Grand-Osage segment was characterized by a wide channel and numerous braided reaches with sand bars and islands (Figure 2). *Schmudde* [1963] estimated that under natural conditions the river was 2-3 times wider than after channel stabilization. Channel migration under natural conditions was sufficient to rework as much

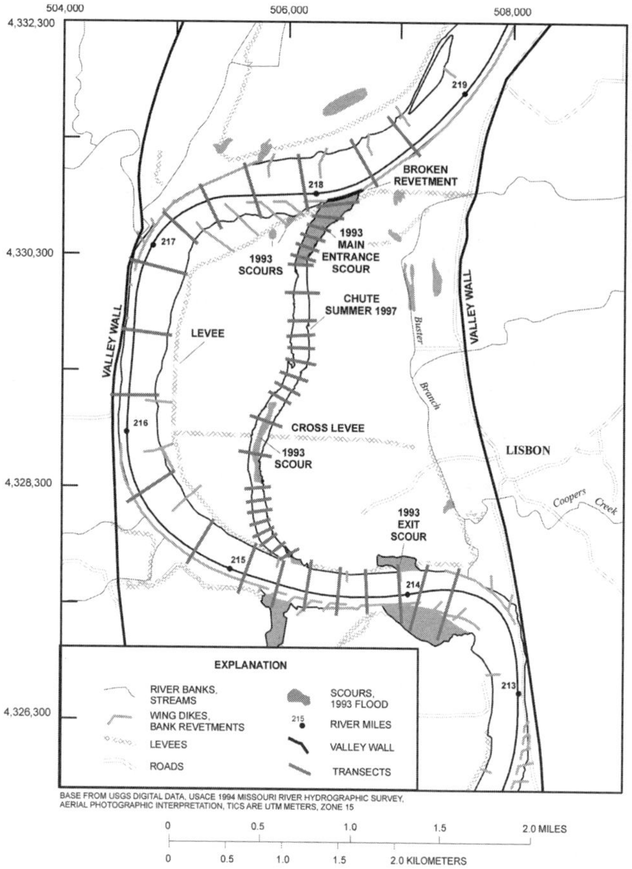

Figure 5. Map of the Lisbon Bottom area, levees, bank revetments, wing dikes, scours created by the 1993 flood, and the chute (as mapped in Summer 1997).

as 1/3 of the flood plain of the Missouri River in approximately 50 years [*Schmudde*, 1963].

Engineering alterations to improve navigability of the Missouri River began as early as the 1830's. Snagging (clearing of woody debris) and channel clearing to improve navigability for keelboats and steamboats were systematic and intensive from 1885-1910 [*Galat et al.*, 1996]. On 1879 project maps [*Missouri River Commission*, 1879], Lisbon Bottom land cover is depicted as dense willows, sandbars, and a few small fields. During the interval 1879-1920, the mainstem channel of the Missouri River moved approximately 1570 m (5,150 ft) across the downstream one third of Lisbon Bottom (Figure 2), leaving the town of Lisbon without river access.

Intensive channel structuring began after authorization of the Navigation and Bank Stability Project by the Flood Control Act of 1944 [*Galat et al.*, 1996]. In general, alterations followed a sequence of clearing of woody debris, construction of wing dikes to narrow the channel, construction of bank revetments to stabilize banks, and construction of levees to prevent inundation of urban, industrial, and agricultural lands [*Ferrell*, 1993; *Ferrell*, 1996; *Galat et al.*, 1996]. Navigation maps from 1954 [*U.S. Army Corps of Engineers*, 1954] show the channel has been largely confined to its present (June 2000) form and position. No levees were mapped along the main channel at Lisbon Bottom in 1954; a former landowner recollected that levees along the channel were not built until the mid-1980s [*Bill Lay*, pers. commun., 1997]. By 1993, Lisbon Bottom was completely leveed along the left bank downstream as far as Coopers Creek (Figure 5). In April of 1993 the levee was inspected and qualified for the U.S. Army Corps of Engineers maintenance program. The levee was built to a height that would protect Lisbon Bottom up to a stage of 9.75 m (32.0 ft) on the Glasgow gage, or approximately 186.2 m (611 ft) above sea level at river mile 218. Based on extrapolation from the streamgage at Boonville this is equivalent to protection from flood stages of 2-5 year recurrence interval. Revetments, wing dikes, and levees as they existed in 1997 are shown in figure 5.

4. THE FLOOD OF 1993

The flood of 1993 broke the Lisbon Bottom levee in multiple places. The exact processes by which the levee broke are unknown. Similar levees along the Missouri River broke by a variety of mechanisms, including overtopping and erosion, liquefaction by seepage water, and lateral erosion from the channel side [*Schalk* and *Jacobson*, 1997]. During the flood of 1993, the peak daily mean flow at Boonville was 20,400 cms (cubic meters per second; 721,000 cfs [cubic feet per second]) on July 30 (Figure 6). This exceeded the estimated 500-year flow at Boonville of 19,810 cms (700,000 cfs) [*U.S. Army Corps of Engineers*, 1994].

The 1993 flood left one large scour at the upstream margin of Lisbon Bottom and three smaller levee breaks and scours (Figures 5, 7). Also, the flow breached the cross levee in numerous places, and at least five exit scours developed along the downstream margin. Bathymetric maps of the main entrance scour (Figure 5) documented 7 m maximum depth in 1994 [*Galat* and others; 1997]. None of the levee breaks were repaired after the 1993 flood, so subsequent floods of smaller magnitude were allowed to flow through the levees (Figure 6).

Lisbon Bottom has been flooded multiple times since July 30, 1993. Sixteen discrete floods in excess of 5,600 cms (198,000 cfs) occurred from August 1993 to March 2000 (Figure 6). This discharge is estimated to be the 2-year flood for this segment of the Lower Missouri River

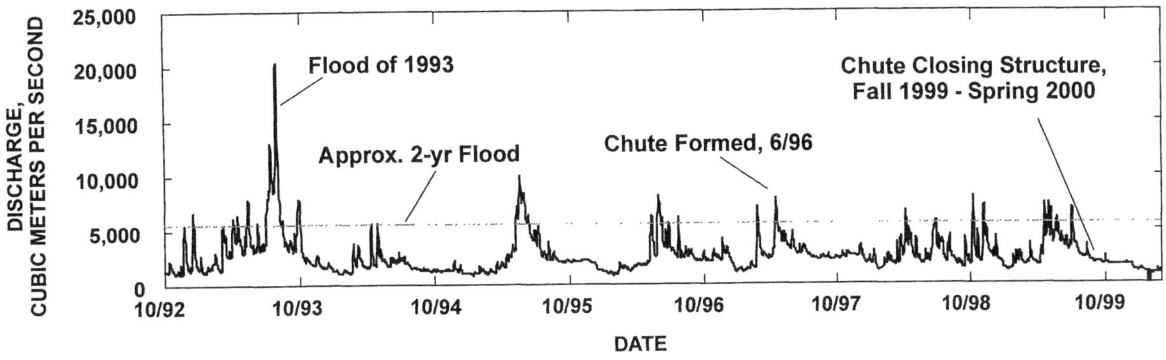

Figure 6. Hydrograph from the Missouri River at Boonville, Missouri, October 1993, to March 2000. The approximate discharge for the 2-year flood is indicated (U.S. Army Corps of Engineers, written communication, 1997).

[*U.S. Army Corps of Engineers, Kansas City District*, written communication, 1997] and would be approximately bankfull in the absence of engineered levees.

The largest daily mean discharge recorded at Boonville, Missouri, between September 1993 and January 1998 was 10,020 cms (354,000 cfs) in May 1995. This flood inundated at least 80% of Lisbon Bottom and resulted in substantial reworking of sand deposits emplaced by the 1993 flood, but it did not change the basic architecture of the scours. The most geomorphically effective flood occurred during late May-June 1996 with a peak flow of approximately 8,200 cms (290,000 cfs). This flood was sufficiently erosive to connect the upstream entrance scour with the non-connected scour at the interior levee and a small exit scour near RM 214.7, thereby creating the chute shown in Figure 5.

5. GEOMORPHIC EVOLUTION AND PHYSICAL HABITAT IN THE LISBON BOTTOM CHUTE

From May-June 1996 to October 1999, the chute evolved with minimal human intervention. During this time, it carried 2-20% of the discharge of the Missouri River, depending on total discharge and the state of repairs of the upstream revetment. In December 2, 1997, the chute was carrying approximately 20% of the total Missouri River discharge (736 cms of 3,560 cms); in May 1998, the chute was carrying 4.8% (85 cms of 1,750 cms); In September 1998, the chute carried 1.7% (28 cms of 1,657 cms).

Since formation in May-June 1996, the chute has incised into flood-plain deposits, widened, migrated laterally, and increased in complexity (Figure 8). Initially, the chute was narrow and bordered by high banks as much as 5 m above low-water levels. As the chute developed a thalweg in the lower half it began to migrate laterally, creating several prominent cutbanks (Figure 8). Consequently, the sinuosity increased and slope decreased (table 2). Initially, the chute slope was approximately equal to the slope of the flood plain, about twice that of the navigation channel. The notches in the revetment at the upstream end of the chute performed as a hydraulic control, and therefore determined the chute slope. The notches were periodically repaired to different specifications 1996 - 1999, resulting in non-systematic variation of the slope. In June 1999, the revetment was sealed in anticipation of construction of a notched hydraulic control structure.

The rate of widening of the chute (measured as average bankfull channel width) was initially rapid, but slowed markedly after June 1997 (Figure 9A). When it was formed in 1996, the chute was relatively straight and had a trapezoidal cross section. By January 1998, the chute had developed three distinct bends and complex channel structure (Figure 8). In the upstream half, the chute is characterized by mid-channel bars and a braided appearance (Figure 10A). The downstream half has developed point bars, an incipient flood plain within the high banks, and a meandering thalweg (Figure 10B).

To quantify and compare the resulting channel morphology with a historical reference, we assembled a dataset of channel cross sections from 1920 navigation charts of the Lower Missouri River [*War Department*, 1920], (Figure 11). Eleven side channels in the vicinity of Lisbon Bottom were selected, and width:depth ratios were calculated, using the average depth of the cross section. The bathymetry was originally mapped in July 1920, prior to systematic USGS discharge data, so the flow duration is unknown; however, the large areas of sand bars on these maps indicate a general low-flow condition.

Width:depth ratios for cross sections in the Lisbon Bottom chute change markedly with discharge. Because flows for all measurements were within the high banks, larger discharges tended to have smaller ratios. Notwithstanding

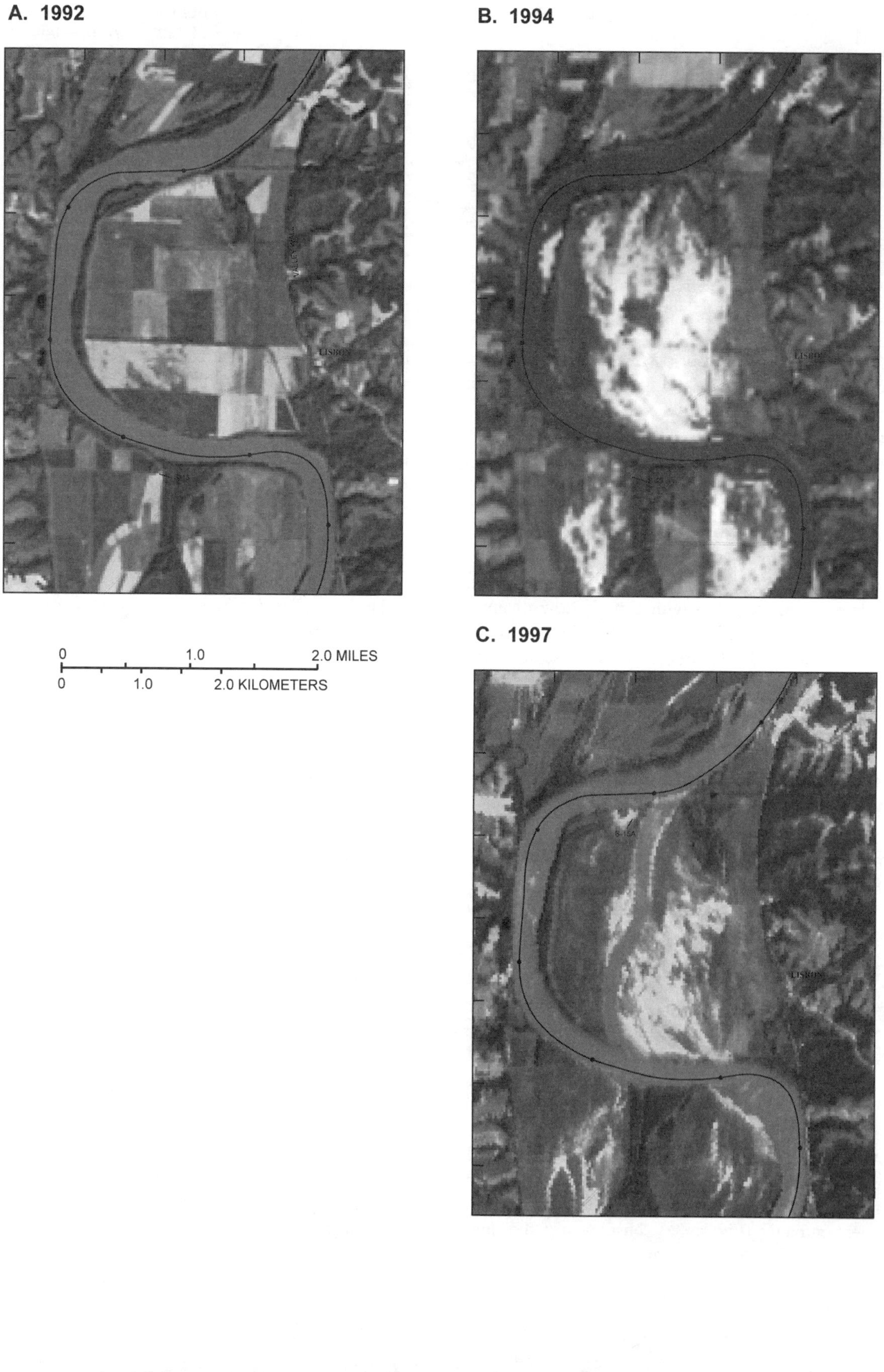
A. 1992
B. 1994
C. 1997
0
1.0
2.0 MILES
0
1.0
2.0 KILOMETERS
LISBON

higher discharge in June 1999 compared to December 1997, width:depth ratios increased in the upper half of the chute, a trend consistent with channel widening. In September 1998, lateral and mid-channel bars were just beginning to emerge in the chute at approximately 2,300 cms in the main channel (measured at Boonville, Missouri). At this discharge, width:depth ratios in the chute were substantially larger. In the upper half of the chute, width:depth ratios were within the range of historical values, indicating a correspondence to the historical reference condition for channel geometry (transects 5-15, Figure 9B).

Velocities and depths in the chute and neighboring navigation channel were assessed at two discharges in December 1997 and May 1998 by ADCP survey (Figures, 5, 12). Distributions of depth and velocity show the broad variation of physical habitat that exists in the navigation channel, primarily resulting from wing dikes that cause constrictions and recirculating eddies. The chute has a relatively high percentage of shallow and slow water; the relative contribution of shallow, slow water increases with decreasing discharge.

Early in the evolution of the chute, the margins of the channel at the upstream one half were underlain by cohesive silty clay and clay beds. These units were the remains of the topstratum of the sedimentary sequence that the chute had cut through. As the chute widened and incised through the cohesive units, the bed material became coarser and dominated by sand. A systematic survey of the chute with a bed classification instrument in June 1999 indicated variation in bed material along the chute that was consistent with the transition from braided to meandering planform. The braided, upstream half was dominated by gravelly sand and sand, whereas the downstream half was characterized by a sand channel, sandy point bars, and mud margins (Figure 13). Gravelly sand substrate at the upstream end of the chute is considered an especially valued substrate because of the relative scarcity of this coarse substrate habitat in shallow, low-velocity areas of the Lower Missouri River (Figure 14). The median bed-material particle diameter on this part of the Lower Missouri River is typically fine to medium sand, in the range of 0.24 – 0.55 mm [*U.S. Army Corps of Engineers,* Unpub. data].

6. BIOLOGICAL RESPONSES TO HABITAT EVOLUTION

Biological responses to the evolving chute habitats vary by trophic level. Assessment of benthic invertebrate communities indicates that the relatively mobile, sandy substrate of the chute supports a community with relatively few taxa [*Poulton et al.,* 1999]: only 4 taxa of macroinvertebrates were collected in sandy substrates within the chute, as compared to 78 taxa associated with rock revetments in the main channel, and 59 taxa associated with mucky sediments in recirculating eddies downstream of wing dikes. Organic snags in the chute, however, had 30 taxa of macroinvertebrates [*Poulton et al.,* 1999] indicating that accumulation of large woody debris in the chute may be necessary to achieve high secondary productivity.

The fish community of the chute has been shown to be quite diverse [*U.S. Fish and Wildlife Service*, 1999]. Collections in 1997-1999 found 64 species of fish in the chute, compared to 26 species in the adjacent navigation channel; 91 species of fish are known to inhabit the Lower Missouri River. Of the 64 species collected in the chute, one was the federally listed pallid sturgeon (*Scaphirhynchus albus*), and four others were species of concern (sicklefin chub *(Macrhybopsis meeki),* sturgeon chub *(Macrhybopsis gelida*), plains minnow (*Hybognathus placitus*), and blue sucker (*Cycleptus elongates*); U.S. *Fish and Wildlife Service*, [1999]). In contrast, waterbird taxa richness in the chute area decreased from 50 to 37 species in the 1.5 years after the chute was formed. This decrease was attributed to replacement of shallow, lentic habitats that existed in the original scour, with lotic habitats in the chute [*Helmers et al.,* 1999].

Vegetative succession on the margins of the chute has been minor because of the steep banks and unstable bars. Only in the Spring of 2000 were substantial areas of vegetation growth observed on incipient flood plain and lateral bars adjacent to the high banks. These communities are mainly grasses and forbs, but they have minor woody components represented by seedling cottonwood (*Populus deltoides*) and willows (*Salix spp.*). Cottonwood-willow com-

Figure 7. Images of Lisbon Bottom showing extensive changes to land cover from 1992 to 1997. Darker gray tones are indicative of vegetation cover. The bright white in (B) and (C) represents sand deposits. From October 1994 to August 1997, much of Lisbon Bottom became more vegetated, despite the floods of May 1995 and June 1996. (A) Pre-flood Thematic Mapper image September 24, 1992, Boonville discharge = 1,870 cms. (B) SPOT image October 11, 1994, Boonville discharge = 1,320 cms. (C) SPOT image, August 23, 1997, Boonville discharge = 2,580 cms. Images courtesy of Earth and Planetary Remote Sensing Laboratory, Washington University, St. Louis. Location is shown on Figure 5.

munities are expected to be persistent in frequently scoured sites like the chute bars [*Mazourek et al.,* 1999].

7. DISCUSSION – UNDERSTANDING GAINED AT LISBON BOTTOM

The chute at Lisbon Bottom is a unique experiment in rehabilitation of large-rivers, because of the extent to which fluvial processes are being allowed to create and maintain physical habitats. The relatively passive approach has allowed a high degree of connection between flows in the river, and flows in the chute and adjacent wetlands. The connection allows for transmission of flood pulses into the chute, with associated exchanges of nutrients, large woody debris, and sediment. Relatively unhindered flows have resulted in continued sediment transport in the chute and resultant dynamic habitats. Hydrologic connection and dynamic habitats generally are considered ecologically beneficial in large rivers [*Junk et al.,* 1989; *Gore* and *Shields*, 1995; *Galat et al.,* 1998; *Sparks et al.,* 1998].

The chute has widened and migrated laterally since formation in 1996. The upstream half of the chute has achieved a wide, shallow, braided channel, and appears to be evolving toward a historical reference condition. The downstream meandering portion has a well-defined thalweg and lateral bars that contribute additional diversity in depth, velocity, and substrate. The chute is presently (June 2000) providing depths, velocities, and substrate that are relatively lacking in the adjacent navigation channel.

The future stability and physical characteristics of the chute are a concern because of potential impacts on river navigation and biological characteristics of Lisbon Bottom. A decreasing rate of widening may indicate adjustment toward an equilibrium channel form. Increasing sinuosity, decreasing slope, and increasing hydraulic radius also may combine to increase frictional energy losses in the chute, and to promote channel stability.

Frictional energy losses and habitat characteristics will be determined largely by whether the chute evolves toward a braided or a meandering planform. Discrimination between braided and meandering channels is frequently described as a slope-discharge threshold [*Leopold and Wolman*, 1957], in which discharge is assessed as a bankfull value. Bankfull discharge is problematic to determine for the chute because it is not clear if channel geometry has adjusted to define bankfull dimensions. Nonetheless, a flood during June 2000 may have been a reasonable approximation of near bankfull flow. This flood produced flow of 200 cms (7,100 cfs) in the chute from 3,700 cms (133,000 cfs) total in the channel. The flow in the chute just reached the top of the depositional lateral bars that were beginning to be stabilized by vegetation; hence, this discharge appears to be close to the concept of bankfull, although it was considerably less than the estimated 2-year flow in the main channel (5,600 cms; U.S. Army Corps of Engineers, Kansas City District, Unpub. data). Figure 15 shows the range of slope and bankfull discharge that might

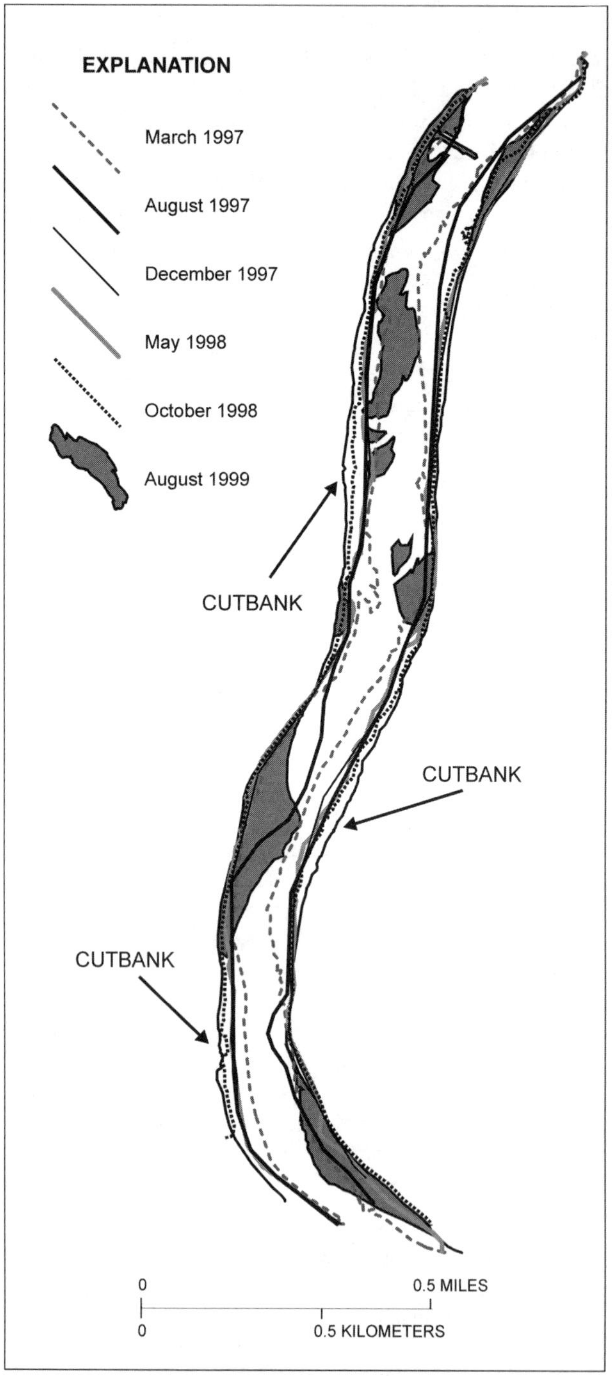

Figure 8. Maps of chute banklines from March 1997 to August 1999, and low-water bars August 1999.

Table 2. Evolution of slope and sinuosity in the Lisbon Bottom chute, 1996 - 2000

[Sinuosity is the ratio of length of the thalweg to straight-line distance; slope of the navigation channel is 0.00016]

	Jun-96	Mar-97	Oct-98	Jun-99	Jun-00
Chute Slope	0.00026	0.00024	0.00023	NA	0.00009
Chute Sinuosity	1.10	1.19	1.24	1.13	1.13
Ratio of Slopes, Chute: Navigation Channel	1.60	1.47	1.42	NA	0.56

NA: Data not applicable because of repairs to upstream revetment.

apply to the Lisbon Bottom chute using discharges from 200 cms to 560 cms (assuming 10% of main channel, 2-year flood in the chute), and with slopes ranging from the actual chute slope to the down-valley slope of the bottom land. These parameter values define an area almost entirely within the meandering portion of the [*Leopold and Wolman*, 1957] threshold relationship.

Another approach to evaluating channel pattern stability fields was proposed by *van den Berg* [1995] based on unit stream power and median bed material particle size. Computed values for unit stream power at bankfull stage in the chute range 6-35 N/m*s; with median bed-material size of 0.24 – 0.55 mm [*U.S. Army Corps of Engineers*, Unpub. data]; these values also plot well within the single-thread meandering field defined by *van den Berg* [1995].

In a re-examination of the braided-meandering transition, *Carson* [1984] argued that the fundamental prerequisite for braiding was high bed-material load, rather than a threshold combination of slope and discharge. In this analysis, slope is considered a dependent variable adjusted to discharge and sediment availability. It would follow that the factors that ultimately control sediment transport into the chute will be critical to determining future channel planform

Sediment load also may define, over the long term, whether the chute fills in with sediment, is maintained as a secondary channel, or will tend toward pirating the main channel of the Missouri River. In an analysis of river avulsion processes, *Slingerland and Smith* [1998] showed that stability of a side-channel chute depended on the ratio of the chute slope to the main channel slope, the ratio of the height of the lip of the chute to flow depth in the main channel, and particle size of the moving bed layer in the main channel. These factors determine the balance between sediment flux through the chute and sediment flux down the main channel. For medium-sand-sized bed-material in the range typical for this part of the Missouri, and a ratio of chute slope to main channel slope of 1.4 – 1.6 (table 2), the *Slingerland and Smith* [1998] analysis is relatively insensitive to the lip geometry and would predict a sustainable side-channel chute ("equilibrium crevasse") rather than sedimentation or avulsion. A somewhat finer particle size (0.1 mm), however, would result in sedimentation of the chute because greater sediment concentration in the water column in the main channel would allow greater transport into the chute. Although prediction from the *Slingerland and Smith* [1998] analysis would not be justified without

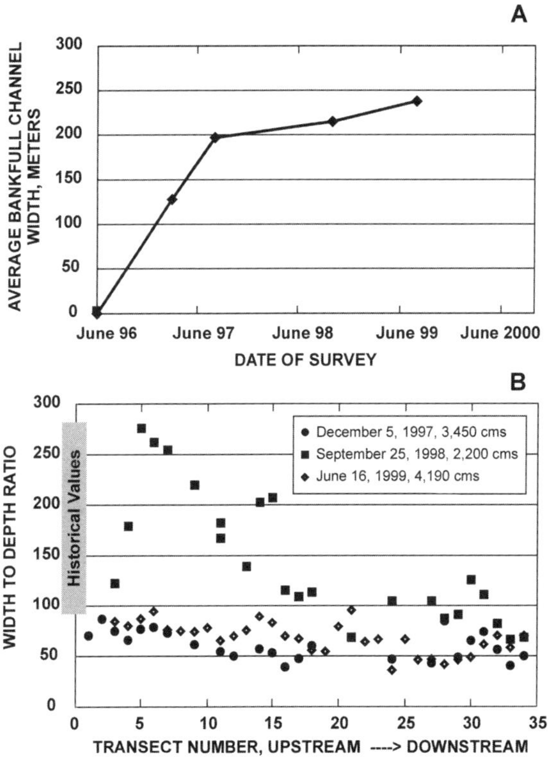

Figure 9. (A) Average chute channel width plotted over time. This channel width is a bankfull channel width mapped from DGPS surveys of the high bank line. (B) Width:depth ratios calculated from bathymetric data from monitoring transects, upstream to downstream in chute. Discharge at Boonville, Missouri, on December 5, 1997 was 3,450 cms; on September 25, 1998, 2,200 cms; on June 16, 1999, 4,190 cms. The range of historical values is from 1920 bathymetric data for side-channel chutes in vicinity of Lisbon Bottom.

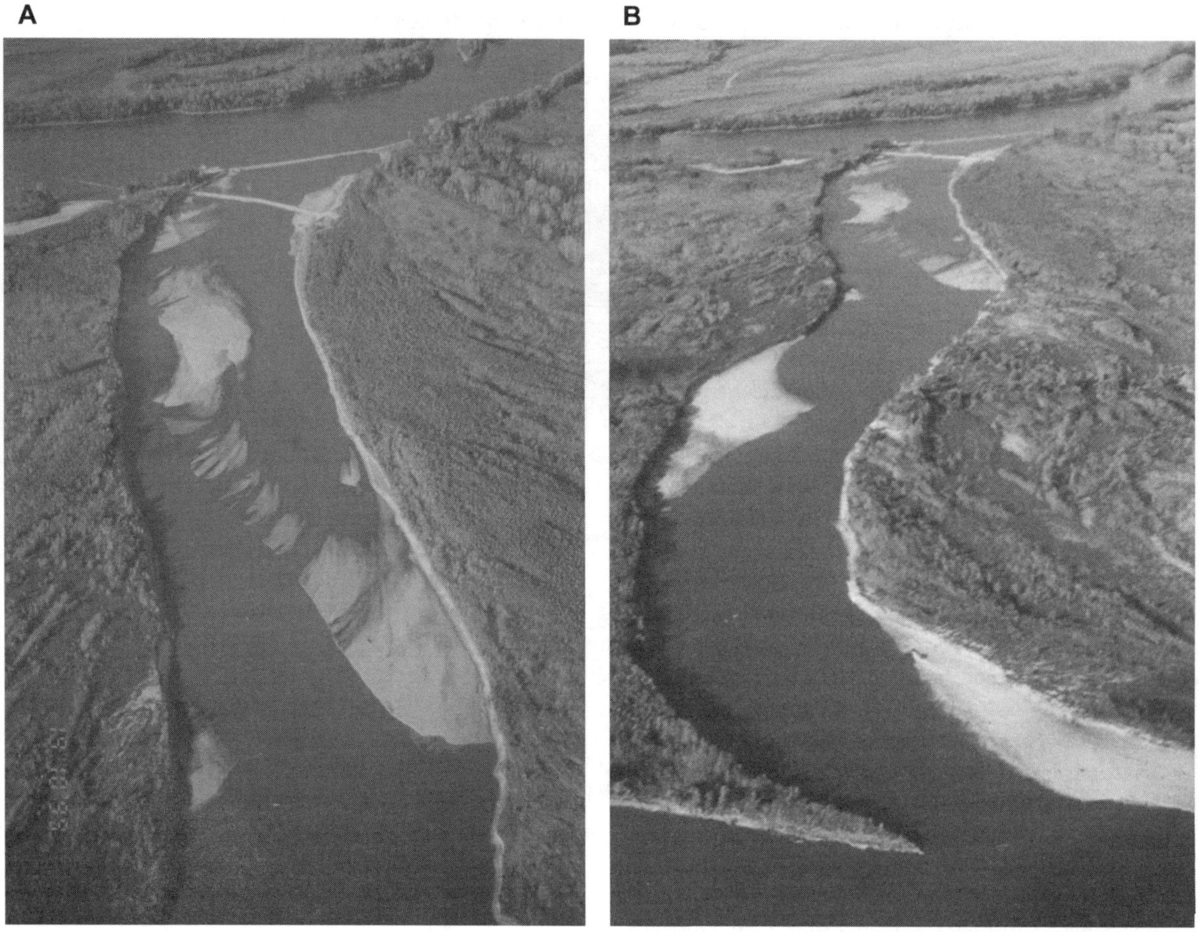

Figure 10. Low-altitude, oblique aerial photographs of the Lisbon Bottom Chute, October 1999. (A) Upper, braided one-half of the chute, looking upstream toward revetment and hydraulic inlet structure (under construction). (B) Lower, meandering one-half of the chute, looking upstream; the navigation channel appears in the foreground and background.

much greater attention to details of model formulation and validation for this site, the analysis supports the idea that long-term evolution of the chute will be dependent on the interplay of sediment load, sediment particle-size distribution, and chute geometry.

Plan-form geometry is an additional potential factor in ultimate chute stability. In a review of 20 channel cutoffs in the southeastern U.S., *Shields and Abt* [1989] found that the rate of channel filling in old, cutoff channels increased with increasing cumulative discharge in the cutoffs, increasing sediment concentration in the main channel, and decreasing sine of the angle between the main channel and the cutoff channels. The smaller the planform angle of approach, the greater potential for sediment transport into the cutoff channel. With an approach angle of about 50°, the Lisbon Bottom chute geometry would not be expected to produce extremely fast or slow filling rates.

Designers of secondary channels on the Rhine River in the Netherlands have concluded that secondary channel systems are inherently unstable over the long term, and will tend either to fill up with sediment or pirate the main channel [*Bareneveld et al.,* 1994; *Schropp*, 1995]. Designs for secondary channels on the Rhine aim to keep all sediment from entering the secondary channel to prevent sedimentation, although it is recognized that low sediment transport in the secondary channel increases the chance that harmful aggradation will occur in the main channel and may lead to excessive incision of the secondary channel [*Schropp*, 1995]. *Barenevel et al.* [1994] argue that careful modeling of discharge and sediment transport can help design a balance of channel dimensions and water/sediment distribution. However, such designs are believed to achieve a secondary channel that would be in equilibrium for no more than several years, after which dredging of the secondary channel would be necessary. The disagreement between the Rhine design experience and the theoretical analysis of *Slingerland and Smith* [1998] indicates the need for empirical documentation of field-scale experiments.

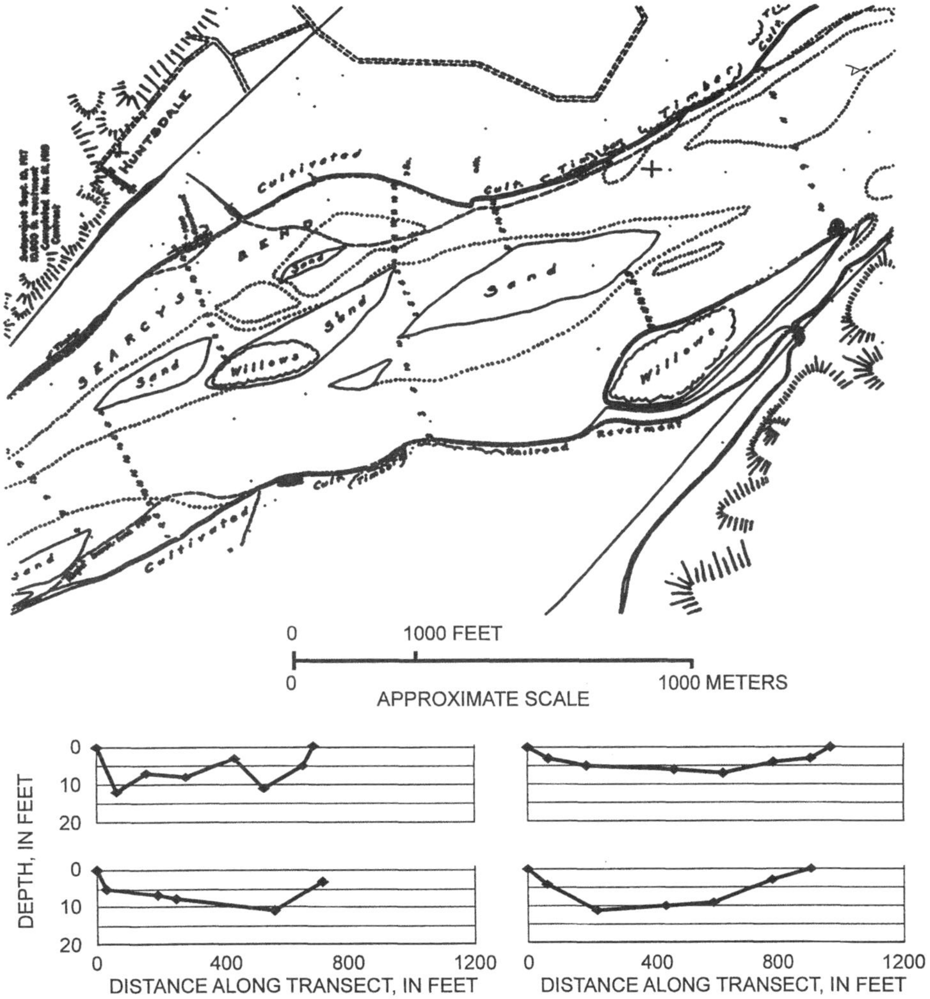

Figure 11. Excerpt from Lower Missouri River navigation charts from 1920 [*War Department,* 1920], showing an example of data used for determining bathymetry for historic side-channel chutes, and four examples of 1920 side-channel chute cross sections.

Although the evolution of the Lisbon Bottom chute was minimally affected by management in 1996-1999, its future will be more controlled. In June 1999, a grade-control structure was installed across the chute approximately 450 m upstream from the downstream end. The design for the grade control structure called for rocks to be keyed into the banks and emplaced into a trench in the channel bed, so it would not affect flow or impede boat and fish passage [*U.S. Army Corps of Engineers*, 1998]. Beginning in autumn 1999 and extending through May 2000, a notched hydraulic control structure was constructed approximately 270 m downstream from the revetment at the upstream end of the chute (Figure 10A). This structure and the revetment (Figures 5, 10A) are notched to allow flow through the structure 95% of the time. The hydraulic structure also was designed to allow an increasing percentage of total flow with increasing discharge, in an attempt to ensure that habitat forming flows continued in the chute. In June 2000, discharges in the chute were 3-6% of the total river discharge when the river was flowing 2,000-3,300 cms, compared to as much as 20% before the structure was in place. Hence, future evolution of the Lisbon Bottom chute may be very different from that which occurred in 1996-1999. Notwithstanding that the water and sediment fluxes into the chute, and bed degradation are now controlled, the Lisbon Bottom chute is unique in the Lower Missouri River in that it retains over 2.5 km of unstructured, naturally evolving side-channel habitats.

Channel widening and migration of the Lisbon Bottom chute in 1996-1999 occurred during a period characterized by abnormally high flows, including eight floods in excess of 2-year recurrence (Figure 6). These high and relatively

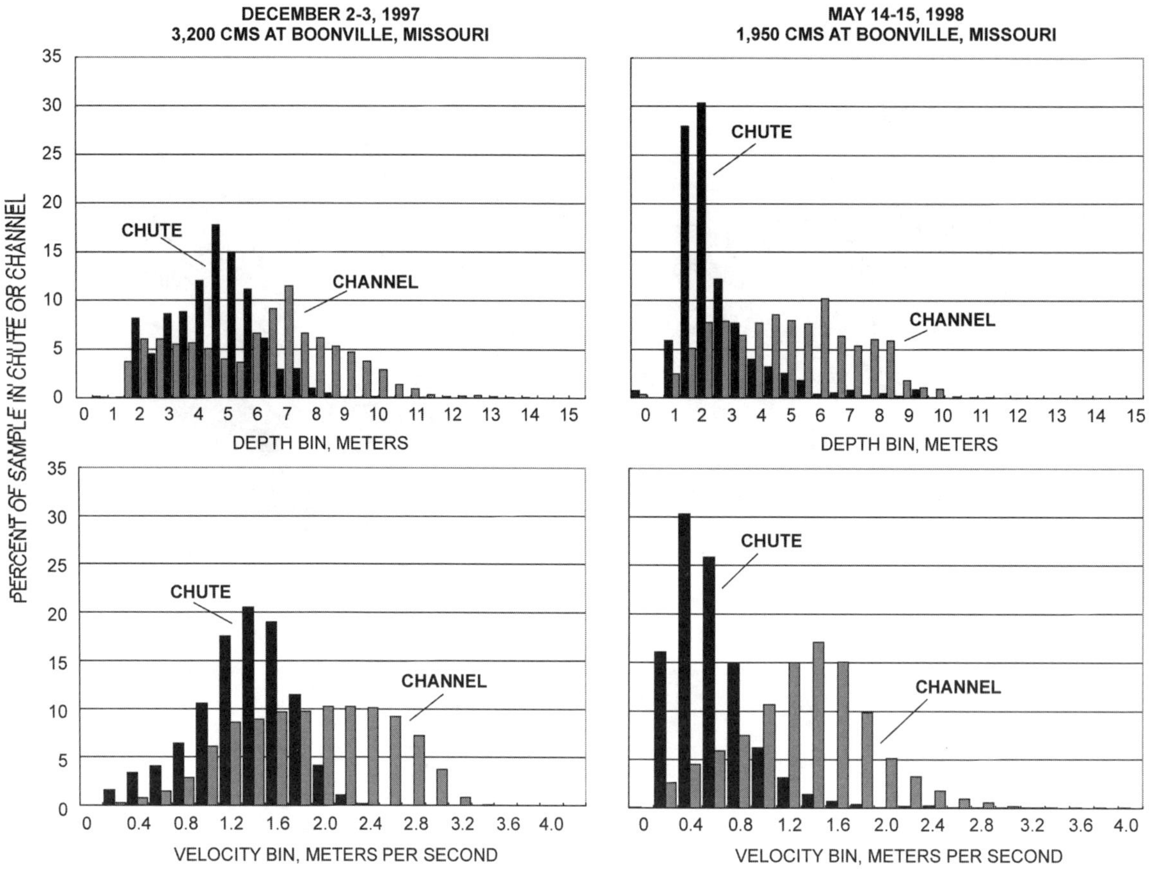

Figure 12. Histograms of samples of depths and velocities in the Lisbon Bottom Chute and Missouri River navigation channel, for two discharges, expressed as percent of sample in each area. Velocity data were collected with a 600 kHz acoustic Doppler current profiler using identical setups in the chute and navigation channel. Velocity data are total point magnitudes throughout the water column. Sample sizes range from 2,100 to 14,250 points.

unrestrained discharges allowed fluvial processes to shape the chute, and to form bars and incipient flood plain within the high banks. These new depositional surfaces (Figure 8) provide a range of elevations that can be inundated by a range of discharges, thereby substantially increasing habitat diversity and aquatic-terrestrial connections. Relatively unrestrained fluvial processes in 1996-1999 created habitat that will be more biologically effective at future, controlled discharges than if the period of habitat formation had not been allowed. Engineered side-channel rehabilitation projects may benefit from designs that similarly allow time for unrestrained habitat-forming floods to create natural features before ultimate stabilization.

Passive rehabilitation projects like Lisbon Bottom generally are less costly than more engineered or intensive approaches. For example, the cost of design and construction of minimal stabilization of the 3.3 km Lisbon Bottom chute has been estimated at $800,000 whereas cost of an engineered 2.9 km side-channel chute near Overton, Missouri has been estimated at $2.4 million. The Overton chute includes upstream and downstream hydraulic control structures, multiple grade-control structures, and substantial riprapped bank revetment to protect an adjacent interstate highway [*U.S. Army Corps of Engineers*, unpub. design]. While the cost of a rehabilitation project is relatively simple to assess, a complete performance evaluation would need to include an integrated assessment of ecological benefits relative to construction and maintenance costs. Quantifying integrated ecological benefits of rehabilitation projects remains a particular challenge.

8. SUMMARY AND CONCLUSIONS

The chute at Lisbon Bottom on the Lower Missouri River has provided a unique opportunity to document the effects of fluvial processes in passive rehabilitation. The opportunity began with the flood of 1993, estimated to have a 500-year recurrence interval. This flood opened the flood plain to fluvial processes and allowed a much smaller flood to carve a natural side-channel chute. The chute

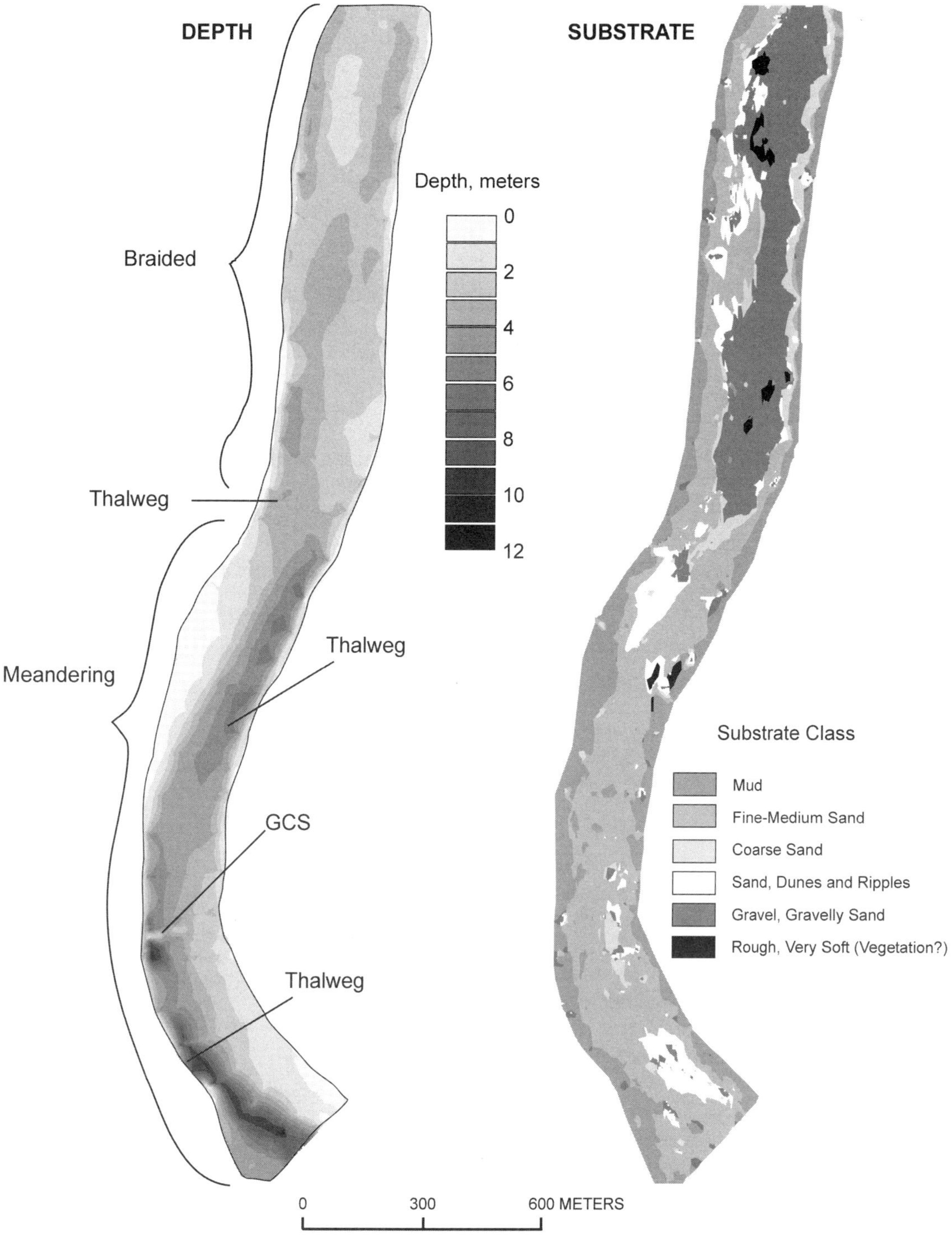

Figure 13. Depth and substrate maps of the Lisbon Bottom chute, June 1999. GCS - Grade Control Structure.

channel morphology has evolved toward an historical analog and is now providing diverse habitats that are in short supply in the engineered navigation channel. The short-term biological benefits have been mixed: the chute has fewer waterbird and benthic macroinvertebrate taxa than adjacent habitats, but greater numbers of fish species compared to the navigation channel.

Long-term biological responses of the chute will depend in part on the physical evolution of habitats, as well as other ecological interactions such as predator/prey rela-

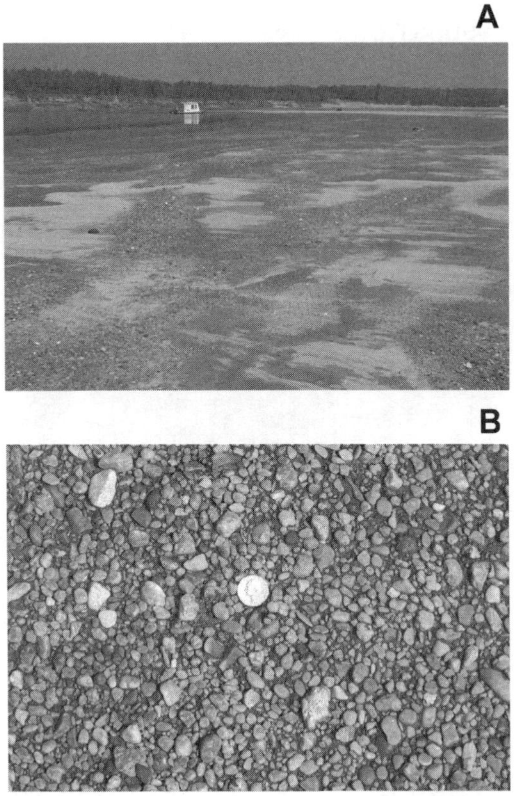

Figure 14. (A) Photograph of patches of sand and sandy gravel on upstream bar, Lisbon Bottom chute. (B) Close-up photograph of well-sorted medium to coarse gravel. Quarter shown in (B) is 24 mm across.

tions, effects of invasive species, altered nutrient fluxes, and changes in water quality. Although the chute is presently (June 2000) characterized by braided and meandering sections, the slope and discharge plot within the stability field for a meandering channel. In addition, loss of sediment supply from the main channel would be expected to encourage meandering rather than braiding in the future. Whether the chute is a persistent feature or will tend to fill with sediment is dependent primarily on factors that control relative fluxes of sediment into the chute and down the main channel. Theory and design experience disagree, however, in their predictions, indicating the need for continued monitoring.

In addition, we anticipate that future biological responses to habitat evolution will depend substantially on large woody debris dynamics and formation of the chute's own flood plain. Large woody debris is relatively rare in the chute as of June 2000. Accumulation and retention would provide fine-scale hydraulic complexity, cover for fish, and stable substrates for benthic macroinvertebrates. Accumulation will be dependent on how effective the chute is in gathering debris transported by the main channel and, as cottonwood-willow communities age, accumulation will depend in part on lateral erosion rates that deliver large woody debris directly to the chute. Retention of large woody debris will depend on discharges and velocities in the chute, and whether marginal areas of shallow depth and slow velocity develop to trap the debris.

When initially formed, the chute was incised within a flood plain so the bank heights were commensurate with a much larger river. The high banks effectively disconnected the chute from the adjacent flood plain, and decreased potential for exchange of nutrients, sediment, invertebrates, and fish. During 1996-1999 the chute migrated laterally and began to form bars and its own incipient flood plain. These fluvial features now provide a range of elevations that can be inundated by a range of discharges, thereby substantially increasing habitat diversity and aquatic-terrestrial connections. As the chute continues to migrate laterally and this flood plain develops, there will be greater opportunity for migration of species from channel to flood plain, for deposition of fine sediments in backwaters, and for retention of large woody debris. Macroinvertebrate production should also increase in the chute's vegetated flood plain. A key factor in habitat availability in the Lisbon Bottom chute was the four years of relatively high discharges and minimal management that allowed fluvial processes to begin the process of creating diverse habitat.

The future of the chute and its physical and biological characteristics are not predictable with confidence. The fluvial processes that create physical habitat can require long time intervals, and the ultimate characteristics of the chute may not be apparent for decades. Biotic responses to the physical habitat template can also require long time intervals as communities and populations adjust to physical habitat. Moreover, unpredictable sequences of hydrologic events or complex biotic interactions may alter trends and produce unexpected results. Continued monitoring and

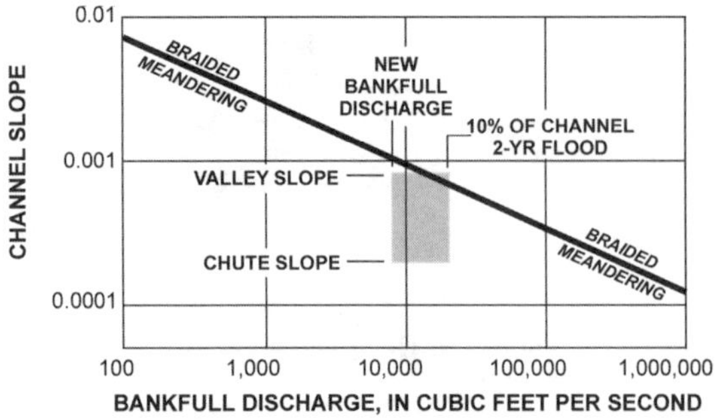

Figure 15. Stability threshold between braided and meandering stream pattern, based on *Leopold and Wolman* [1957]. Shaded area shows range of conditions for the Lisbon Bottom chute.

evaluation of the evolution of the Lisbon Bottom chute will be necessary to assess long-term ecological benefits of passive rehabilitation in this setting.

Acknowledgments. The authors acknowledge help in collecting field data from: Henry Hauck, Scott Southern, Larry Lumpkin, Ellen Ehrhardt, Aaron Delonay, and Maria Panfil. We especially acknowledge help from Ray Arvidson and Curt Niebur for access to remotely sensed imagery. The ideas presented here have benefited from informed discussions with David Galat, Duane Chapman, Barry Poulton, Jim Fairchild, Jim Milligan, and Todd Tillinger. We thank the Big Muddy National Fish and Wildlife Refuge (J.C. Bryant and Tom Bell) and the U.S. Fish and Wildlife Service for logistical support and access to the Lisbon Bottom site.

REFERENCES

Barneveld, H.J., R.L.J. Nieuwkamer, and G.J. Klaasen, Secondary channels: opportunities for and limitations of habitat restoration, *Water Science and Technology*, *29*, 335-345, 1994.

Burke, T. D., and J.W. Robinson, River structure modifications to provide habitat diversity, *The Mitigation Symposium*, Fort Collins, CO, July 16 1979-July 20 1979, pp. 556-561, 1979.

Cals, M.J.R., R. Postma, A.D. Buijse, and E.C.L. Marteijn, Habitat restoration along the River Rhine in The Netherlands: putting ideas into practices, *Aquatic Conservation: Marine and Freshwater Ecosystems*, *8*, 71-79, 1998.

Carson, M.A., The meandering-braided river threshold; a reappraisal, *Journal of Hydrology*, *73*, 315-334, 1984.

Federal Interagency Stream Restoration Working Group, *Stream Corridor Restoration; Principles, Processes, Practices*, Washington, D.C., USDA, Natural Resources Conservation Service, Variously paginated, 1998.

Ferrell, J., *Big Dam Era - A Legislative History of the Pick-Sloan Missouri River Basin Program*, Omaha, U.S. Army Corps of Engineers, 228 pp., 1993.

Ferrell, J, *Soundings - 100 Years of the Missouri River Navigation Project*, 171 pp., Omaha, U.S. Army Corps of Engineers, 1996.

Forrest, M. H., and R. Ettema, Sediment imbalance in rivers: simulation possibilities and problems, *Proceedings of the Symposium on Restoration Planning for the Rivers of the Mississippi River Ecosystem, Rapid City, South Dakota, September 14 1992-September 17 1992,* edited by L.W. Hesse, C.B. Stalnaker, N.G. Benson, and J.R. Zuboy, pp. 415-425, 1993.

Funk, J. L., and J.W. Robinson, *Changes in the channel of the lower Missouri River and effects on fish and wildlife*, Jefferson City, Missouri, Missouri Department of Conservation, 52 pp., 1974.

Galat, D.L., L.H. Fredrickson, D.D. Humburg, K.J. Bataille, J.R. Bodie, J. Dohrenwend, G.T. Gelwicks, J.E. Havel, D.L. Helmers, J.B. Hooker, J.R. Jones, M.F. Knowlton, J. F. Kubisiak, J. Mazourek, A.C. McColpin, R.B. Renken, and R.D. Semlitsch, Flooding to restore connectivity of regulated, large-river wetlands, *Bioscience 48*, 721-733, 1998.

Galat, D.L., J.F. Kubisiak, J.B. Hooker, and L.M. Sowa, Geomorphology, distribution, and connectivity of Lower Missouri River flood plain waterbodies scoured by the flood of 1993, *Verh. Internat. Verein. Limnol.*, *26*, 869-878, 1997.

Galat, D. L., J.W. Robinson, and L.W. Hesse, Restoring aquatic resources to the Lower Missouri River: Issues and initiatives, *Overview of the River-Floodplain Ecology in the Upper Mississippi River Basin*, edited by D.L. Galat and A.G. Frazier, Washington, DC, US Govt. Printing Office, pp. 49-71, 1996.

Gore, J.A., and F.D. Shields, Can large rivers be restored? *Bioscience, 45*, 142-152, 1995.

Harberg, M. C., J.I. Remus, S.C. Rothe, J. Becic, J., and L.W. Hesse, Restoration planning for an abandoned Missouri River chute, *Proceedings of the Symposium on Restoration Planning for the Rivers of the Mississippi River Ecosystem, Rapid City, South Dakota, September 14 1992-September 17 1992,* edited by L.W. Hesse, C.B. Stalnaker, N.G. Benson, and J.R. Zuboy, pp. 360-371, 1993.

Helmers, D. L., D.D. Humburg, A.C. McColpin, E. Ehrhardt, J.A. Vogel, K.J. Bataille, and L.H. Fredrickson, Waterbirds, *Initial Biotic Survey of Lisbon Bottoms, Big Muddy National Fish and Wildlife Refuge, U.S. Geological Survey, Biological Science Report, USGS/BRD/BSR - 2000-0001,* edited by D.D. Humburg, and V.J. Burke, pp. 55-62, 1999.

Hesse, L.W., and W. Sheets, The Missouri River hydrosystem, *Fisheries, 18*. 5-14, 1993.

Jacobson, R.B. and M.S. Laustrup, 2000, Habitat assessment for pallid sturgeon overwintering surveys, Lower Missouri River, *U.S. Geological Survey file report*, http://www.cerc.usgs.gov/pdf_docs/micra_whole.pdf, 2000.

Junk, W.J., P.B. Bayley, and R.E. Sparks, The flood pulse concept in river-floodplain systems, *Special Publication Canadian Fisheries and Aquatic Sciences, 106*, 110-127, 1989.

Klein, J.P., G. Maire, and F. Exinger, The restoration of former channels in the Rhine alluvial forest: the example of the Offendorf nature reserve (Alsace, France*), Water Science and Technology, 29*, 301-305, 1994.

Latka, D.C., J. Nesteler, and L.W. Hesse, Restoring physical habitat in the Missouri River: A historical perspective, *Proceedings of the Symposium on Restoration Planning for the Rivers of the Mississippi River Ecosystem, Rapid City, South Dakota, September 14 1992-September 17 1992,* edited by L.W. Hesse, C.B. Stalnaker, N.G. Benson, J.R. Zuboy, pp. 350-359, 1993.

Leopold, L. B., and M.G. Wolman, River channel patterns: braided, meandering, and straight, *U.S. Geological Survey Prof. Pap. No. 282-B*, 39 pp., 1957.

Lubinski, K. S., and S. Gutreuter, Ecological information and habitat rehabilitation on the upper Mississippi River, *Proceedings of the Symposium on Restoration Planning for the Rivers of the Mississippi River Ecosystem, Rapid City, South Dakota, September 14 1992-September 17 1992,* edited by L.W. Hesse, C.B. Stalnaker, N.G. Benson, J.R. Zuboy, pp. 87-100, 1993.

Mazourek, J., D. Martin, D.D. Humburg, and L.H. Fredrickson, Post-flood vegetation communities, *Initial Biotic Survey of Lisbon Bottoms, Big Muddy National Fish and Wildlife Refuge, U.S. Geological Survey, Biological Science Report,*

USGS/BRD/BSR - 2000-0001, edited by D.D. Humburg and V.J. Burke, pp. 21-27, 1999.

Missouri River Commission, *Missouri River Commission Survey Maps - 1879,* Missouri River Commission, 1:63,360 scale, 1879.

Morlock, S.E., Evaluation of acoustic Doppler current profiler measurements of river discharge, *U.S. Geological Survey Water-Resources Investigations Report 95-4218*, 37 pp., 1996.

Plafkin, J. L., M.T. Barbour, K.D. Porter, and R.M. Hughes, *Rapid bioassessment protocols for use in streams and rivers,* U.S. Environmental Protection Agency, EPA/444/4-89-001, 176 pp., 1989.

Poulton, B.C., J.F. Fairchild, M.L. Wildhaber, Benthic macroinvertebrates associated with specific habitats and substrates in the Lower Missouri River (abstract), *Proceedings of the North American Benthological Society, 1999 Meeting, Duluth, Minnesota,* 1999.

Rothe, S., Restoration of diversity to channelized Missouri River (abstract), *Proceedings, Society for Ecological Restoration Annual Meeting, Seattle, WS,* Online document http://www.ser.org/library/seattle/seattleindex.html, 1995.

Rukavina, N.A., Substrate mapping in the Great Lakes nearshore one with a Roxann acoustic sea-bed classification system, *Proceedings, Canadian Coastal Conference, Guelph, Ontario,* edited by M.G. Skafel, pp. 338-349, 1997.

Schalk, G. K., and R.B. Jacobson, Scour, sedimentation, and sediment characteristics at six levee-break sites in Missouri from the 1993 Missouri River flood, *U.S. Geological Survey Water-Resources Investigations Report 97-4110*, 72 pp., 1997.

Schlosser, I. J., A conceptual framework for fish communities in small warmwater streams, *Community and Evolutionary Ecology of North American Stream Fishes.* 269-299, Norman, University of Oklahoma Press, pp. 17-24, 1987.

Schmudde, T.H., Some aspects of the Lower Missouri River flood plain, *Annals of the Association of American Geographers, 53*, 60-73, 1963.

Schropp, M.H.I., Principles of designing secondary channels along the River Rhine for the benefit of ecological restoration, *Water Science and Technology, 31*, 379-382, 1995.

Shields, F. D., Jr., Environmental guidelines for dike fields, *River meandering: proceedings of the conference Rivers '83 sponsored the Waterway, Port, Coastal and Ocean Division of the American Society of Civil Engineers, New Orleans, LA*, edited by C.M. Elliott, pp. 430-442, 1984.

Slingerland, R., and N.D. Smith, Necessary conditions for a meandering-river avulsion, *Geology, 26,* 435-438, 1998.

Sparks, R.E., J.C. Nelson, and Y.Yin, Naturalization of the flood regime in regulated rivers, *Bioscience, 48*, 706-720, 1998.

U.S. Army Corps of Engineers, *1954 Missouri River Hydrographic Survey – Rulo, Nebraska to the Mouth,* 1:4800 scale, 1954.

U.S. Army Corps of Engineers, *The Great Flood of 1993 Post-Flood Report - Upper Mississippi River and Lower Missouri River Basins - Main Report,* U.S. Army Corps of Engineers, 60 pp., 1994.

U.S. Army Corps of Engineers, *Public Notice,* Kansas City, Missouri, U.S. Army Corps of Engineers, Kansas City District, 7 pp., 1998.

U.S. Fish and Wildlife Service, *Big Muddy National Fish and Wildlife Refuge, unpublished file report,* Fort Snelling, Minnesota, U.S. Fish and Wildlife Service, 22 pp., 1999.

U.S. Government, *Water Resources Development Act of 1986 (P.L. 99-662)*, Washington, D.C., 1986.

van den Berg, J.H., Prediction of the alluvial channel pattern of perennial rivers, *Geomorphology, 12,* 259-279, 1995.

War Department, *Missouri River from the Mouth to Quidaro Bend,* Washington, D.C., War Department, scale 1:12,000, 1920.

Robert B. Jacobson and Mark S. Laustrup, U.S. Geological Survey, 4200 New Haven Road, Columbia, Mo 65201.

Michael D. Chapman, U.S. Army Corps of Engineers, 700 Federal Building, 601 E. 12th Street, Kansas City, Missouri 64106

Mitigating Boatwake-Induced Erosion Along the Kenai River, Alaska

Joseph M. Dorava

Water Resources Department, Vierbicher Associates Inc., Reedsburg, Wisconsin

Increased sedimentation and loss of streamside cover associated with streambank erosion have been identified as key factors decreasing salmon returns to important anadromous fish rivers. Boatwake-induced streambank erosion has been implicated in delivery of sediment to the in-channel habitats and with the destruction of riparian cover along the Kenai River in south central Alaska, an economically important salmon river generating as much as $78 million annually in state-wide benefits. Boatwake-induced erosion generally can be reduced by either decreasing the size and number of boatwakes striking the banks or by reinforcing the banks to more effectively resist the erosion. Recent controlled boat experiments have identified several opportunities to mitigate boatwake-induced erosion along the Kenai River. Data from these boat experiments indicate a substantial decrease in the weight of bank material eroded as a result of increasing the distance between the boat and the bank and decreasing passenger load. The data also indicate that protecting streambanks with spruce trees cabled to the bank or bio-logs attached to the bank can reduce streambank erosion induced by boatwakes. Challenges for resource managers implementing bank erosion mitigation strategies include understanding the erosion control method's effectiveness, its public acceptance, and its enforceability.

INTRODUCTION

North American Pacific Salmon use freshwater streams to spawn and support initial growth of their offspring. Juvenile salmon typically spend from 0.5 to 2 years rearing in freshwater, and their rearing success depends heavily on cool, clear, and clean water with adequate depths, navigable water velocities, ample food supplies, and abundant cover to protect them from predators [*Groot and Margolis*, 1991]. These important juvenile salmon habitat features are commonly found along the shoreline margins of rivers, especially those with characteristics similar to the Kenai River in south central Alaska [*Bendock and Bingham*, 1988].

Geomorphic Processes and Riverine Habitat
Water Science and Application Volume 4, pages 217-226

Excessive sedimentation and habitat losses have been identified as important contributing factors in the decline of the salmon population in the Pacific Northwest [*Beschta*, 1989; *Bjorn*, 1969; *Meehan and Swanston*, 1977]. Fish habitat provided by streamside vegetation, overhanging banks, clear, cool water, and appropriately sized substrate can be altered or destroyed by encroaching streamside development and excessive river use. For example, development along the shoreline and boating in the river can alter near-shore water depths and velocities and increase bank erosion [*Dorava*, 1995; *Dorava and Moore*, 1997]. These activities can reduce the amount and the quality of juvenile salmon habitat, which in turn reduces salmon populations. In gravel-bed rivers such as the Kenai River, bank erosion can lead to deposition of fine-grained material on the streambed and suffocation of salmon eggs prior to emergence [*Beschta*, 1989]. Excessive sedimentation can also reduce the ability of juvenile salmon to find food, as water clarity is degraded [*Groot and Margolis*,

1991]. Increased turbidity also reduces light penetration and thereby decreases the zone of primary productivity in the river. In addition, as streams become wider and shallower as a result of unnaturally high rates of bank erosion, streamside and in-stream cover habitat can be reduced, leading to increased predation of juvenile salmon and high water temperatures [*Estes and Kuntz,* 1986; *Liepitz,* 1994].

Wakes generated by boats have been recognized as a contributing cause of streambank erosion by many investigators [*Alaska Department of Natural Resources,* 1986; *Barrick,* 1984; *Bhowmik and Demissie,* 1982, 1983; *Bhowmik and others,* 1982; *Bradbury and others,* 1995; *Bush,* 1988; *Camfield and others,* 1980; *Garrad and Hey,* 1987; *Hagerty,* 1989; *Jaakson,* 1988; *Johnson,* 1994; *Klingeman and others,* 1990; *Limerinos and Smith,* 1975; *Nanson and others,* 1994; *Reckendorf,* 1989; *Reckendorf and Saele,* 1991; *Scholer,* 1974; and *Von Krusenstierna,* 1990]. Along the Kenai River, where commonly more that 300,000 angler days of sport fishing occur annually [*Liepitz,* 1994], the greatest amount of bank erosion has been documented in areas with the heaviest boat activity, suggesting that the two processes are linked [*Dorava and Moore,* 1997]. Streamside development and boatwakes have been observed to degrade fish habitat by altering water velocities and reducing available cover by increasing erosion of susceptible streambanks along the Kenai River [*Dorava,* 1995; *Dorava and Moore,* 1997]. Because it is highly desirable for some people to live on the water, streamside development has been increasing considerably along the Kenai River and this has contributed to increased boat use on the river.

In many anadromous fish streams and especially along the Kenai River where the salmon fishery generates more than $78 million annually for Alaska's economy, management of streamside development and in-river uses such as boat traffic can be used to mitigate the effects of habitat loss. Developing mitigation strategies requires an understanding of the tradeoffs between protecting a resource while restricting public access to it. Furthermore, an understanding of the benefits of various mitigation practices, such as protecting the banks from erosion or reducing erosive actions against the streambank, will make it easier for resource managers to knowledgeably select effective alternatives for protecting important juvenile salmon habitat.

Purpose and Scope

Opportunities to more effectively protect the Kenai River salmon fishery by reducing the loss of rearing habitat resulting from powerboat operations are evaluated using data from two studies where powerboats were operated under controlled conditions. Data describing the maximum boatwake height and amount of bank erosion resulting from various boat operating conditions are presented for consideration in the development of regulations to protect the fishery. Several bank protection techniques are also compared for their effectiveness at reducing the height of boatwakes striking the bank and subsequently reducing the erosion of bank material. These data provide resource managers an assessment of the benefits to riverine habitat from alternative mitigation strategies. Once the effectiveness of a specific erosion control technique is understood, the next challenge for resource managers is determining if the public will accept an increase in regulation, and whether the new regulation is enforceable.

Recently completed experiments along the Kenai River evaluated the effectiveness of bank protection and boatwake reduction techniques. Data from these experiments present alternative opportunities to reduce juvenile salmon habitat losses by either managing boat operating characteristics or by protecting the banks from erosion with commonly used stabilization techniques. Data describing the effectiveness of various erosion control techniques and an understanding of geomorphic processes on the Kenai River can be used by resource managers to develop reliable and effective mitigation strategies that will protect riverine habitat valuable to the economically important salmon fishery.

Background

Within the sparsely developed 5,700-km^2 watershed of the Kenai River, residents, fishermen, and natural resource managers have recognized the economic importance of the salmon fishery. Several harvest restrictions and habitat mitigation measures have been used to protect the fishery. For example, currently only two chinook salmon greater than 0.50 meters in length can be taken annually from the river. In addition, powerboat use is forbidden in some areas of the upper river, and where powerboat use is allowed on the river downstream from Kenai Lake, the motor size has been limited to 35 horsepower. Furthermore, powerboat passenger loads are restricted to six or less, and commercially operated professional fishing guides are constrained to a 12-hour window of operation between 6 am and 6 pm. These management efforts are designed to provide safe equitable access to the fishery while mitigating the effects of over harvesting and habitat loss.

EXPERIMENTAL DESIGN AND DATA

During an initial controlled boat experiment in August 1996, three types of boats passed by a wake-recording

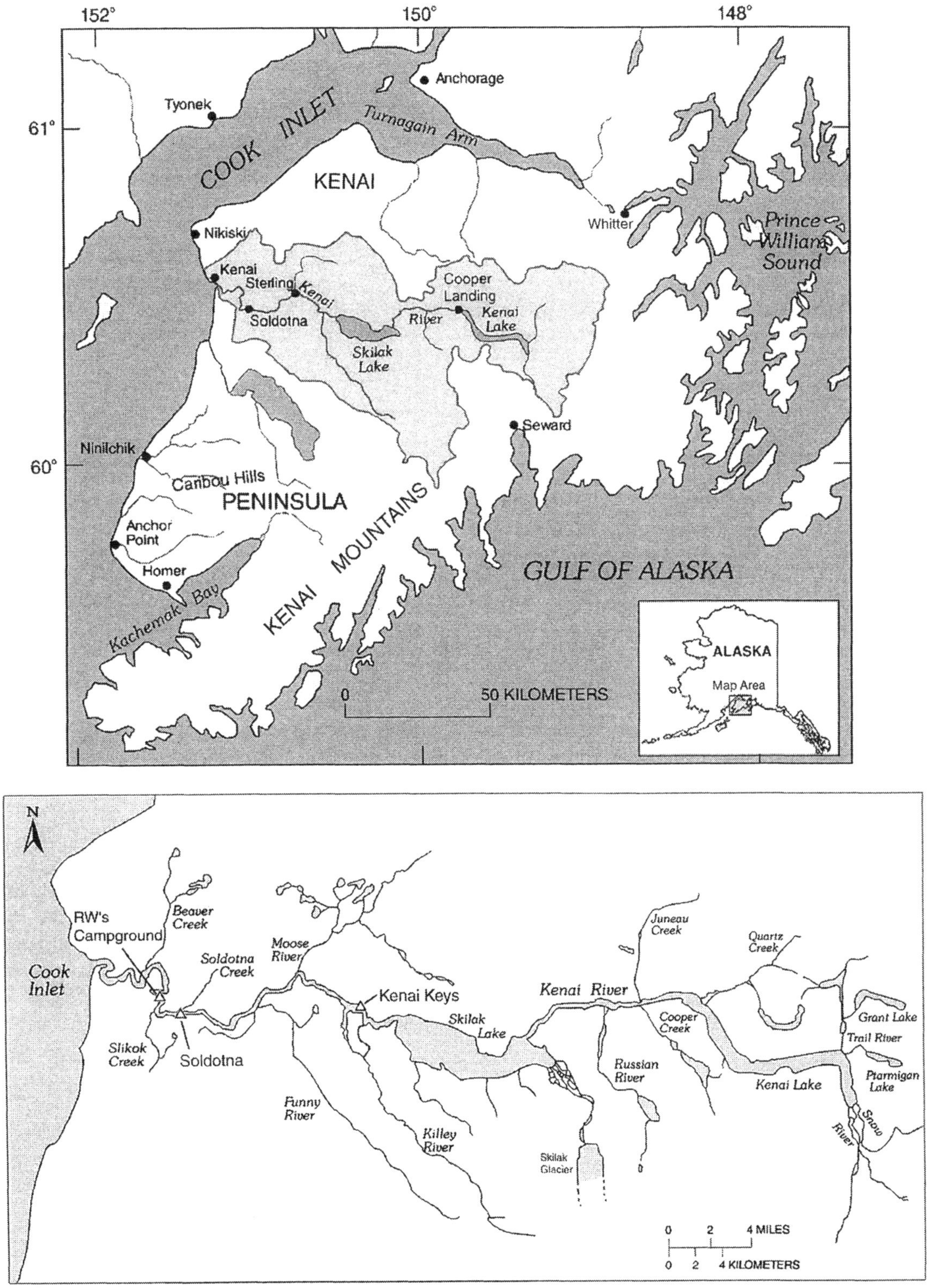

Figure 1. Map showing study sites (triangle) and important local geographic features of the Kenai River watershed.

Table 1. Boatwake heights recorded at the Kenai Keys on the Kenai River, August 27-28, 1996

Type of boat and motor	Percent of channel width	Average of two maximum wake heights measured while boat passed upstream and downstream (data in meters)			
		1 passenger	2 passengers	4 passengers	6 passengers
Flat-bottomed, 6-meter long 35 horsepower 4-cycle outboard	Next to bank	0.15	0.17	0.20	0.28
	25	0.08	0.12	0.12	0.15
	50	0.08	0.08	0.08	0.12
Semi-V, 6-meter long 35 horsepower 4-cycle outboard	Next to bank	0.16	0.17	0.23	0.08
	25	0.14	0.13	0.16	0.19
	50	0.12	0.11	0.13	0.17
Inflatable, 4.9-meter long 30 horsepower 2-cycle outboard	Next to bank	no data	0.14	0.13	no data
	25	no data	0.09	0.11	no data
	50	no data	0.07	0.06	no data

gage and an erosion-monitoring site near the Kenai Keys about 8 km downstream from Kenai Lake (Figure 1). Each boat passed by these two nearby data-collection sites carrying different passenger loads and traveling at a different distance from the bank. Data depicting the maximum boatwake height were recorded by an ink pen connected to a float that made a trace on a rapidly rotating paper chart [*Dorava and Moore*, 1997]. Material eroded from the bank was collected in a baking pan pinned to the stream bottom adjacent to the bank and later dried and weighed and identified as swash load. Maximum boatwake height and swash load data were collected and averaged for an upstream and a downstream pass of each boat operated under each condition of distance and passenger loading (Tables 1 and 2). Although the number of samples for each operating condition limits this data set, the original intention was to qualitatively or semi-quantitatively compare alternative powerboat operating characteristics. Using these data, there appears to be a greater reduction in boatwake height from moving boats toward the mid-channel sailing line than from reducing passenger loads (Table 1). Further examination of the data, including all the measured boatwake heights and swash loads especially the maximum values of these data (Figure 2), indicates that there is a rapid increase in bank erosion when maximum boatwake heights increase from about 0.15 to 0.20 meter. The rapid increase in the swash loads depicts conditions where the bank-erosion rate is very susceptible to small increases in boatwake height. This condition indicates that large boatwakes may be producing most of the bank erosion.

The results of this initial experiment present opportunities for managers to regulate powerboat operation and reduce the loss of salmon habitat. However, managers must also consider the public sensitivity to regulations, especially any regulations that potentially restrict equitable access to the resource. For example, if similar boatwake height and bank-loss reductions can be accomplished by moving powerboats farther off shore compared to additional limitations on allowable passenger loading, public access to the fishery would be essentially unaffected.

A second set of controlled powerboat experiments evaluated the effectiveness of various streambank protection measures. These additional experiments used similar boatwake-height and bank-erosion data-collection techniques as the first experiment, but also estimated the amount of cover habitat provided by each bank protection measure. During this experiment a single boat type (flat bottom) was operated with a set passenger load (two passengers), and in a defined sailing line (25 percent of the distance across the channel). The boat repetitively passed by a wake-recording gage and an erosion-monitoring site in the lower Kenai River near RW's Campground (Figure 1). Data were collected during 20 boat passes with no bank protection in place and then again during 20 passes with two various bank protection measures (bio-logs and cabled spruce trees) in place.

Results of the second controlled boatwake experiment indicate that boatwake attenuation and erosion control are similar for bio-log or cabled spruce tree bank protection measures (Figures 3 and 4). In fact, using a two-sample t-

Table 2. Swash load samples collected at the Kenai Keys on the Kenai River, August 27-28, 1996

Type of boat and motor	Percent of channel width	Sediment weight collected while boat passed upstream and downstream (Data in grams per meter of bank)			
		1 passenger	2 passengers	4 passengers	6 passengers
Flat-bottomed, 6-meter long 35 horsepower 4-cycle outboard	Next to bank	283	730	2309	2011
	25	194	179	313	834
	50	45	343	238	104
Semi-V, 6-meter long 35 horse-power 4-cycle outboard	Next to bank	1028	1281	1370	864
	25	447	119	no data	no data
	50	253	no data	194	no data
Inflatable, 4.9- meter long 30 horsepower 2-cycle outboard	Next to bank	no data	136	134	no data
	25	no data	45	30	no data
	50	no data	30	74	no data

test (∝ = 0.05) [*Zar*, 1984] no significant difference was detected in the maximum boatwake height or weight of material eroded from the bank for the bio-log or spruce tree tests. However, both significantly (∝ = 0.05) [*Zar*, 1984] reduced the maximum boatwake height and the weight of material eroded from the bank, compared to the unprotected bank. Data from other studies also indicate that spruce trees provide about three times more cover habitat for juvenile salmon than bio-logs [*Dorava*, 1999]. Furthermore, in another similar experiment near a site in the city of Soldotna about 8 km upstream, when an unprotected natural bank was subjected to the same boatwakes as an adjacent bank that had been stabilized with multiple erosion control measures, including bio-logs, cabled spruce trees, and growing willow shoots, about 3.7 times as much sediment was eroded from the unprotected bank (Figure 5). These data help streamside property owners and resource managers select appropriate stabilization techniques for their riverbanks depending on the availability of materials and the type and amount of boat traffic passing by their site.

DISCUSSION OF RESULTS

The record of average maximum boatwake heights and the corresponding swash-load values from the first experiment were converted to their logarithmic equivalents and the transformed exponential function was analyzed using linear regression (Figure 2). This transformed linear relation was then used for predicting swash-load val-

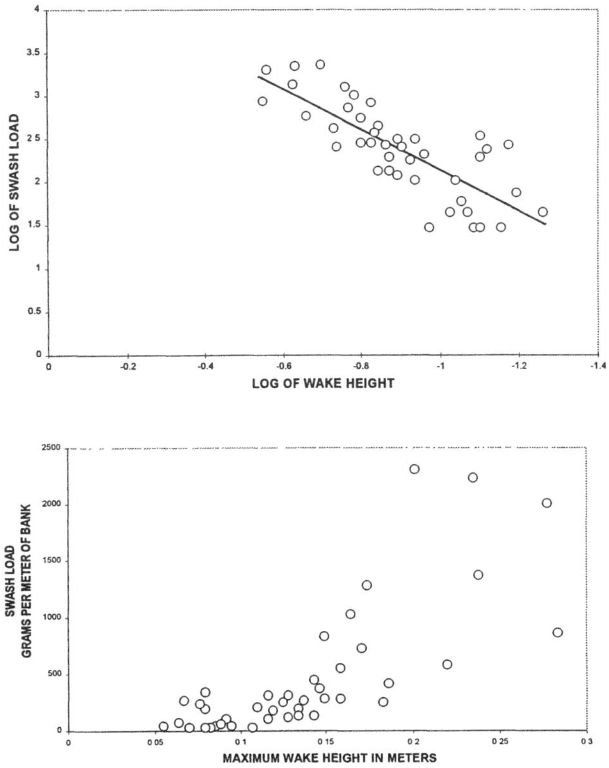

Figure 2. Wake height and swash load data collected during a controlled boat experiment on the Kenai River near the Kenai Keys (lower figure). Log transformed wake height and swash load data with a line indicating the least squares, linear regression (upper figure).

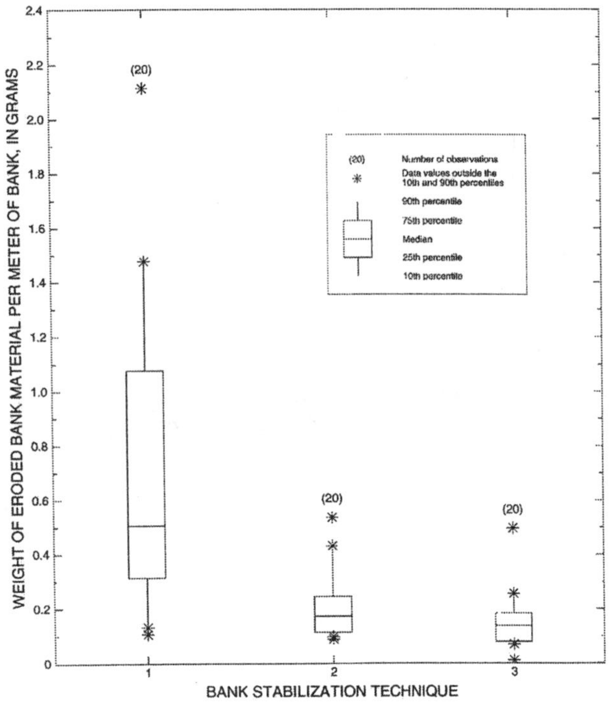

Figure 3. Box plots comparing wake height data collected during a controlled boat experiment on the Kenai River near RW's Campground. Bank stabilization technique 1 depicts the natural unprotected bank; technique 2, the bank protected by cabled spruce trees; and technique 3, the bank protected by a bio-log attached to the streambed.

ues at the Kenai Keys study site resulting from boatwake heights that were not directly measured but were within the range of the data measured and environmental conditions present during the experiment. Using this predictive relation, 12,123 boatwakes recorded at the Kenai Keys study site during 1996 were estimated to produce about 2.1 meters of lateral bank erosion along the approximately 1-meter-high bank. This prediction assumes that all the boatwakes were 0.14 meter in maximum height, the average calculated for boatwake heights at this site (Table 3), and that the bank material had a uniform composition and weight. This estimate of lateral bank erosion is greater than twice the 0.8 meter of bank erosion measured at this site (Table 4). This over- prediction results most likely because the effects of boatwakes on the bank change with different environmental conditions, such as water level in the river, moisture content in the bank, and composition of the bank. Additionally, the swash-load value prediction assumes that all the material collected in the sample pan originated from the streambank, and because some material collected in the pan was most likely resuspended streambed material, the bank-erosion predictions are inherently overestimated.

The predictive relation is, however, a useful tool to compare the relative effects of different boatwake heights. For example, if maximum boatwake heights are reduced by some mitigation strategy to 0.07 meter, or about one-half of the average maximum boatwake height recorded at this site, the estimated annual bank erosion would be reduced to less than 0.43 meter. According to the results of this experiment, however, obtaining this small (0.07 meter) boatwake height is impractical, attainable only if the boat is an inflatable type, traveling in the middle of the river channel, with four or fewer passengers (Table 1). As another example, the predicted bank erosion can be reduced by about one-half when the maximum boatwake height is reduced by only about one-third from 0.14 to 0.10 meter. This lower boatwake height is possible to achieve by numerous combinations of boats and operating characteristics (Table 1).

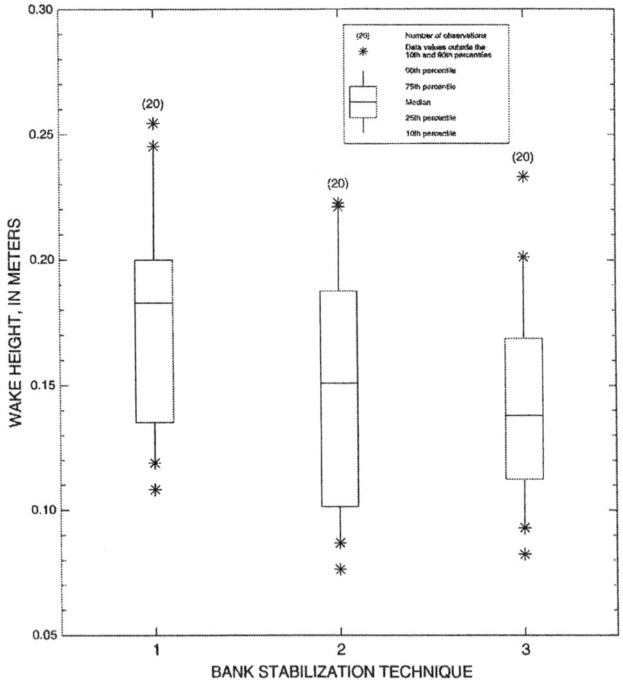

Figure 4. Box plots comparing the weight of material eroded from the banks during a controlled boat experiment on the Kenai River near RW's Campground. Bank stabilization technique 1 depicts the natural unprotected bank; technique 2, the bank protected by cabled spruce trees; and technique 3, the bank protected by a bio-log attached to the streambed.

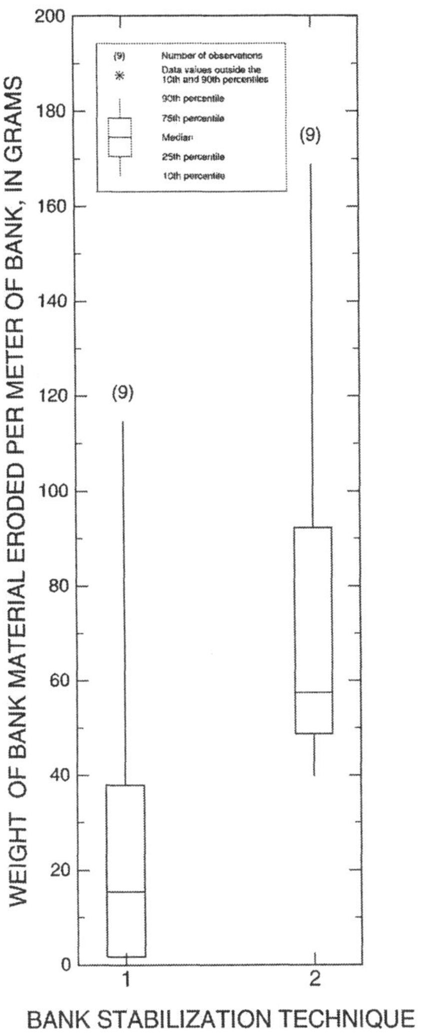

Figure 5. Box plots comparing the weight of material eroded from the banks during a controlled boat experiment on the Kenai River near the City of Soldotna. Bank stabilization technique 1 depicts the bank protected by a series of bio-engineered practices including cabled spruce trees, in combination with bio-logs attached to the streambed, and willow sprouts growing through the bio-logs; and technique 2, depicts the natural unprotected bank.

DEVELOPING MITIGATION STRATEGIES

Opportunities and Benefits

Approximately 14 percent of the soil types in a half-mile-wide corridor along the Kenai River between Cook Inlet and Skilak Lake have been characterized by the Natural Resources Conservation Service (NRCS) as being easily eroded [*Lehner*, 1994]. Although numerous erosion processes and soil characteristics contribute to this classification by NRCS, the implication is that these soils are susceptible to boatwake-induced erosion. Additionally, about 48 percent of the banks of the Kenai River downstream from Skilak Lake have previously been characterized as having moderate or high relative sensitivity to streamside development [*Scott*, 1982]. Included in *Scott's* [1982] characterization were the potential effects of constructing canals, groins, bank protection, and gravel mining. Additionally, *Scott* [1982] recognized the potential for increased river uses, such as boatwakes, to be a significant source of suspended sediment. Boatwakes can generate a substantial source of fine-grain size suspended sediment that can clog spawning gravel, and, at the same time, boatwakes can destroy streamside vegetation that provides protective cover for juvenile salmon. Understanding active geomorphic processes along the river can provide guidance to regulators who develop mitigation measures for land use or in-river uses that are site-specifically designed. For example, highly erosive areas might be restricted to non-motorized boat uses or require specific bank protection measures to be installed.

Mitigating the effects of boatwakes is possible by two primary methods:

(1) reducing the size and number of boatwakes, and

(2) interrupting the bank erosion process.

For example, if altering boat-operating regulations can reduce the number and/or size of boatwakes, not as much bank would erode. In addition, if the streambank face could be reinforced or protected from boatwakes by some device or mechanism this also would reduce erosion. The problem with some commonly used streambank stabilization techniques, such as riprap or timber bulkheads, however, is that they do not provide productive salmon habitat. Thus it is important to consider the additional benefits to the fishery when bank protection techniques, such as cabled spruce logs or bio-logs, are used because they provide valuable cover habitat.

Challenges and Consequences

Any practice that controls either the generation of boatwakes or the erosion of streambanks by boatwakes, must consider the socio-economic and socio-political aspects of the practice, as well as its documented performance. For example, the data collected during the experiments indicate that limiting the number of passengers allowed in a boat reduces the size of boatwakes generated, but this mitigation strategy may also result in a commensurate increase in the number of boats on the river, which already has a limited carrying capacity [*Whittaker and Shelby*, 1993]. Con-

Table 3. Maximum wake heights recorded at three sites on the Kenai River,1996

Site	Date	Number of wakes	Maximum wake height, (meters) Range	Average
RW's Campground	July 12-16	3,550	0.03-0.20	0.11
Soldotna	July 5-9	224	0.05-0.21	0.09
Kenai Keys	July 19-23	2,200	0.05-0.37	0.14

sidering these secondary effects of a mitigation strategy is important, but detailed analysis of these secondary effects is beyond the scope of the presently available data.

Developing mitigation strategies that reduce the size of wakes generated by boats was the primary focus of the first controlled boat experiment. Boatwake-height reduction has been recognized as an effective measure for bank erosion mitigation [*Nanson and others*, 1994; *Von Krusenstierna,* 1990]. Along the Connecticut River in New Hampshire, for example, regulations have created a boatwake-free zone within about 50 meters of the bank, which indirectly controls boat-operating conditions [*Barry Keith, New Hampshire Fish and Game, oral communication,* 1997]. Data from the Kenai River experiments demonstrate several opportunities for mitigating boatwake heights by regulating boat hull design, passenger load, or the boats operating position in the river. Maximum boatwake height was generally reduced more by moving a boat across the river channel from a position next to the streambank to one at mid-channel than by decreasing passenger loading from six to one (Table 1).

Substantial reductions in bank erosion are evident in the data when boatwake heights are less than about 0.20 to 0.15 meter (Figure 2). The experimental data indicate that obtaining boatwake heights in this range or lower was possible by several combinations of distance, passenger loading, and boat type (Tables 1 and 2). The experimental data also generally support a potential reduction in bank erosion could be obtained by simply operating powerboats near the middle of the river or with fewer passengers.

Because the Kenai River fishery is so popular and economically important it will be a challenge to implement mitigation strategies that restrict public access to the fishery. Furthermore, the mitigation opportunities identified during the experiments should only be implemented after consideration of their ability to reduce streambank erosion, as well as their enforceability and acceptability.

The data analyses presented here demonstrate the effectiveness of various boatwake mitigation strategies developed during experiments with various boat operating characteristics and bank protection measures. The documented effectiveness of these erosion mitigation opportunities should be one consideration and the potential consequences resulting from their enforcement and their public acceptability are two other serious considerations. Resource managers must also consider the possibility that additional untested mitigation strategies, such as restricting boat speed, further limiting outboard motor horsepower size, or implementing additional spatial or temporal restrictions on powerboat use, may also provide substantial reductions in boatwake induced erosion and provide improved protection of riverine habitat.

Table 4. Comparison of maximum bank-loss and boat passes at three sites on the Kenai River, 1996

Site	Bank loss (meters)	No. boats passing each site during periods of bank loss
	Motorized segment	
RW's Campground	1.14	22,008
Soldotna	0.43	2,770
Kenai Keys	0.80	12,123

Acknowledgments. Portions of the research described in this paper were supported by the Alaska Department of Fish and Game, U.S Geological Survey, and Vierbicher Associates Inc. Thoughtful guidance was provided during the research by Christopher Estes, Kevin Scott, Liska Snyder, Steve Maynord, Chris Waythomas, and Gary Liepitz. Lance Trasky and Janet Curan provided insightful comments on an earlier version of the paper.

REFERENCES

Alaska Department of Natural Resources, Kenai River comprehensive management plan, Alaska Department of Natural Resources report prepared in cooperation with the Kenai Peninsula Borough, 384 p., 1986.

Barrick, L. S., Kenai River bank erosion study, Alaska Department of Fish and Game, Division of Fisheries Rehabilitation Enhancement, and Development, Report Number 41, 82 p., 1984.

Beschta, R. L., The intrusion of fine sediments into a stable gravel bed, *J. of the Fisheries Research Board of Canada*, v. 36, no. 2, p. 204-210, 1989.

Bhowmik, N. G., and M. Demissie, Waves generated by river traffic, Conference on Applying Research to Hydraulics Practice, American Society of Civil Engineers Hydraulic Division, *Proceedings*, p. 179-187, 1982.

Bhowmik, N. G., and M. Demissie, Bank erosion by waves, Conference on Frontiers in Hydraulic Engineering, American Society of Civil Engineers Hydraulics Division, Shen H.T. ed., Proceedings, p. 195-200, 1983.

Bhowmik, N. G., M. Demissie, and C. Y. Guo, Waves generated by river traffic and wind on the Illinois and Mississippi Rivers, Univ. of Illinois, *Water Resour. Res.* Rep. No. 167, 90 p., 1982.

Bjorn, T. C., Embryo survival and emergence studies, Idaho Fish and Game Department Job Compliance Report, Project F-49-R-7, 11 p., 1969.

Bradbury, J., P. Cullen, G. Dixon, and M. Pemberton, Monitoring and management of streambank erosion and natural revegetation on the lower Gordon River, Tasmanian Wilderness World Heritage Area, Australia, *Environmental Management* v. 19, no. 2, p. 259-272, 1995.

Bush, J. E., Relative physical impacts of jet boats prop boats and canoes in an Ozark stream, Missouri Department of Conservation, Final Report, 9 p., 1988.

Camfield, F. E., R. E. L. Ray, and J. W. Eckert, The possible impact of vessel wakes on bank erosion, U.S. Coast Guard, Office of Research and Development, Final Report CG-W-1-80, 192 p., 1980.

Dorava, J. M., Effectiveness of streambank stabilization techniques along the Kenai River, Alaska, U.S. Geol. Surv. *Water-Resources Investigations Report* 99-4156, 21 p., 1999.

Dorava, J. M., Hydraulic characteristics near streamside structures along the Kenai River, Alaska, U.S. *Geol. Surv. Water-Resources Investigations Report* 95-4226, 41 p., 1995.

Dorava, J. M., and G. W. Moore, Effects of boatwakes on streambank erosion, Kenai River, Alaska, *U.S. Geol. Surv. Water-Resources Investigations Report* 97-4105, 84 p., 1997.

Dorava, J. M. and K. M. Scott, Role of glaciers and glacial deposits in the Kenai River watershed and the implications for aquatic habitat, in Gray J.S., and Riehle J.M., eds. *U.S. Geological Survey Prof. Paper* 1595, p. 3-11, 1998.

Garrad, P. N., and R. D. Hey, Boat traffic, sediment resuspension and turbidity in a broadland river, *J. of Hydrology*, v. 95, p. 289-297, 1987.

Hagerty, D. J., Ohio River bank erosion-traffic effects, American Society of Civil Engineers, *J. of Waterway, Port, Coastal and Ocean Engineering*, v. 115, no. 3, p. 404-408, 1989.

Jaakson, R., River recreation boating impacts, American Society of Civil Engineers, *Journal of Waterway, Port, Coastal and Ocean Engineering*, v. 114, no. 3, p. 363-367, 1988.

Johnson, S., Recreational boating impact investigations, Upper Mississippi River system, Pool 4, Red Wing, Minnesota, Report by the Minnesota Department of Natural Resources for the National Biological Survey, Environmental Management Technical Center, Report EMTC 94-S004, 48 p. + appendixes, 1994.

Klingeman, P. C., H. Matin, and C. C. Huang, Investigation of motorboat-induced streambank erosion on the Lower Deschutes River, Oregon State Univ., Water Resour. Res, Institute Report, 67 p. + appendixes, 1990.

Lehner, D., Kenai River cooperative river basin study, U.S. Department of Agriculture, Soil Conservation Service, Technical Report, (variously paged), 1994.

Liepitz, G. S., An assessment of the cumulative impacts of development and human uses on fish habitat in the Kenai River, Alaska Department of Fish and Game Technical Report No. 94-6, 63 p., 1994.

Limerinos, J. T., and W. Smith, Evaluation of causes of levee erosion in the Sacramento-San Joaquin Delta, California, *U.S. Geol. Survey Water-Resources Investigations* 28-74, 53 p., 1975.

Meehan, W. R., and D. N. Swanston, Effects of gravel morphology on fine sediment accumulation and survival of incubating salmon eggs, *U.S. Forest Service Research Paper* PNW-220, 16 p., 1977.

Nanson, G. C., A. Von Krusenstierna, E. A. Bryant, and M. R. Renilson, Experimental measurements of riverbank erosion caused by boat-generated waves on the Gordon River, Tasmania, *Regulated Rivers Research and Management,* v. 9. p. 1-14, 1994.

Reckendorf, F., Kenai River streambank erosion, Special report, Portland, Ore., Soil Conservation Service, West National Technical Center, 57 p., 1989.

Reckendorf F., and L. Saele, City of Soldotna, Alaska, Kenai River bank inventory report July 1991, U.S. Department of Agriculture, Soil Conservation Service Report (variously paged), 1991.

Scholer, H. A., The effects of speedboat activities on riverbanks, chap. 5 in *Geomorphology of New South Wales coastal rivers,*

Univ. of New South Wales Water Research Laboratory Report No. 139, p. 103-108, 1974.

Scott, K. M., Alaska, *U.S. Geol. Surv. Prof. Paper* 1235, 35 p., 1982.

Von Krusenstierna, A., Riverbank erosion by boat-generated waves on the lower Gordon River, Tasmania. The Univ. of Wollongong, Australia, M.S. thesis, 136 p., 1990.

Whittaker, D., and B. Shelby, Kenai River carrying capacity study, Important findings and implications for management. Alaska Department of Natural Resources, Division of Parks and Outdoor Recreation Report, 52 p., 1993.

Zar, J. H., *Biostatistical analysis (2nd edition),* New Jersey, Prentice Hall, 718 p., 1984.

Joseph M. Dorava, Water Resources Department, Vierbicher Associates Inc., 400 Viking Drive, P.O. Box 379, Reedsburg WI 53959.

Importance of Archival and Process Studies to Characterizing Pre-Settlement Riverine Geomorphic Processes and Habitat in the Puget Lowland

Brian D. Collins and David R. Montgomery

Department of Geological Sciences, University of Washington, Seattle, Washington

The synergy of archival investigations and field studies of relatively undisturbed rivers allows quantitative, process-based, river-specific description of the historic (pre-European settlement) riverine landscape and salmonid habitat in the Puget Lowland area, a landscape greatly altered by human activity in the last 150 years. Archival sources can document historic conditions and can suggest, but cannot demonstrate, causality between physical and biological processes and landscape features. Field studies help develop process models, whereas archival studies are necessary to capture intra-region variability in river morphology and process. Capturing this variability helps determine where those causal models apply. Field studies also help illuminate smaller-scale processes that archival sources commonly do not describe, but only archival sources can reveal landscape-scale processes and features that no longer exist due to landscape alteration and fragmentation. In the Puget Sound region, these combined approaches demonstrate that, on a watershed scale, the preponderance of river and wetland habitat was in lowland river valleys. Some rivers had an anastomosing pattern with many floodplain sloughs, in large part because of wood jams. Others had a single-thread meandering pattern, with oxbow lakes and vast floodplain wetlands. Estuarine marshes and channels were also extensive, but the amount and type varied with the rivers' different geologic histories. Wood formerly had a dominant influence on riverine processes at all scales; a true program of river restoration therefore must emphasize restoring floodplain forests, the river's lateral erosion, and wood jams. The former abundance and diversity of lowland habitats also suggests the need to place greater emphasis on lowland restoration, and to consider such restored habitats as potential watershed refugia.

INTRODUCTION

Restoring and protecting habitats in Puget Lowland rivers on which endangered Pacific salmonids and other species depend, requires answering these questions: (1) Where was habitat historically, and how did that distribution differ from today?, and (2) What were the geomorphic processes that created the habitat, and how do those processes differ today? Answering these questions is a first step in understanding the quantitative and qualitative manner in which the last 150 years of settlement has changed the region's riverine landscape and the types, function, and amounts of aquatic habitat. Such an understanding is necessary for de-

Geomorphic Processes and Riverine Habitat
Water Science and Application Volume 4, pages 227-243

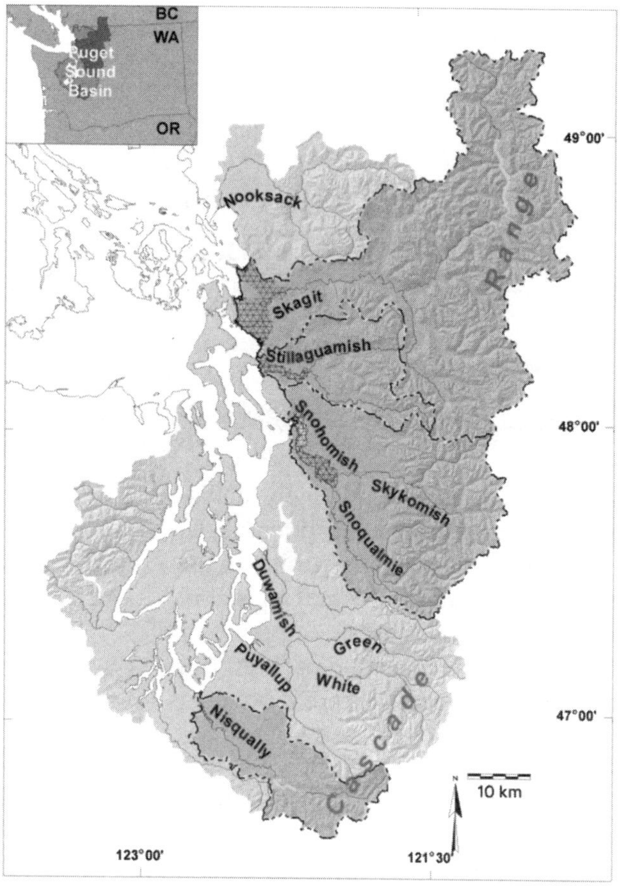

Figure 1. Locations of Puget Lowland rivers and study sites (hatched pattern) on the Nisqually, Snohomish, and Stillaguamish rivers.

veloping a strategy to restore lost or damaged habitats, and to protect remaining habitats.

Answering these questions is difficult in landscapes that have been greatly modified or degraded by human actions. Such is the case in the Puget Sound basin, where dramatic changes to the riverine landscape have occurred. Beginning in the 1850s, settlers diked and drained estuarine and floodplain marshes, channels, and sloughs; cleared channels of wood jams and other obstructions; deforested valley bottoms; and diked, dredged, and leveed rivers [for overview, see *Sedell and Luchessa,* 1981]. As a consequence, the present-day function and appearance of the riverine landscape differs radically, and the amount and quality of salmonid habitat is considerably reduced [e.g., *Beechie et al.,* 1994]. Because of the extent of recent landscape change, a conceptual understanding of the historic landscape's structure and function requires a combination of archival research and contemporary field studies in remnant natural areas that can provide at least local analogs to historic conditions. In the Puget Sound basin, the lower Nisqually River is the only major river that provides a suitable analog, in having significant areas with natural banks and a mature floodplain forest [for description see *Collins and Montgomery,* in press].

Archival and field approaches are complementary and can be productively used iteratively. First, while archival studies can show associations between landscape variables, field studies allow the investigation of causal relationships. In particular, field studies help generate models of historic river and habitat dynamics. However, because geologic history and other factors cause considerable regional variation in river morphology and dynamics, it is not possible to generalize from such models without using archival sources to constrain the applicability of a model to a given river. Second, field studies emphasize processes and features at a smaller scale, which may not be well documented in the archival record. On the other hand, archival sources allow the study of landscape-scale processes, for which no remaining analogs may exist, owing to landscape fragmentation.

This paper describes the application of archival approaches and contemporary field studies in an on-going study of Puget Lowland rivers. The study goals are to understand the process dynamics that created the historic riverine landscape and habitat, and to develop quantitative, process-based, and river-specific templates for river and habitat restoration, and more broadly, provide context for river management strategies.

STUDY AREAS

The Puget Lowland is a topographic low between Washington's Olympic Mountains to the west and Cascade Range to the east (Figure 1). Repeated Pleistocene advances of the Puget Lobe of the Cordilleran ice sheet deposited a thick fill of glacial sediments (the "great Lowland fill" [*Booth,* 1994]). Glaciers sculpted the fill into forms including north-south trending streamlined hills, and subglacial-runoff carved channels into the fill. Major rivers draining the eastern Puget Lowland include the Nooksack, Skagit, Stillaguamish, Snohomish, Duwamish, Puyallup, and Nisqually (Figure 1). Their gradients decline rapidly as they flow across the lowland toward Puget Sound. Some rivers incised relatively narrow (1- to 3-km wide) valleys into the Pleistocene glacial fill (e.g., the Nisqually, Stillaguamish, and upper Nooksack rivers). In other cases, subglacial runoff carved lower-gradient, wider valleys (3- to 5-km wide) in which the rivers flow (e.g., the Puyallup, Duwamish, Snohomish, Snoqualmie, and lower Nooksack rivers). Voluminous estuarine sedimentation from postglacial erosion and volcanism created extensive spreading deltas (e.g., the Skagit River) or confined, tidal-influenced valley bottoms (e.g., the Duwamish and Snohomish rivers). Data presented in this paper are primarily from the Nisqually, Snohomish, Stillaguamish, and Skagit rivers (Figure 1).

The Nisqually River begins at the terminus of the Nisqually Glacier in Mount Rainier National Park (Figure 1) and drains an 1,890 km^2 area. The LaGrande and Alder hydroelectric projects dam the river at kilometers 68 and 71, respectively. Downstream of these dams, the river passes through primarily agricultural and rural land, then traverses or borders the Fort Lewis military reservation, between river kilometers 4 and 31. The Nisqually River delta at Puget Sound is largely within the Nisqually National Wildlife Reserve. In the study reach, between river kilometers 7.4 and 18.7, the river incised a valley approximately 70 m deep into the general land surface created by the last glaciation. The forested valley bottom is between 0.6 and 1.1 km wide, and includes numerous floodplain channels. The channel's average bankfull width is 100 m, the gradient ranges between 0.002 and 0.005, and it is upstream of tidal influence. The reach is unique in the Puget Lowland in that it has a valley-bottom floodplain with a forest exhibiting mature characteristics, and no artificial levees or bank protection. An historical investigation (G. Walter, Nisqually Indian Tribe, Natural Resources Department, oral communication, 2000) indicates that early logging was confined to removal of the larger conifers, which were cut into bolts and floated to a downstream mill. The United States Army annexed both sides of the river during World War I, and no logging is believed to have taken place since. Limited land clearing exists on private land in the lower river kilometer on the left bank (south) side.

The Skykomish and Snoqualmie rivers join to form the Snohomish River (drainage area 4,640 km^2), which flows westward 34 river kilometers to Possession Sound at the city of Everett (Figure 1). The gradient of the Snohomish River declines downstream from 0.0004 to 0.00006. Several major sloughs (Steamboat, Ebey, and Union sloughs) split from the main river, which is dredged to maintain its navigability in its lowest 5 river kilometers. Because of the river's low gradient, tidal backwater extends to about kilometer 29.

The North and South forks of the Stillaguamish River (1,770 km^2 drainage area) join at river kilometer 29 at the town of Arlington. From there the river flows westward to Port Susan and Skagit Bay. The Stillaguamish is steeper than the Snohomish; its gradient declines from 0.0009 to 0.0003. The Stillaguamish splits at kilometer 18 and rejoins at kilometer 10, which is the approximate upper limit of tidal influence. It then flows into Port Susan through Hat Slough. Before about 1906 at the head of Hat Slough, the river flowed instead through what is now a minor tidal slough past the town of Stanwood, into Port Susan through South Pass, and also northward into Skagit Bay through West Pass [*U. S. Army Corps of Engineers,* 1929]. Levees and bank armor are extensive on the Stillaguamish and Snohomish rivers, and agriculture dominates both valley bottoms. Riparian forests are generally either absent or narrow and dominated by small deciduous trees.

The Skagit River drains 8,270 km^2 of mountainous terrain in British Columbia and Washington's North Cascade Range, as it descends to its delta in the Puget Lowland. Downstream of the town of Mount Vernon, about 15 river kilometers from the river's mouth, the river splits into two major distributaries, the North Fork and South Fork. The river has built an extensive Holocene delta. Immediately following glaciation, it is likely that glacial-era sediments in the drainage basin eroded rapidly [*Church and Slaymaker, 1989*; *Benda et al.,* 1992; *Beechie et al,* this volume] and deltaic sedimentation and progradation were rapid. Volcanic events have occurred at Glacier Peak at least nine times within the past 5,500 years [*Beget*, 1982], and deposits from Glacier Peak volcano probably account for a significant portion of the delta. The much smaller Samish River flows on the northern part of the greater Skagit River delta.

PRIMARY ARCHIVAL SOURCES

Government Land Survey Records

The General Land Office (GLO) conducted its cadastral survey of the Puget Lowland between about 1850 and 1880, which in nearly all river valleys was prior to widespread building of sea or river dikes and prior to stream clearing and floodplain logging. Thus the GLO records are an essential source for characterizing pre-European settlement riverine conditions. Field notes include several types of information on natural vegetation, which botanists have used in numerous studies since at least the 1920s [e.g., *Sears,* 1925] to reconstruct pre-settlement vegetation cover [for reviews see *Whitney*, 1996, *Whitney and DeCant,* 2001 ; for recent examples see *Galatowitsch*, 1990; *Nelson et al.,* 1998, and *Radeloff et al.,* 1999]. The same information is also useful for characterizing riparian and valley bottom forests, including the size and species of recruitable wood [*Collins et al.,* in press], and for mapping and characterizing riverine wetlands [*North and Tevarsham,* 1984] and changes to channel widths [*Knox,* 1977].

"Bearing" or "witness" tree records from GLO field notes are one important data source. At reference points at the corners of mile-square sections and half way between corners ("quarter section" points), surveyors measured the distance and the compass direction to several nearby trees. At section corners, surveyors were instructed to identify four witness trees, and two at quarter section boundaries. If there were no trees nearby, surveyors built a mound of earth. In their field notes, surveyors recorded the diameter and common name of each witness tree, and the distance and bearing to it. In addition to these regularly-spaced points, surveyors also established "meander corner points" at the banks of navigable rivers and sloughs where section lines intersected them and documented two bearing trees. These meander points allow us to separately characterize streamside trees from valley-bottom trees.

Notes on bearing trees are useful for characterizing the size and species frequency of riparian and floodplain trees. Because instructions issued by the GLO evolved over time [see *White,* 1991 for a compilation of instructions given to federal land surveyors], to account for potential biases it is important to interpret field notes for a given study area in light of instructions given to surveyors in that area and at that time. For example, the published criteria used to select trees varied with time, which in turn might differ from actual practice [see *Collins and Montgomery*, in press for details on interpreting field notes in the Puget Lowland]. One important bias in characterizing species frequency and size results from bearing trees being greater than 7.5 cm in diameter. This means that bearing tree records under-represent smaller-diameter species (e.g., vine maple [*Acer circinatum*] and willow *[Salix spp.*]). On the other hand, bearing tree records would accurately characterize species frequency based on basal area (the percent of the sum of cross-sectional area of all trees accounted for by the cross-sectional area of any one species). To evaluate this hypothesis, in the Nisqually River valley we relocated survey points for which bearing trees were recorded by land surveyors in 1873 [see *Collins and Montgomery*, in press]. On Indian Reservations, points were established at 1/16-corners, including within sections [*White,* 1991]. At each of 26 points, we established bearing trees following our interpretation of the instructions to surveyors in effect for the original survey. In addition, we also recorded species and diameter of all trees larger than 0.01 m in diameter in a 314-m^2 plot centered at the survey point. Comparing the bearing tree data to the plot data confirmed that bearing trees can provide a good estimate of species abundance by basal area.

Measured distances from survey points to bearing trees are also useful for mapping the spacing or density of tree cover; for example, in delineating savanna areas [*Radeloff et al.,* 1999] or wetland types. In addition to recording witness trees, surveyors were instructed to make extensive notes of land and water features they encountered, including major changes to vegetative community, streams and marshes, and the width of all "water objects." Springs, lakes and ponds and their depths, the timber and undergrowth, bottomlands, visual signs of seasonal water inundation, and improvements were also to be noted along section lines. In practice, the completeness of this information varies from surveyor to surveyor, but it provides important secondary data for interpreting the historic landscape, particularly in describing wetlands. For example, the date at which surveyors' observations of water depth were made and their notes on indicators of seasonal water depths makes it possible to characterize summer and winter water depths in wetlands.

The maps GLO surveyors drew from their field notes reflect field data along section boundaries where surveyors walked; except where major channels were meandered, mapping in the interior of sections cannot be considered accurate. In addition to channels, the GLO map information includes wetlands, which in many cases show indeterminate boundaries within the interior of sections, line fragments showing the edge of forested or cultivated fields, and some in-channel wood accumulations.

Other Archival Sources

The U. S. Coast and Geodetic Survey (USC&GS) charted the coastline and coastal rivers, generally to the upstream limit of navigation, often a few tens of kilometers inland. Charts also show the adjoining vegetation type, such as forest, salt marsh, freshwater marsh, and cultivated fields. In eastern Puget Sound, some early, less-detailed charts were made in the 1850s, with more detailed and accurate charts made in the late 1870s to late 1880s at a scale of 1:10,000 or 1:20,000. These charts were in most cases created after widespread tidewater diking. These charts are the basis for earlier estimates of estuarine wetlands in Puget Sound [*Bortleson et al.,* 1980], which do not take into account wetland areas diked prior to the earliest USC&GS mapping.

During the period the USC&GS charted the coastline, the predecessor to the U. S. Army Corps of Engineers (USACOE) began in 1876 filing annual reports on investigations of rivers in western Washington [*Annual Reports of the Chief of Engineers, U.S. War Department;* hereafter abbreviated *U. S. War Department*]. These descriptions of river conditions generally highlighted wood, which often completely blocked rivers to steamboat navigation. Beginning in 1880, engineers began clearing this wood; by the end of the decade, they engaged in a regular program of "snagging" that continues to this day. Annual reports on snagging and other river improvement programs in most years include the number and size of snags removed. Other useful sources of historical information include various government reports [e.g. *Nesbit et al.,* 1885; *Mangum et al.,* 1909]; settlers accounts; contemporary histories [e.g., *Interstate Publishing Company,* 1906]; photographs (the earliest useful photographs we have located are from the 1880s); aerial photos (beginning in about 1930); and U. S. Geological Survey (USGS) topographic maps, starting in the 1890s.

BASIN-WIDE DISTRIBUTION OF RIVERINE HABITATS

Within the watersheds draining the western Cascade Range, the majority of channel and wetland area accessible to salmonids was in the larger rivers and floodplains of Puget Lowland [*Sedell and Luchessa,* 1981; *Beechie et al.,* 1994; this volume]. This is known primarily from archival research into lowland habitats, in combination with contemporary field studies on headwater tributaries.

The historic abundance of habitat in lowland floodplains and deltas can only be determined from archival materials, because so little of these habitats remain. For example, in the Snohomish River valley in approximately 1860, prior to significant landscape modifications by settlers, the majority of land area was either channel or wetland (Plate 1A). The GLO records and USC&GS charts are the primary sources for this interpretation. By allowing sometimes spatially-fragmented archival information to be generalized by topography or elevation, contemporary DEMs created from LIDAR or from detailed topographic surveys help increase spatial resolution (see the section "Floodplain Habitats" later in this paper for more detail on methods for Plates 1 and 2). Vast floodplain wetlands and extensive estuarine marshes accounted for nearly two-thirds (62%) of the valley bottom. By the 19th century's end, much of this wetland had been diked, ditched, and drained (Plate 1B); by the middle of the 20th century, only patches remained (Plate 1C). In the Skagit-Samish delta and Stillaguamish estuary, extensive wetlands (Plate 2A) were diked even sooner than in the Snohomish River estuary. Dikes had isolated the vast majority of wetlands by the end of the 19th century (Plate 2B), and extensive distributary channels in the Skagit estuary were closed off by the middle of the 20th century (Plate 2C).

Although the overall character of these three adjacent estuaries is similar, the amount and type of habitat in each differed significantly, because of differences in landforms and geologic histories. Estuarine wetland, including abundant blind-tidal and distributary channels, dominated deltas of the Skagit-Samish river systems (Figures 2A and 2D; in Figure 2, estuarine channels with characteristics of blind-

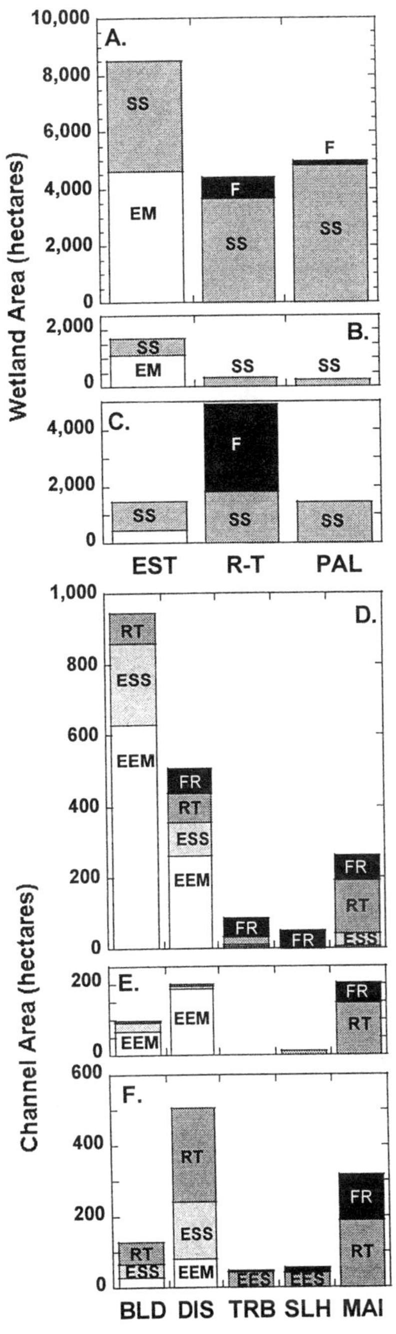

Figure 2. (A, B, C): Historic (~1860) wetlands in the (A) Skagit-Samish delta; (B) Stillaguamish estuary; (C) Snohomish River valley. Wetland type: EST: estuarine; R-T: riverine-tidal; PAL: palustrine. Vegetation type: EM: emergent; SS: scrub-shrub; F; forested. (D, E, F): Area of channels, within different zones defined by vegetation and hydrology, in the (D) Skagit-Samish delta; (E) Stillaguamish estuary; (F) Snohomish River valley. Channel type: BLD: blind tidal; DIS: distributary; TRB: tributary; SLH: slough; MAI: mainstem. Habitat types: EEM: estuarine emergent; ESS: estuarine scrub-shrub; RT: riverine-tidal; FR: freshwater.

tidal channels but also receiving some distributary flow are combined with blind channels). This dominance presumably reflects the spreading-delta geometry, which creates a large amount of land in contact with saltwater, and the low-gradient of a large delta, which leads to extensive intertidal inundation. The Snohomish River, by contrast, has a large area of freshwater distributary and blind channel (Figure 2F) and extensive riverine-tidal wetlands (Figure 2C). (We use "riverine-tidal" [*Cowardin et al.*, 1985] to refer to wetlands created by tidal backwater effects). This abundance presumably reflects the very low gradient of the Snohomish River valley, which was eroded by Pleistocene subglacial runoff. In addition, because of confining valley walls, the absence of a divergent delta reduces the amount of estuarine channel relative to the Samish-Skagit delta. Finally, the lower Stillaguamish River, which flows in a relatively steep and confined valley, is dominated by a relatively small amount of estuarine wetland and channels (Figures 2B and 2E). Differences in these three adjacent estuaries demonstrate the important role of archival sources in characterizing the abundance and variation of aquatic habitats, and that archival investigation of each is necessary to characterize specific distributions and quantities of habitats and associated controlling variables.

RIVER MORPHOLOGY AND PROCESSES

Archival sources provide snap-shot-in-time descriptions of river morphology and the variables that influence it, but causality between associated elements can only be inferred. Field studies help develop causal models, and archival studies, in turn, help to assess the adequacy of these models, and their applicability to different rivers. Because nearly all rivers have been modified and isolated from their floodplains, archival materials are important for inferring river pattern and the nature of river-floodplain connections.

River Pattern and Floodplain Interaction

Maps by the GLO and earliest USGS topographic maps show that many rivers in the Puget Lowland currently having a single channel had an anastomosing pattern historically with multiple main channels and abundant floodplain sloughs. The Nisqually River at present (and historically) exemplifies this pattern (Figure 3A); field and aerial photo analyses show that wood jams are instrumental to creating this pattern. Jams promote and regulate channel avulsion, according to interpretations of sequential aerial photographs. For example, in the Nisqually study reach, jams can divert flow or plug channels, switching flow from one channel to another. Wood jams also appear responsible for maintaining floodplain sloughs. For example, in the 11-km-long study reach of the Nisqually River, 18 floodplain sloughs receive water from the main river during low-flow. Each of these floodplain channels has a wood jam associated with its inlet; Figure 3A shows some of these sloughs and associated jams in part of the study reach. In each case, the jam acts to regulate the amount of flow into the channel, which is typically a former main channel, and prevents the main flow from reoccupying the channel.

The river pattern of the lower Stillaguamish River on 19th and early 20th century GLO and topographic maps was similar to that of the (current and historic) Nisqually River. In the Stillaguamish, GLO field notes and annual reports by the Army Engineers [*U. S. War Department,* 1881] describe or map wood jams associated with many floodplain sloughs. Beginning in the 1920s and primarily in the 1930s, levees were constructed along the Stillaguamish, and floodplain sloughs were disconnected from the river and in many cases drained. At present, the river has a single channel, except for one remaining island, and no floodplain sloughs (Figure 3B). Floodplain channels present on maps from 1930 and 1941, but which now exist only as topographic depressions or ditches unconnected to the river, are mapped in Figure 3B with dashed lines to illustrate the historic river pattern and the Stillaguamish's historic similarity to the current pattern of the Nisqually River.

Early maps show that not all Puget Lowland rivers had a multiple-channel, branching pattern. For example, the lower-gradient Snohomish River appears typical of a second type of river plan form characterized primarily by a single, meandering channel. Early maps of the Snohomish show oxbow lakes typical of meandering river systems (Plate 1A), landforms absent from the valleys of anastomosing rivers. The pattern of the Nisqually River suggests that the meandering process defines the extent of the migration belt in anastomosing rivers, but that within the area defined by the meander train, jam-caused avulsion is frequent. The switching of channels creates a network of main channel branches and floodplain sloughs. By contrast, meandering rivers such as the Snohomish apparently migrated slowly enough to create vast floodplain marshes (e.g., Plate 1A) and were characterized by meander-cut-off avulsion typically associated with meandering rivers. Such primarily single-thread, meandering rivers appear to have been typical of river valleys created by late-Pleistocene glacial meltwater channels; the Snoqualmie, Duwamish, and Green rivers are also examples. Wood jams in meandering rivers may not have served the channel-splitting and channel-switching functions to the same extent as in anastomosing rivers. However, historic maps [e.g., the Skagit River; *USACOE*, 1898] suggest that jams maintained inlets to floodplain channels in both types of rivers.

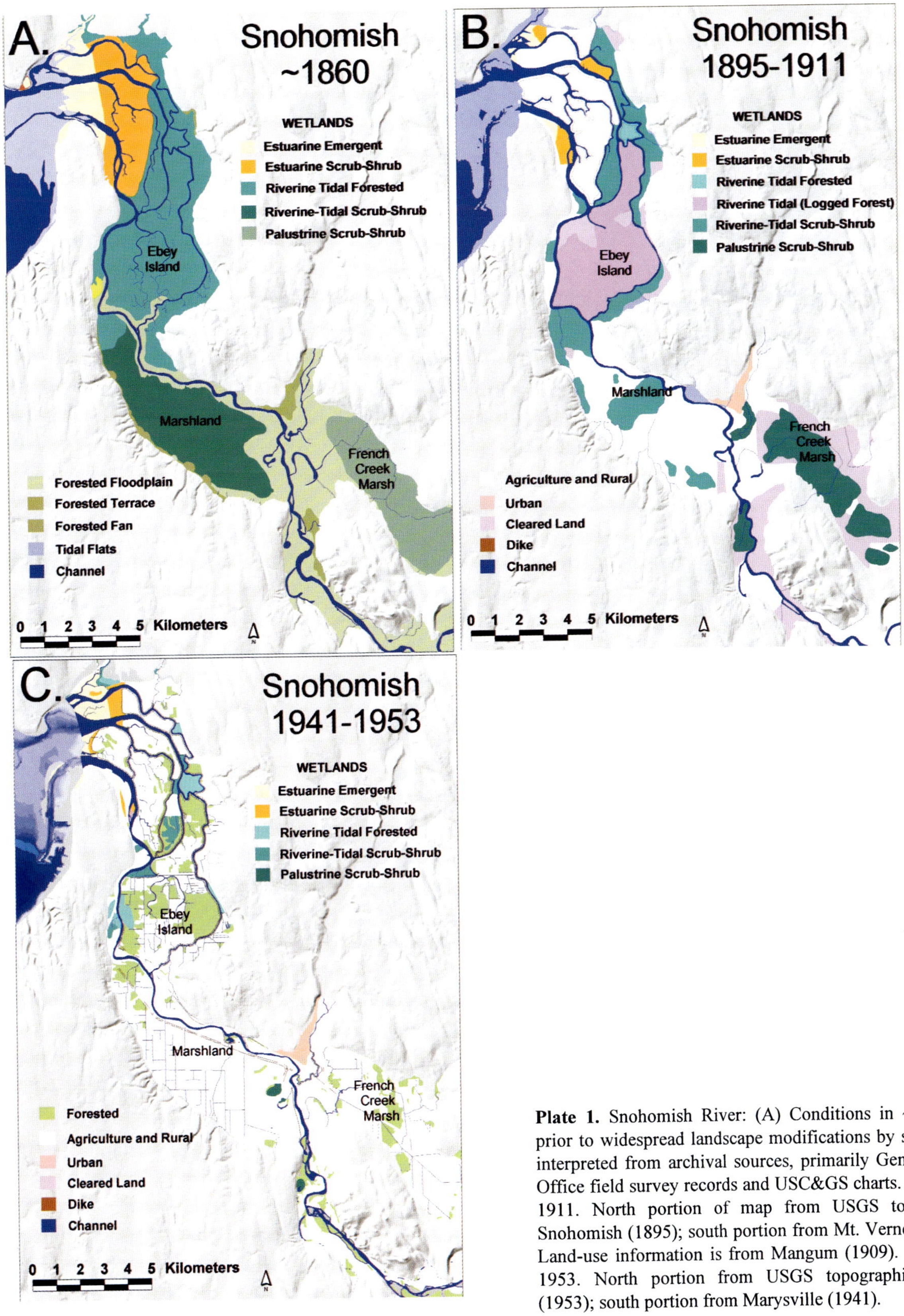

Plate 1. Snohomish River: (A) Conditions in ~1860, or prior to widespread landscape modifications by settlers, as interpreted from archival sources, primarily General Land Office field survey records and USC&GS charts. (B) 1895-1911. North portion of map from USGS topographic Snohomish (1895); south portion from Mt. Vernon (1911). Land-use information is from Mangum (1909). (C) 1941-1953. North portion from USGS topographic Everett (1953); south portion from Marysville (1941).

Figure 3. (A) The Nisqually River, from 1999 aerial photographs. Those floodplain channels having flowing water in summer 2000, and which are obscured by tree cover on the aerial photographs, are shown with gray, and were mapped from field work in 1998 and 2000 onto 1999 1:12,000-scale ortho-photographs. Jams shown with black. "J" indicates a jam that is associated with the inlet to a floodplain slough. (B) The Stillaguamish River, from 1990 aerial photographs. Dashed lines indicate relict floodplain sloughs that are no longer present, but were shown on 1930 and 1941 maps. Flow in both panels is from right to left.

While field studies help to develop models of river-floodplain dynamics, regional variation revealed by archival sources shows there is not a single model for all rivers. The archival perspective helps to indicate which model is appropriate for a particular river.

Floodplain Habitats

Interpretations of historic habitats on the Skagit, Stillaguamish, and Snohomish deltas illustrate approaches to mapping floodplain habitats from archival materials. Identifying relict sloughs that have been disconnected from river or tidal access by dikes is relatively straightforward. The GLO maps, USC&GS maps, USGS topographic maps, and early aerial photographs are primary sources of information. However, mapping and characterizing wetlands requires additional analysis of historic sources.

In cases where USC&GS charts were made prior to widespread diking, as in the Snohomish River estuary, wetland extent is easily determined. However, it is also useful to delineate different wetland types, because different vegetation types in wetlands are surrogates for differing habitat conditions such as salinity and hydrologic regime [*Hayman et al.,* 1996], and the amount of channel area differs with wetland type. Map symbols distinguishing freshwater wetlands from salt marsh, and forested versus scrub-shrub and emergent vegetation, are subtle or ambiguous on the Snohomish USC&GS map. Information on the species and spacing of trees from the GLO notes supplement the USC&GS mapping for delineating estuarine wetlands in the Snohomish. Moreover, on the Skagit and Stillaguamish deltas, which are more typical in having been diked prior to USC&GS mapping, GLO notes are the primary information source. In these cases we used the boundary between forested and cultivated areas mapped by the USC&GS as a first approximation of the landward limit of scrub-shrub estuarine wetland, assuming that settlers would have preferred to homestead in emergent and scrub-shrub areas instead of clearing adjacent dense forest. Observations by surveyors in the GLO notes confirm this as the boundary between forest and estuarine wetland (frequently referred to as "tidal prairie"). A digital elevation model made from detailed mapping by the Army Corps of Engineers allowed generalization by elevation, providing finer-scale spatial detail. The mapped extent of estuarine wetland on the Skagit delta is also confirmed by the width of remaining, modern emergent zone (approximately 1km) and the local land slope.

The distance to bearing trees from GLO survey points helps differentiate scrub-shrub wetland from adjacent forested wetlands having more-closely-spaced trees (Plate 3A). Locations where GLO surveyors built mounds because no woody vegetation was nearby also delineate emergent from scrub-shrub estuarine wetlands (Plate 3A). The distribution and frequency of tree species is also useful for delineating wetland types. For example, witness tree records for the Snohomish estuary show "yew" (western yew, *Taxus brevifolia)* and "juniper" (common juniper, *juniperus commonis*, or Rocky Mountain juniper, *Juniperus scopulorum*) only in the estuarine scrub-shrub and riverine-tidal forested wetland area (Plate 3B). In the same valley, pine (probably shore pine, *Pinus contorta,* or western white pine, *Pinus strobus*) clusters in the riverine-tidal scrub-shrub and palustrine scrub-shrub wetlands. Spruce (Sitka spruce, *Picea sitchensis*), while widely distributed, are common in the estuarine wetlands, and species such as "maple" and "cottonwood" (bigleaf maple, *Acer macrophyllum,* black cottonwood, *Populus trichocarpa*) occur only in the freshwater portions of the valley (Plate 3B). Tree diameters also show characteristic patterns useful for delineating habitats (Plate 3C). Large-diameter trees are limited to the six species shown in Plate 3C. Spruce occurs conspicuously as the only large tree in the estuarine-scrub-shrub and riverine-tidal-forested wetlands. Cedars (western redcedar, *Thuja plicata*) are the largest trees, and occur almost exclusively in the riparian and floodplain areas of the freshwater and upper half of the tidal-freshwater areas of the valley (Plate 3C). These areas are also characterized by a diversity of large-diameter species (including *Acer macrophyllum* and *Populus trichocarpa*), unlike elsewhere. Finally, the riverine-tidal scrub-shrub and palustrine scrub-shrub wetlands have few large trees of any species.

Water depths recorded by GLO surveyors are an additional indicator useful for mapping and characterizing wetlands and their aquatic habitat. For example, the "Marshland" area was a 2,000-ha riverine-tidal scrub-shrub wetland on the Snohomish River floodplain (Plates 1A and 3). Based on bearing trees and general descriptions in the GLO notes, Marshland appears to have been a patchwork of scattered, sparse-tree-covered areas, willow-hardhack *(Spiraea douglasii)* shrub thickets, and open marsh. Much of Marshland was flooded with a few feet of water at the time of the GLO survey in February, 1871 (Plate 3A) Of 23 survey points, the field notes indicate the depth of water at 11 points, where depth averaged 0.67 m; at an additional two points the water was too deep for access. Five additional points were too wet to build a mound, and three more points were described as "swamp." Thus, between 13 and 21 of 23 points had standing water. The water at three points was greater than a meter in depth, and the two points too deep to access were presumably deeper, meaning that at least five points were deeper than 1 m. Most points were described as "subject to overflow" to depths greater than the water that was present at the time of the survey. The

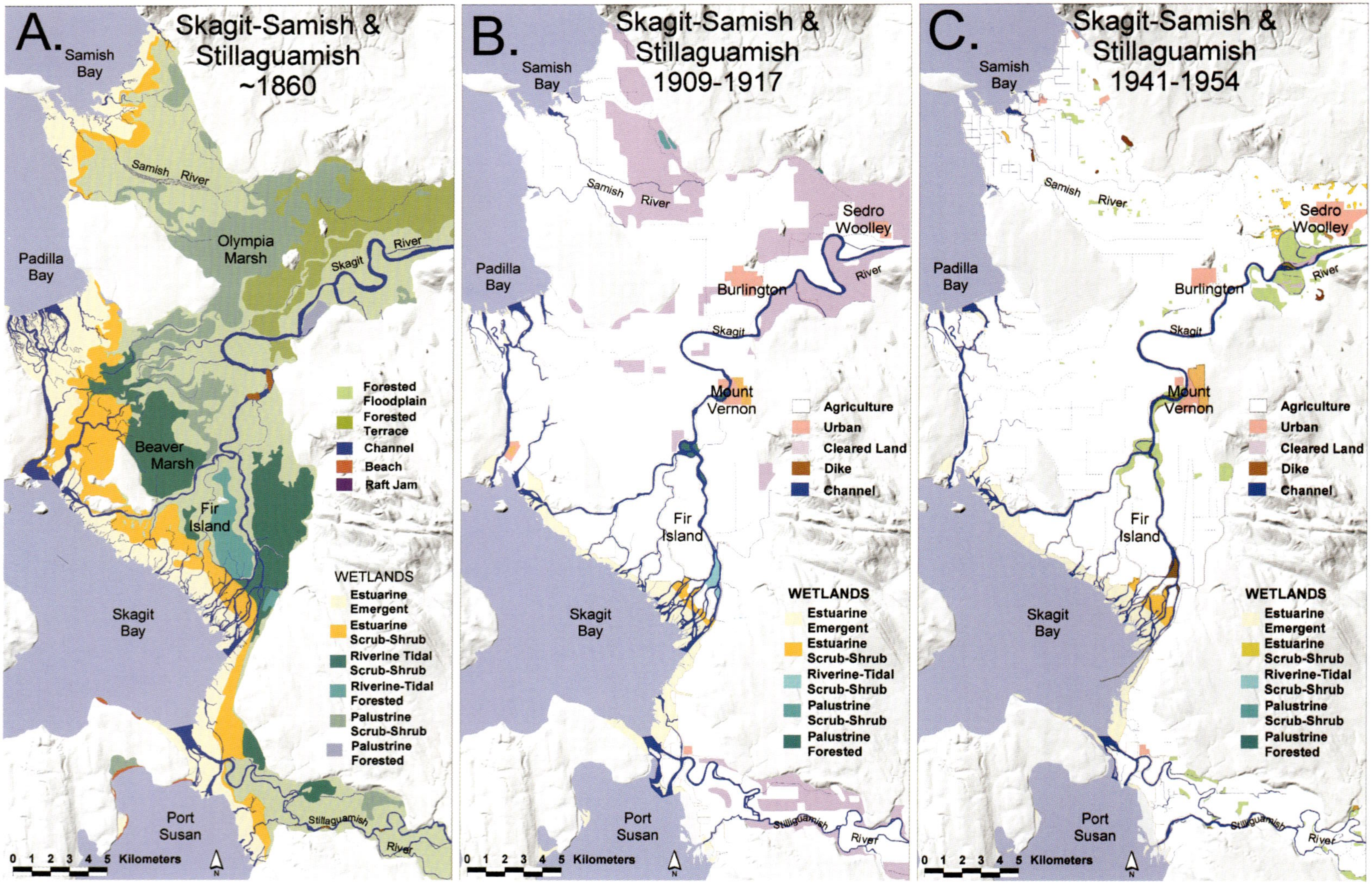

Plate 2. Skagit-Samish Delta and lower Stillaguamish River: (A) Conditions in ~1860, or prior to widespread landscape modifications by settlers, as interpreted from archival sources, primarily General Land Office field survey records and USC&GS charts. (B) 1911-1917, from USGS topographic maps Samish (1917) and Mt. Vernon (1911). Land-use information is from Mangum (1909). (C) 1941-1954. From: USGS topographic maps Samish (1954), Mt. Vernon (1943), Stanwood (1943), Marysville (1941), and 1937 aerial photographs.

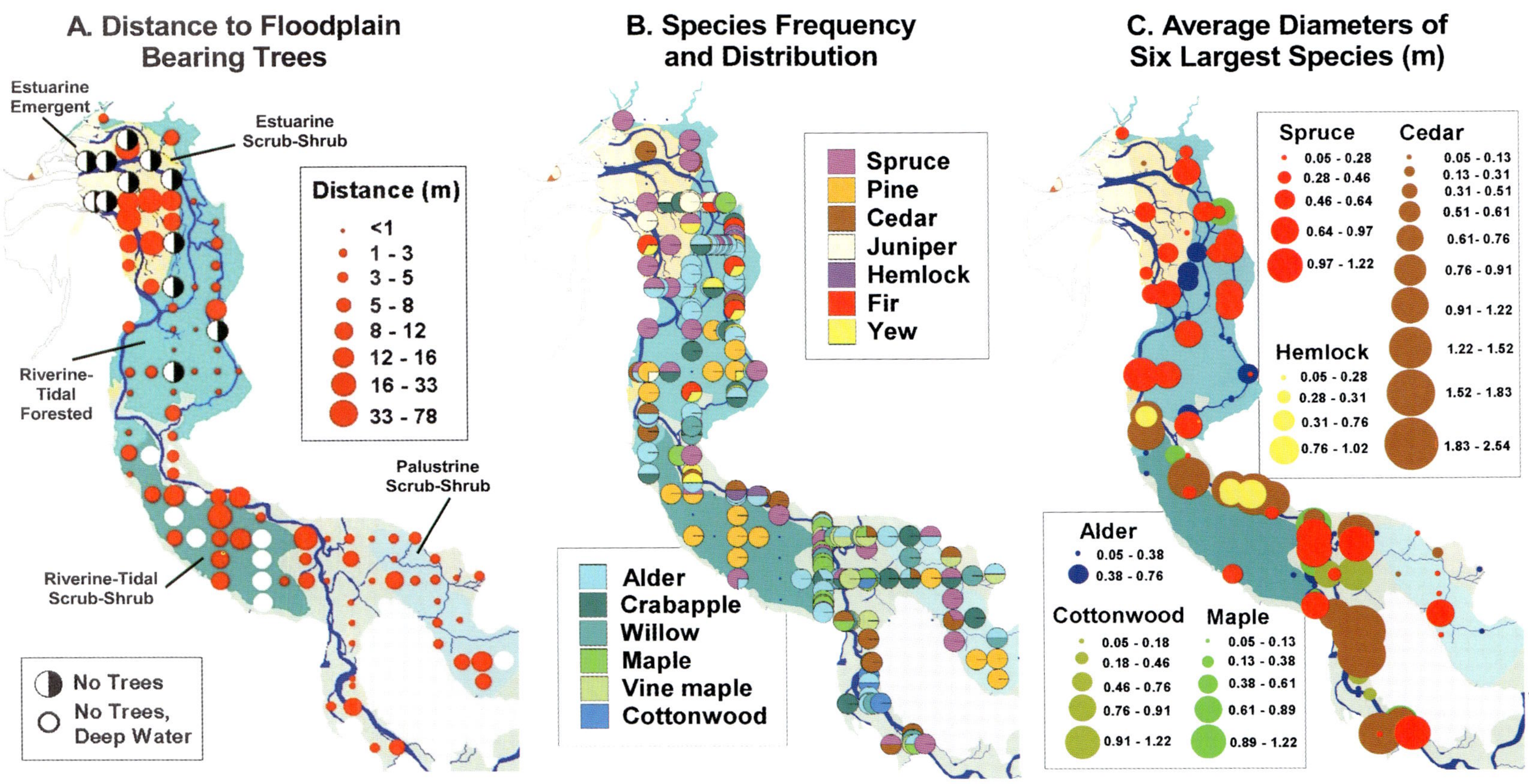

Plate 3. Snohomish River valley. (A) Average distance to "valley bottom" trees in 1873 GLO notes. (B) Distribution and frequency of tree species. (C) Average diameter of six largest-diameter species.

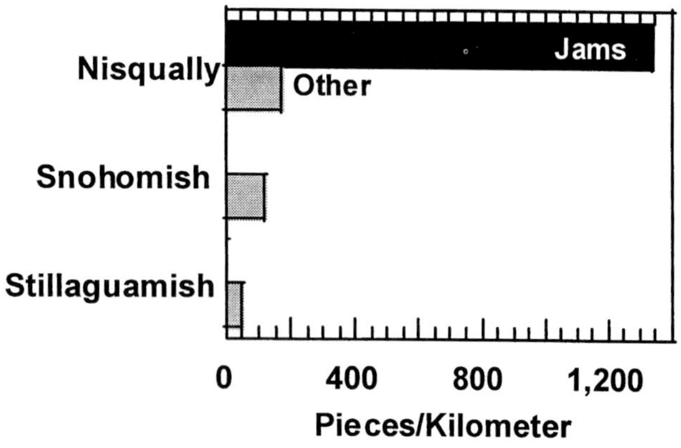

Figure 4. (A) Distribution of wood pieces in the sampled reach of the Nisqually River in 1998-1999, (B) the Snohomish River (C) and Stillaguamish River study reaches.

published map shows "subject to overflow 2 to 6 ft" (0.6 to 1.8 m). The extent of summer inundation is not known directly, but might be indicated by the widespread absence of trees and the prevalence of scrub willows. Additionally, the soils in about one-fifth of the area of Marshland are mapped as the Mukilteo Muck [*Debose and Klungland,* 1983]; organic soils developed under sedges and rushes. Earlier soils mapping [*Mangum,* 1909] describes most of the Marshland area as "muck and peat."

The GLO map shows a second large marsh in the Snohomish River valley, the "French Creek Marsh," as 1,400 ha on the north side of the valley, upstream of the town of Snohomish (Plate 1A). This more densely vegetated marsh had a different hydrologic regime and source of flooding than Marshland. It was surveyed in late July and August of 1866; because July and August is in the region's dry season, the recorded water depths can reasonably be taken as summer, minimum-water-level conditions. Several descriptions indicate that beaver dams flooded the area. For example, descriptions include the following: "...swampy and generally overflowed to the depth of 12 inches in consequence of beaver dams...", and "...mostly overflowed now to the depth of from 4 to 8 inches...liable to annual inundation of 36 inches...." Overall, the GLO notes suggest the marsh experienced shallow summer flooding (less than 0.3 m) from beaver dams, and periodic winter inundation of about 1 m. It appears that most of the marsh was flooded by beaver dams in the summer. Soils in about one-third of the marsh (31%) are mapped as Mukilteo Muck [*Debose and Klungland,* 1983], which as indicated previously is a very deep soil formed "in organic material derived dominantly from sedges."

In-Channel Wood

Our approach to investigating historic functions of wood in rivers was to collect field data from the Nisqually River and similar data from the Snohomish and Stillaguamish rivers. We also used archival sources to determine that field data from the Nisqually is a reasonable surrogate for historic conditions [*Collins et al.,* in press]. Archival materials additionally provide information on wood accumulations for which there are no longer existing analogs, as well as information on geographic variation in wood characteristics.

We inventoried by boat in the three rivers in September-October 1998 [for detail see *Collins et al.*, in press]. The Nisqually River had much more wood per channel width than the other two rivers—approximately 8 and 21 times more than the Snohomish and Stillaguamish, respectively (Figure 4). Wood in jams accounts for most of the difference; excluding jams from the Nisqually wood count, the amount was comparable to the other two rivers. We suspect that few jams occur in the Snohomish and Stillaguamish rivers for two reasons. One is the absence of long, large-diameter pieces with rootballs, which in the Nisqually River act as key pieces that initiate and stabilize jams (Figure 5A). The large wood pieces having rootballs

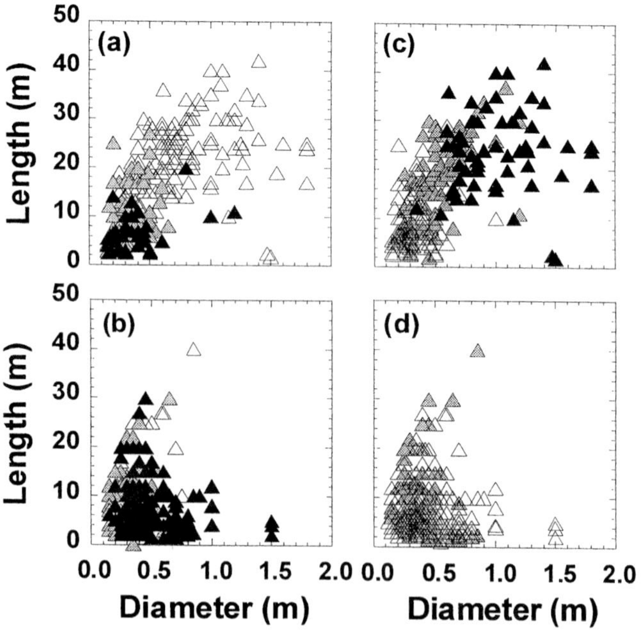

Figure 5. Wood dimensions in the Nisqually (A, C) and Snohomish (B, D) rivers, 1998. (A, B): solid symbol: no rootball; gray symbol: rootball present; hollow symbol: key piece with rootball. (C, D): solid symbol: recently recruited; gray symbol: intermediate age; hollow symbol: older.

are no longer available to the Snohomish and Stillaguamish rivers, because they have lacked mature riparian forests for more than a century [e.g., *Plummer et al.,* 1902]. Another reason is the two rivers recruit far less wood than the Nisqually, because the leveed rivers cannot erode their floodplain, which also generally lacks a riparian forest. The presence of two upstream dams on the Nisqually River makes that river's accumulation of wood all the more striking, and points to the importance of local wood recruitment. In contrast, neither the Stillaguamish nor Snohomish is dammed, and thus have no limit on wood transport from upstream.

Very little recently recruited wood is found in the Stillaguamish and Snohomish rivers, compared to the Nisqually (Figure 5B). Most older wood in the Stillaguamish and Snohomish is decay-resistant cedar, presumably relict from before forests were cleared a century ago. This reflects, in part, a decrease in wood recruitment from historic conditions. Also, without jams, the rivers retain far less wood. The lack of retention is reinforced by recently recruited wood being small in diameter and readily transported.

To supplement these field data, one quantitative indicator of historic wood abundance in regional rivers, and its change through time, is the record of river snagging by the Army Corps of Engineers (Figure 6). Nearly all snagging occurred after huge raft jams had been cleared, and much of the riparian forest had already been cut down, so the number of snags recorded would be less than the number that would have been present under pristine conditions. Nonetheless, snags were abundant; for example, 30,000 snags were removed from the lower Skagit River between 1898 and 1908. Between 1880 and 1980, 150,000 snags were removed from five rivers, including the Stillaguamish and Snohomish, with more than one-half of these from the Skagit. A diminishing rate of snag removal after 1900 (evident in Figure 6) reflects the decline in recruitment of wood large enough to lodge in the riverbed and remain stable. This in turn presumably reflects the effects of riparian logging (and particularly the removal of very large trees from the valley-bottom forest), leveeing, and bank protection. The much greater removal rate from the Skagit River than the Stillaguamish and Snohomish implies a large inter-river variation in wood load.

We know of two quantitative measures of wood size in Puget Lowland rivers. First, snag-boat captains' records include the cumulative length of wood removed from Puget Sound rivers and the largest and smallest diameter of snags. These records indicate that very large pieces were represented in the wood load—the annual maximum snag diameter between 1889 and 1909 ranged from 3.6 to 5.3 m [*U. S. War Department*, 1889-1909]. These great diameters are confirmed by engineers' observations [e.g., *U. S. War*

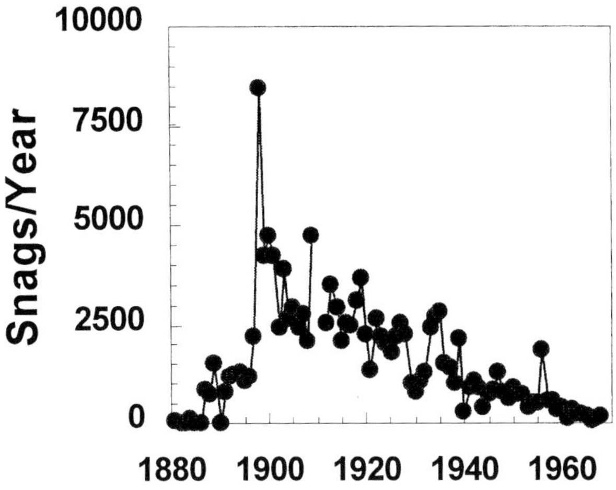

Figure 6. Snags removed per year in Puget Sound rivers. The accounting year runs from July 1 to June 30.

Department, 1895]. The second, and more extensive data source, is from the GLO field notes. Bearing trees recorded on floodplains represent the size of wood that would over a period of time be recruited to the river. Records include a significant number of large-diameter trees, including many individuals of at least six species large enough to form key pieces in jams (Figure 7). For example, in the Nisqually, Stillaguamish, and Snohomish rivers valley bottoms, the largest 40%, 20%, and 30% of bearing trees were greater than 0.5 m in diameter, and 15%, 7%, and 7% were larger than 1.0 m in diameter. Trees were smaller in immediate streamside areas. Comparing these tree size distributions to the minimum diameter of wood that form key pieces in the Nisqually, about 40% of witness trees on the Nisqually valley bottom and 30% of stream-side witness trees were large enough to potentially form key pieces.

None of the largest wood accumulations that existed historically remain to be studied in the field. Kilometers-long raft jams formerly plugged channels, sometimes for hundreds of years. For example, the Mount Vernon Jam had existed for at least a century. A pioneer had learned from the Indians that its surface supported live trees two to three feet in diameter [*Interstate Publishing Company,* 1906, p. 106]. The jam was packed solidly enough that it could be crossed "at almost any point." The jam was described as 9 m deep, consisting of "from five to eight tiers of logs, which generally ranged from three to eight feet in diameter" [*Interstate Publishing Company,* 1906]. Beneath the previously described Mt. Vernon raft jam were in some places "furious cataracts," and in others "deep black pools filled with fish" [*Interstate Publishing Company*, 1906]. The river was as deep as 7 m below the jam at the lowest water stage. Extensive distributary-plugging drift jams

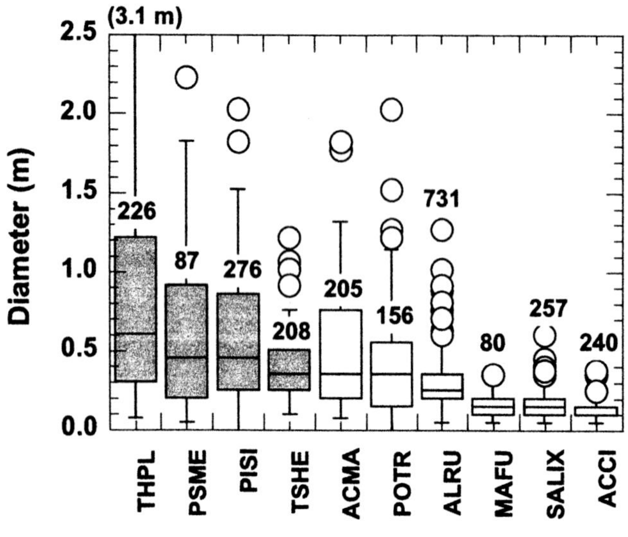

Figure 7. Diameters of bearing trees in the valley bottoms of the Snohomish, Stillaguamish, and lower Skagit rivers, from General Land Office field notes. Numbers are sample size. THPL = western redcedar *(Thuja plicata);* PSME = Douglas fir *(Pseudotsuga menziesii);* PISI = Sitka spruce *(Picea sitchensis);* TSHE = western hemlock *(Tsuga heterophylla);* ACMA = bigleaf maple *(Acer macrophyllum);* POTR = black cottonwood *(Populus trichocharpa);* ALRU = red alder *(Alnus rubra);* MAFU = Pacific crabapple *(Malus fusca);* SALIX = willow spp. *(Salix spp.);* ACCI= vine maple *(Acer circinatum).* Each box encloses 50% of the data with the median value displayed as a line. The lines extending from the top and bottom of each box mark the minimum and maximum values, excepting outliers, or points with values greater than the inner quartile plus 1.5 times the inner two quartiles. One western redcedar with a diameter of 3.1 m plots outside the graph.

were integral to channel dynamics on deltas [*Collins et al.*, in press]. Contemporary accounts and map evidence (Plate 2A) suggest the dominant role of the Mt. Vernon raft jam in causing landscape-scale flooding patterns on the lower Skagit River [see *Collins et al.*, in press]. In the absence of contemporary field examples, hydraulic modeling would be necessary to establish the effects of such wood accumulations.

Wood and Pools

Comparing pools in the Nisqually River with other rivers from which wood has been removed, levees constructed, and floodplain forests cut down, is the best available indication of how pool characteristics have changed in Puget Lowland rivers. In autumn 1998, we measured 85 pools in the Nisqually River study reach, a pool spacing equivalent to 1.4 channel widths (CW) per pool. Wood was the dominant factor forming 61% of pools, including 26% associated with mapped, stable jams [see Figure 10 in *Collins et al.*, in press]. This finding is similar to that of *Abbe and Montgomery* [1996, Figure 3] who found wood forming 70% of observed pools in a 25-km-long reach of the Queets River in Olympic National Park. In the Nisqually, pools associated with jams were considerably deeper than other pools, the mean depth being three times greater than free-formed pools. Jam-associated pools were twice as deep as pools formed by individual pieces having attached rootballs, augmented by wood, or formed by banks.

In the Stillaguamish River, we measured 84 pools in three sample reaches. In contrast to the Nisqually, the pool spacing ranged between 3 and 5 CW/pool in the three reaches, or two to three times less frequent than in the Nisqually. Only one-ninth (11%) of the pools in the Stillaguamish were formed by wood. More than one-half formed along riprap-armored banks. These pools also were deep; however, these pools lacked cover, and provided considerably less habitat value than pools associated with wood [e.g. *Beamer and Henderson*, 1998]. In the Snohomish River, we measured 18 pools in an 8.5-km-long reach, beginning at the confluence of the Snoqualmie and Skykomish rivers. Wood created only one relatively shallow pool. Riprap banks were the most common pool-forming factor, and also formed the deepest pools. One riprap-bank-formed pool was 7 m deep. The pool spacing of 3 CW/pool was twice that of the Nisqually River, indicating one-half as many pools. Comparing pool data from the Snohomish and Stillaguamish rivers to the Nisqually indicates the transition from a freely migrating river with mature floodplain forest to a leveed river with little riparian recruitment reduced the number of pools by two to three times.

APPLICATION TO RIVER MANAGEMENT AND RESTORATION PLANNING

The historic abundance of lowland habitat, and the greater losses in habitat that have occurred in lowland areas [*Beechie et al.*, 1994; *Beechie et al.*, this volume] indicate that habitat restoration must include lowland floodplains and rivers. Moreover, it is possible that these mostly vanished or badly degraded habitats, if restored or rehabilitated, could more logically serve as refugia for salmonids than disturbance-prone headwater areas, as commonly argued in conservation planning [e.g., *Frissell et al.*, 1993; *Doppelt et al.*, 1993]. Restored or rehabilitated lowland floodplain habitats might be appropriately considered as future refugia not only because of their historic quantitative importance, but also because they are buffered from upstream watershed disturbances (e.g., large sediment and water influxes). The potential to recoup these inherently buffered habitats also challenges current thinking about the sequencing of watershed restoration measures. Watershed restoration professionals often emphasize the need to first restore headwater processes, such as reducing erosion associated with logging roads, prior to undertaking restoration projects downstream. However, this may not be the only rule of thumb applying to large watersheds having low-gradient, wide-floodplain rivers in the lower part of the watershed, as in the eastern Puget Sound area.

To a large extent this habitat was created by a dynamic river-floodplain connection. Many factors could complicate reconnecting historic floodplain habitats with the river. For example, river incision would complicate reconnecting floodplain sloughs. Comparing channel bed surveys in the Stillaguamish River in the vicinity of several floodplain sloughs currently proposed for reconnection shows that the river downcut by 1 to 2 m between 1929 to 1937 (before the sloughs were disconnected) and 1991. Elsewhere, many former floodplain sloughs and distributaries have been ditched or filled, in some cases with toxic materials, complicating efforts to restore flow into them.

Archival and field studies together indicate that wood was formerly a dominant influence on riverine processes and habitats at all spatial and temporal scales [*Collins et al.,* in press]. This contrasts with the current regional models of wood being relatively ineffective in the large rivers [e.g., *Bilby and Bisson,* 1998], a perception that may reflect the cumulative effect of human actions more than the pristine river condition. Based on the central importance of wood, we have suggested a framework for reestablishing floodplain forests, linked to restoring a self-sustaining, dynamic river morphology and habitat [see *Collins and Montgomery,* in press]. The framework includes a focus on developing large trees to create key pieces for jams. In the Nisqually River, fast-growing black cottonwood (*Populus trichocarpa*) commonly form key pieces, arguing for an emphasis on including fast-growing tree species in initial riparian restoration planning giving way to slower-growing, more durable species in a type of "restoration succession" [*Collins and Montgomery*, in press]. Data from the Nisqually River and experience elsewhere suggests that self-sustaining restoration can occur within 50-100 years, much earlier than commonly assumed. Placed key pieces or constructed jams [*Abbe et al.,* 1997] can be important in interim decades.

Analysis of historic meander belts widths can provide general guidelines for the proportion of a valley bottom in which a river migrated or avulsed. The different spatial and temporal scales over which different types of rivers interact with the valley bottom and with floodplain forests have fundamental implications for the intensity and types of land use that are compatible with naturally functioning

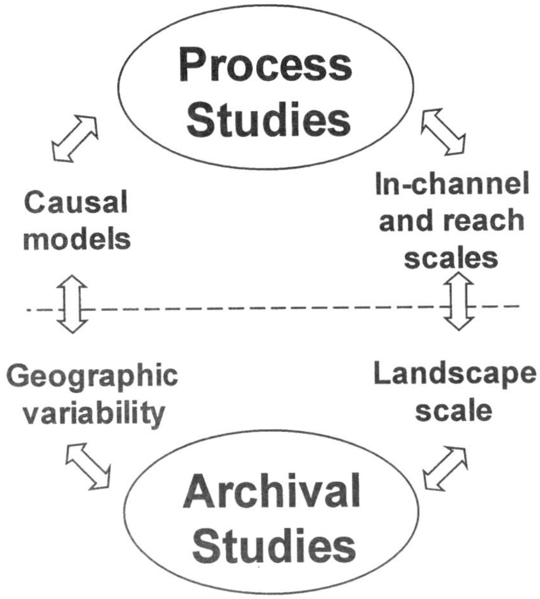

Figure 8. Synergy of archival and field studies in characterizing historic riverine processes and habitats.

lowland rivers. Long-term progress toward restoring Puget Sound rivers will need to incorporate an understanding of the relations between channel pattern, channel width, and the space and time scales in which floodplain forests interact with fluvial and habitat-forming processes. To fully recoup historic habitat, land use and land-use planning must be compatible with a dynamic channel and floodplain.

Field and archival studies can be used to characterize historic river dynamics in an iterative and synergistic way (Figure 8). Archival sources generate hypotheses about landscape processes, and field studies can then generate process models. These conceptual models provide the basis for designing effective river management and habitat restoration and conservation schemes; archival materials can provide the basis for applying them to a given location. Field studies provide more small-scale information and insight into form and process than it is typically possible to discern from archival sources. On the other hand, there is a danger of focusing only on contemporary process studies, and overlooking the landscape-scale features and processes that no longer exist. For example, a focus on a 50-ft-wide streamside buffer on a lowland river would neglect the fact that most of the habitat was historically farther from the river. Without an understanding of the larger (landscape) scale, management efforts can focus on managing the microcosm, instead of addressing structures or processes that exist or operate at a landscape scale.

How to prioritize and order restoration activities within a watershed, and understanding the interactive mechanics of geomorphic processes and riverine habitat are central to effective restoration planning. Historic studies in the Puget Lowland suggest answers to these questions that are contrary to commonly held assumptions about the historic distribution of habitats, the processes that generate them, and how, where, and in what order habitats might best be restored. This example supports the argument for the importance of undertaking historic analysis early in restoration and conservation planning.

Acknowledgements. Funding for work drawn upon for this paper was provided by: the Quaternary Research Center, Puget Sound Regional Synthesis Model (PRISM), and Center for Streamside Studies at the University of Washington; the USDA Forest Service Pacific Northwest Forest and Range Research Station; Bullitt Foundation; Skagit System Cooperative; and Seattle District of the U.S. Army Corps of Engineers. We thank GIS specialists Amir Sheikh and Harvey Greenberg for their assistance.

REFERENCES

Abbe, T. B., and D. R. Montgomery, Large woody debris jams, channel hydraulics and habitat formation in large rivers, *Regulated Rivers Research and Management,* 12: 201-221, 1996.

Abbe, T. B., D. R. Montgomery, and C. Petroff, Design of stable in-channel wood debris structures for bank protection and habitat restoration: an example from the Cowlitz River, WA, pp. 809-816 in: *Proceedings of the Conference on Management of Landscapes Disturbed by Channel Incision*, edited by S. S. Y. Wang, E. J. Langendoen, and F. D. Shields, Jr., University of Mississippi, 1997.

Beechie, T., E. Beamer, and L. Wasserman, Estimating coho salmon rearing habitat and smolt production losses in a large river basin, and implications for habitat restoration, *North American Journal of Fisheries Management* 14: 797-811, 1994.

Beechie, T., B. D. Collins, and G. Pess, Holocene and recent geomorphic processes, land use, and salmonid habitat in two north Puget Sound river basins, in: *Geomorphic processes and riverine habitat*, edited by J. M. Dorava, B. Palcsak, F. Fitzpatrick, and D. R. Montgomery, American Geophysical Union, Washington, D. C., this volume.

Beamer, E. M. and R. A. Henderson, Juvenile salmonid use of natural and hydromodified stream bank habitat in the mainstem Skagit River, northwest Washington, report by Skagit System Cooperative, LaConner, WA, to U. S. Army Corps of Engineers, Seattle, WA, 51 p., 1998.

Beget, J. E., Recent activity at Glacier Peak, *Science* 215: 1389-1390, 1982.

Benda, L., T. J. Beechie, A. Johnson, and W. C. Wissmar, Morphology and evolution of salmonid habitats in a recently deglaciated river basin, Washington State, U. S. A., *Canadian Journal of Fisheries and Aquatic Sciences* 49: 1246-1256, 1992.

Bilby, R. E. and P. A. Bisson, Function and distribution of large woody debris, pp. 324-346 in *River ecology and management: lessons from the Pacific coastal ecoregion*, edited by R. J. Naiman and R. E. Bilby, Springer, NY, 1998.

Booth, D. B., Glaciofluvial infilling and scour of the Puget Lowland, Washington, during ice-sheet glaciation, *Geology* 22: 695-698, 1994.

Bortleson, G. C., M. J. Chrzastowski, and A. K. Helgerson, Historical changes of shoreline and wetland at eleven major deltas in the Puget Sound region, Washington, *U.S. Geological Survey Hydrological Investigations Atlas* HA-617, 1980.

Church, M. and O. Slaymaker, Disequilibrium of Holocene sediment yield in glaciated British Columbia. *Nature* 337: 452-454, 1989.

Collins, B. D., D. R. Montgomery, and A. D. Haas, Historic changes in the distribution and functions of large woody debris in Puget Lowland rivers, *Canadian Journal of Fisheries and Aquatic Sciences*, in press.

Collins, B. D. and D. R. Montgomery, Forest development, log jams and restoration of floodplain rivers in the Puget Lowland, *Restoration Ecology,* in press.

Cowardin, L. M., V. Carter, F. C. Golet, and E. T. LaRoe, Classification of wetlands and deepwater habitats of the United States, U. S. Fish & Wildlife Service, Report FWS/OBS-79/31, 1985.

Debose, A. and M. W. Klungland, Soil survey of Snohomish County Area, Washington, 1983.

Doppelt, B., M. Scurlock, C. Frissell, and J. Karr, *Entering the watershed: A new approach to save America's river ecosystems.* Island Press, Covelo, CA, 1993.

Frissell, C. A., W. J. Liss, and D. Bayles, An integrated, biophysical strategy for ecological restoration of large watersheds, pp. 449-456, in: *Changing roles in water resources management and policy,* edited by N. E. Spangenbery and D. F. Potts, American Water Resources Association, Herndon, VA, 1993.

Galatowitsch, S. M., Using the original land survey notes to reconstruct pre-settlement landscapes in the American West, *Great Basin Naturalist* 50: 181-191, 1990.

Hayman, R. A., E. M Beamer, and R. E. McClure, FY 1995 Skagit River chinook restoration research, Skagit System Cooperative Research Progress Report No. 1, Final Project Performance Report, NWIFC Contract #3311, 1996.

Interstate Publishing Company, *An illustrated history of Skagit and Snohomish Counties,* 1117 p., 1906.

Knox, J. C., Human impacts on Wisconsin stream channels, *Annals of the Association of American Geographers* 67(3): 323-342, 1977.

Mangum, A. W. and Party, *Reconnaissance soil survey of the eastern part of Puget Sound,* U. S. Soils Bureau, 1909.

Nelson, J. C., R. E. Sparks, L. DeHaan, and L. Robinson, Presettlement and contemporary vegetation patterns along two navigation reaches of the upper Mississippi River. pp. 51-60 in *Perspectives on the land use history of North America: a context for understanding our changing environment*, edited by T. D. Sisk, U. S. Geological Survey, Biological Resources Division, Biological Sciences Report USGS/BRD/BSR-1998-0003, 1998.

Nesbit, D. M. with Contributions from U. S. Coast Survey, S. L. Boardman, Eldridge Morse, and others, Tide marshes of the United States, *USDA Miscellaneous Special Report* No. 7, Government Printing Office, Washington, D. C, 1885.

North, M. E. A., and J. M Teversham, The vegetation of the floodplains of the Lower Fraser, Serpentine and Nicomekl Rivers, 1859 to 1890, *Syesis* 17: 47-66, 1984.

Plummer, G. H., F. G. Plummer, and J. H. Rankine, Map of Washington showing classification of lands, Plate 1 in: Gannet, H., The forests of Washington, a revision of estimates. *U. S. Geological Survey Professional Paper* 5, Series H, Forestry, 2, 1902.

Radeloff, V. C., D. J. Mladenoff, H. S. He, and M. S. Boyce, Forest landscape change in the northwestern Wisconsin Pine Barrens from pre-European settlement to the present, *Canadian Journal of Forest Research* 29: 1649-1659, 1999.

Sears, P. B., The natural vegetation of Ohio, *Ohio Journal of Science* 25: 139-149, 1925.

Sedell, J. R., and K. J. Luchessa, Using the historical record as an aid to salmonid habitat enhancement, in: *Acquisition and utilization of aquatic habitat inventory information*, edited by N. B. Arementrout, Proc. of Symp. held 28-30 Oct., 1981, Portland, Oregon. pp. 210-223, American Fisheries Society, 1981.

U. S. Army Corps of Engineers (USACOE), *Stillaguamish River, Wash; letter from the Secretary of War transmitting report from the chief of engineers on Stillaguamish River, Wash., covering navigation, flood control, power development, and irrigation,* Washington, U.S. Government Printing Office, 1929.

U. S. Army Corps of Engineers (USACOE), Index Map of Skagit River, From its Mouth to the Town of Sedro, Washington, 1898.

U. S. War Department. *Annual Reports of the Chief of Engineers, U. S. Army, to the Secretary of War* (continues 1907-1944 as: *Report of the Chief of Engineers, U. S. Army*; 1945-1953 as: *Annual Report of the Chief of Engineers, U. S. Army*; 1954-current year as: *Annual Report of the Chief of Engineers, U. S. Army, on Civil Works Activities*), 1876-1906.

White, C. A., *A history of the rectangular survey system,* U. S. General Printing Office, Washington, D. C., 774 pp., 1991.

Whitney, G. G. J. P. DeCant, Government land office surveys and other early land surveys, pp. 147-172 in *The Historical Ecology Handbook*, edited by D. Egan and E. A. Howell, Island Press Springer, Washington, D.C., 2001.

Whitney, G. G. *From coastal wilderness to fruited plains: A history of environmental change in temperate North America from 1500 to the present,* Cambridge University Press, Cambridge, U.K., *451 pp.*, 1996.

Brian D. Collins and David R. Montgomery, Department of Earth and Space Sciences, Box 351310, University of Washington, Seattle, WA 98195

Section 5

Geomorphology, River Ecology, and Ecosystem Management

Geomorphology, River Ecology, and Ecosystem Management

David R. Montgomery

Department of Earth & Space Sciences, University of Washington, Seattle, Washington

INTRODUCTION

The recent surge of interest in geomorphological influences on river ecology is driven in part by societal desires to reduce or reverse the effects of human actions on aquatic ecosystems. Papers in this volume illustrate the broad range of such interactions and approaches to understanding them. Taken together these studies outline the potential for integrating research on geomorphology and river ecology into watershed management or restoration planning.

The late 20th century paradigm of ecosystem management hypothesizes that resource use can be redesigned to continue without jeopardizing the long-term integrity of natural ecosystems [*Slocombe*, 1993a,b; *Grumbine*, 1990]. But as pointed out almost a decade ago by *Lugwig et al.* [1993], it remains difficult to identify a modern precedent. Developing the management designs likely to favor successful applications of ecosystem management, let alone evaluating the hypothesis that it can live up to its promise, requires careful evaluation of the impacts of human activities, natural events, and their interactions on ecosystems that supply resources and services to human societies. Yet ecological systems present a number of management challenges: they are exceedingly complex, we didn't build them, we don't have their plans, and we don't (and probably can't) fully understand how they work given the interdependencies of their constituents. Hence, development of sustainable management strategies for aquatic ecosystems requires an intimidatingly sophisticated level of knowledge of the spatial context and causal linkages among human actions, watershed processes, channel conditions, and ecosystem response.

Geomorphic Processes and Riverine Habitat
Water Science and Application Volume 4, pages 247-253

Geomorphology encompasses a wide range of processes that influence landforms and landscape dynamics. These processes, together with their interactions with tectonic, hydrologic, and geochemical processes, create habitats and structure the spatial distributions and temporal dynamics of habitats. In turn, habitat characteristics influence ecological systems, which can themselves influence geomorphological processes (Figure 1). Because of these fundamental interconnections, the field of ecology broadly includes the study of organisms and their interactions with each other and their environment. In the past several decades, ecologists have come to increasingly recognize the importance of the "geomorphic template" that can structure ecological processes, habitat characteristics, and their dynamic interactions [*Swanson et al.*, 1988]. Riverine ecosystems have particularly tight coupling to geomorphological processes due to the gravity-driven routing of materials and disturbances down channel systems, and the inability of aquatic organisms to remove themselves from harms way. Consequently, a general framework for generating the scientific understanding necessary to support credible efforts to implement ecosystem management will require integration of geomorphology and river ecology. This volume on geomorphic processes and riverine habitat brings together diverse research efforts and specific case studies to illustrate various approaches to refining an understanding of the interactions among geomorphology, river ecology and people.

REGIONAL NATURE OF RIVER PROCESSES AND ECOLOGY

The studies in this volume also illustrate the strongly regional character of linkages between river processes and aquatic ecology. Quite simply, rivers in different regions have profoundly different character even though river processes are driven by general physical relations that govern the flow of water, sediment transport, and interactions

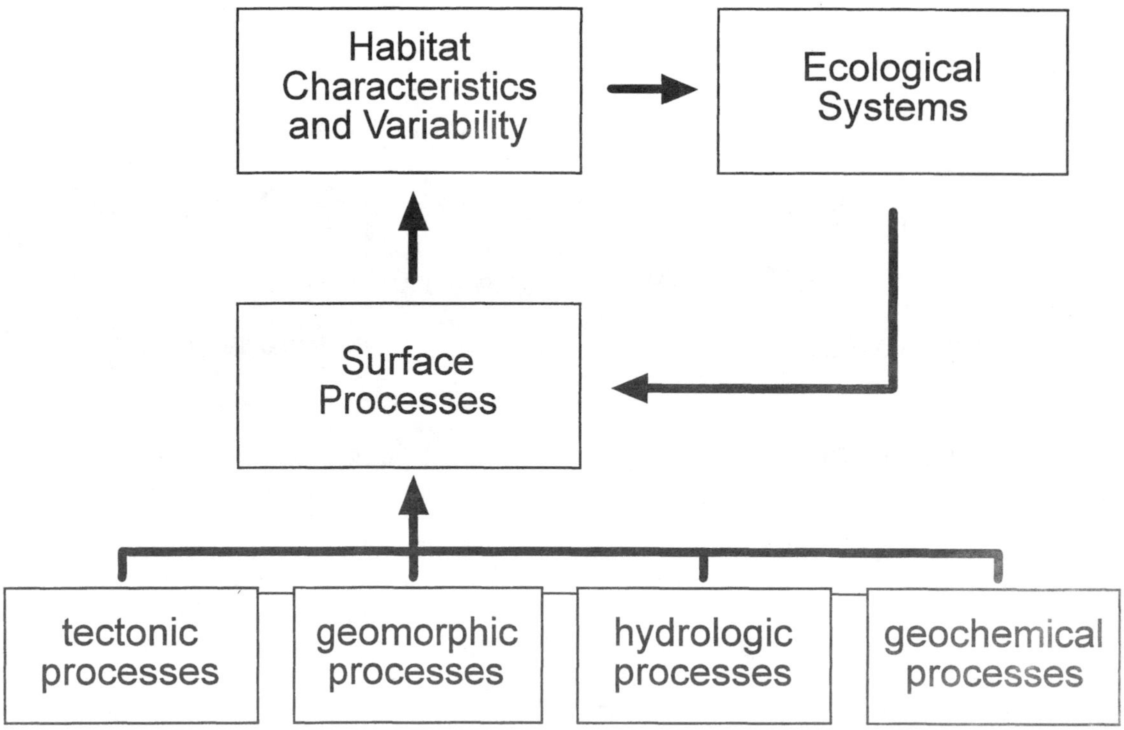

Figure 1. Schematic illustration of the role of surface processes on shaping habitat characteristics and variability and the potential for ecological systems to influence surface processes.

with bed and bank-forming materials. Despite the emphasis typically placed on generalized aspects of rivers, there are important (and interesting) fundamental differences among rivers in different regions of the world. Rivers in the old-growth forests of the Pacific Northwest are loaded with huge logs that profoundly influence channel characteristics, disturbance processes, and stream ecology [*Naiman and Bilby*, 1998]. Rivers draining the flanks of volcanoes can be inundated by extreme sediment loads, such as those at Mount Pinatubo, which even after a decade of monsoonal rainfall remain overloaded with sediment from the massive 1991 eruption [*Montgomery et al.*, 1999]. Rivers in arid regions tend to lack armored beds and therefore have very high bedload transport rates [*Laronne and Reid*, 1993]. Ozark rivers have complex hydrologic relations to local groundwater systems due to the peculiarities of karst terrain [*Jacobson et al.*, this volume]. Oregon Coast Range channels have a strong debris flow control on disturbance regimes and valley bottom landforms [*Lancaster et al.*, this volume], whereas across the Cascade Range in eastern Oregon variations in valley width control disturbance regimes and valley-bottom ecology [*McDowell*, this volume]. In Puget Sound river systems, inherited glacial influences structure valley-bottom and riverine landforms [*Beechie et al.*, this volume; *Collins and Montgomery*, this volume]. More recent historic changes in rates of hillslope processes still affect the character of channels in Puerto Rico [*Larsen and Román*, this volume]. In low-gradient systems interactions with fauna inclined toward hydraulic engineering strongly influences channel processes and characteristics, thereby creating conditions more favorable to themselves [*Pasternack*, this volume]. Australian researchers argue that many concepts developed in North American rivers do not apply down under (*Brizga and Finlayson*, 2000). And the list goes on.

In short, river systems display marvelously rich and varied characteristics, dynamics, and relations to ecological systems in spite of the generality of the underlying physics. In addition to influences set by geological setting and climate history, different types of river systems have different levels of local versus routing controls on channel characteristics and disturbance regimes [*Montgomery*, 1999]. For purposes of river rehabilitation and restoration, an understanding of regional and local influences on fluvial processes can be as important as a solid grasp of the fundamental physics that drive fluvial processes.

PREDICTABILITY OF CHANNEL CONDITIONS

The influence of spatial and temporal variability on channel conditions spans a wide range of scales and imparts an element of inherent unpredictability to channel conditions. In a simplistic worldview, spatially and temporally oriented views of variability in geomorphic processes can be cast as at odds with one another. But both geomorphology and the hydrological sciences have a long history of recognizing the intertwined relationship between both types of variability—perceived dichotomies between spatial and temporal views of hydrogeomorphic influences on riverine ecosystems are forced distinctions given the widespread recognition that both dimensions are central to an understanding of rivers and riverine ecosystems.

Theories of landform organization span a wide range of scales from theories for controls on site or channel-reach scale characteristics and response to the production and routing of sediment through landscapes and ultimately landform evolution. Different types of theories are relevant at each of these scales, and different types and scales of features can be considered as transients at different scales in a hierarchy of landscape theories. An alluvial fan is a transient sediment storage element (a temporary reservoir) at the landform evolution scale, but functions as a long-term sediment sink at the reach scale. Conversely, channel morphology can be transient over decades to centuries at the reach scale but effectively constant over longer time scales more pertinent to landscape evolution. In this manner a landscape shaped by intermittent processes can maintain an approximately steady-state morphology over time scales longer than those governing individual events can.

In addition to the importance of scale, geomorphic and hydrologic variability differ across a landscape. Due to the dispersive effects of channel network routing, low-gradient channels may be inherently less variable, and therefore more predictable, than mountain channels. In other words, the character of the temporal variability in channel properties and processes can be highly dependent upon spatial location in the channel network.

Geomorphologists do not generally embrace the view of landscapes as static systems, but instead recognize that a temporally variable system can maintain relatively stable system characteristics over time scales long enough to integrate across specific transport or disturbance events [*Schumm*, 1977; *Bull*, 1991]. In other words, a coarse-scale steady state can arise from the integration of very dynamic fine-scale processes. Similarly, feedbacks operating in a dynamic system may dampen the propagation of highly variable input signals, as demonstrated by Lancaster et al. [*this volume*] for the Oregon Coast Range. In this area, hillslope lowering rates approximately match basin-wide sediment yields [*Reneau and Dietrich*, 1991] indicating that the range is in an erosional steady-state over long time frames, even though sediment introduction to the channel system is dominated by catastrophic landsliding [*Swanston and Swanson*, 1976]. The key issue is not whether the landscape is either in steady state or exhibiting a stochastic component to sediment transport—these issues are not mutually exclusive given the different time scales over which they operate. Rather, the key issue is how to relate event-driven variability in channel processes and conditions to longer-term patterns and states, and to the dynamics of ecological systems.

LANDSCAPE ELEMENTS AND ASSEMBLAGES: THE BUILDING BLOCKS OF LANDSCAPES

Landscapes are hierarchically organized systems [*Frissell et al.*, 1986] in which landscape elements form general functional units characterized by distinct morphology, landscape position and types of active processes. Perhaps the simplest set of morphologically distinct landscape elements recognizes six basic types: hillslopes, channels, valleys, hollows, floodplains, and fans. More elaborate landform classifications have been developed for a variety of purposes, but this basic list of landscape elements reflects fundamental units—each of which has a number of varieties that influence its characteristics (e.g., bedrock and soil mantled hillslopes, alluvial and debris flow fans, confined and unconfined valleys) and the manner in which the element influences or interacts with upslope, neighboring, or downslope landscape elements. Perhaps the most attractive characteristic of this set of landscape elements is that it mirrors the intuitive general organization that geomorphologists bring to reading a landscape, yet allows for regionalization through different types of landscape elements, variations in their spatial arrangement, and the specific geologic, climatic, and human history of a landscape.

The full complement of landscape elements does not occur in every landscape, and the arrangement and sequencing of landscape elements can vary substantially in different environments, histories, and geological contexts. For example, fine-scale topographic hollows are not well developed in areas where glaciers scoured high-frequency topographic features from the walls of major valleys. Extensive floodplains are rare in unglaciated mountain drainage basins, whereas the distribution of floodplains along rivers in glaciated mountain systems can reflect inherited variations in valley slope. One would not expect to find debris flow fans in gentle topography. Extensive networks of unchanneled valleys (also known as hollows or dambos) characterize extensive areas of southern Africa (*Thomas and Goodie*, 1985). The wide variety of landscapes around the globe illustrates the range of possible types, spatial arrangements, and sequencing of landscape elements. How various landscape elements are arrayed influences how they interact, and a landform assemblage is a distinctive suite or arrangement of various landscape elements.

The location-specific array of landscape elements and the associated disturbance history are key to understanding landscape and channel dynamics in mountain landscapes, and the attendent interactions with and influences on ecological systems. Examples pertinent to understanding the dynamics of riverine ecosystems include the effects of patterns in the sequencing of channel types and valley bottoms on sediment storage and transport processes. In channel networks with alternating confined and unconfined stretches of valley bottom, and, therefore, high and low-gradient channel types, the amplitude of pulses of sediment supply will be damped by repeated deposition at locations where the net local transport capacity declines downstream; i.e., at locations where the local slope decreases and the discharge does not. Differences between colluvial and earthflow-dominated hillslopes illustrate the potential effect of different types of hillslopes on sediment transport and storage processes. Colluvial hillslopes are slopes on which the soil mantle is transported downslope to channels at rates typically ranging from <0.1 to several mm yr^{-1} [*Young*, 1972; *Selby*, 1993]. Sediment movement usually occurs as small events that integrate to approximate a slow gradual "creep" of material downslope that acts as a distributed sediment source for mountain channel networks. Earthflow-dominated hillslopes are either bedrock or soil-mantled hillslopes on which large earthflows are common. These large landslides move episodically at average velocities of 0.1 to >1 m yr^{-1} [*Kelsey*, 1978; *Keefer and Johnson*, 1983; *Iverson and Major*, 1987], and can rapidly introduce large amounts of sediment to specific locations in mountain channel networks. Although position in the channel network influences the effects of landscape assemblages on channel processes, the actual response of a channel also depends on the ability of the channel to transmit the imposed sediment load.

Landscape assemblages influence channels by determining the type, amount, and size of sediment and wood debris supplied to the channel. Several reach-scale examples help to illustrate this point. The sequencing of debris flow fans along a channel network can buffer downstream reaches from headwater disturbances originating in debris flow-prone channels. Debris-flow fans approximately define the downslope limit of debris flow propagation and their distribution through mountain channel networks influences the disturbance regime in headwater channels. A trellis-shaped channel network favors short runout debris flows by forcing deposition at right angle tributary junctions [*Johnson et al.*, 2000], whereas a dendritic channel network favors longer runout debris flows with deposition occurring farther down the channel network at lower slopes [*Ikeya*, 1981; *Takahashi et al.*, 1981]. Unconfined channels with wide valley bottoms can be buffered from the direct influences of hillslope processes, whereas hillslope processes are a strong influence in confined channels with little to no valley bottom (such as along steep, incised inner gorges). In a landscape assemblage composed of bedrock hillslopes, alluvial fans, and wide unconfined valley bottoms, the channels will be relatively buffered from local variability in hillslope processes. In contrast, channel processes will be strongly coupled to hillslope processes in a landscape with earthflow-dominated hillslopes, confined valleys, and no debris fans. Consequently, an understanding of the nature of the building blocks that compose a particular landscape is fundamental to understanding how geomorphological processes function as ecological disturbance processes at the watershed or landscape scale.

IMPLICATIONS FOR ECOSYSTEM MANAGEMENT

The huge range, and inherently limited understanding, of interactions between geomorphological processes and river ecology should make us careful about how confidently we believe we can actively "manage" ecosystems toward desired outcomes. Our simplified understanding of the system may lead to fundamentally erroneous conclusions, as highlighted by Lancaster et al.'s [*this volume*] illustration of how sediment retention by log jams fundamentally alters the system dynamics predicted by models lacking that one key element. An important lesson that follows from Lancaster et al.'s [*this volume*] chapter is that although models of channel and habitat dynamics based on routing of stochastic sediment inputs might be pertinent to understanding the variability of river ecosystems, such models may yield misleading interpretations in a landscape with abundant sediment storage elements. The key message is that we should be cautious about our ability to predict ecosystem response based on simplified models of complex systems.

Many contemporary watershed assessment and management programs include inadequate understanding of environmental systems, inadequate consideration of what should be monitored, inappropriate methods for measuring key variables, and insufficient spatial and temporal scales to capture critical process linkages and contexts. Poor understanding of inter-relations between disturbance agents and channel response often results in collecting extensive data on non-essential variables and neglect of crucial variables. If causal pathways are poorly delineated, then monitoring cannot detect or understand why a system changed in a particular way because causality can be remote in space and time. Lack of process understanding can lead to inappropriate methods or models that are either blind to linkages among processes or that are inadequate to detect probable changes.

A parallel problem with many channel assessment and monitoring programs is the tendency to use similar methods and response indicators in different portions of a watershed. Landscapes need to be stratified into areas dominated by similar processes in order to design and implement effective assessment and monitoring strategies. If stratified too finely, however, local dynamics are hard to interpret because they are largely driven by external processes. If landscape stratification is too coarse, then it may be difficult to delineate causal linkages. Hence the spatial scale over which an assessment or monitoring program is applied strongly influences its usefulness for ecosystem management. Monitoring strategies applied, for example, across large ecoregions lack the spatial context necessary to interpret causality and generate information that is likely to be uninterpretable unless coupled to the history and spatial context of the watershed from within which it is drawn. Such an approach cannot capture the disturbance history, time trends, and the spatial context of the place needed to develop well-founded management or policy recommendations.

Another significant problem for long-term monitoring programs is that measurements are not acquired at appropriate temporal scales. Most processes do not proceed at a steady rate through time; it is far more common for watershed and ecological processes to be characterized by long periods of inactivity interspersed with short bursts of activity in which a landslide fills the channel, a flood scours large wood from the channel, and so on. Only if the resulting signal and impact are long lasting relative to the monitoring interval can these processes be measured. Current monitoring programs, however, too often rely on measurements when and where field personnel can conveniently arrive at the field site, rather than consideration of how the system works coupled to thoughtful experimental design.

Effective monitoring must involve measuring things that may change, reflect identifiable processes, and may be used to interpret channel response in the particular landscape in question. This highlights the need to develop and test practical monitoring indicators that are scientifically based and sensitive to change, and will differ from place to place. For channel networks this involves identifying key "currencies" traded back and forth between a channel and its watershed: sediment, organic debris, water, energy and nutrients. Production, recruitment, and transport dynamics, as well as contemporary inventories, need to be assessed and examined in relation to life histories of pertinent organisms. This task can be simplified by applying knowledge/hypotheses or conceptual models about how the local system works, but in general over-parameterized models are not useful because of the difficulty determining or constraining them in practical applications. Hence, landscape management strategies founded upon documented linkages between geomorphological processes and ecological systems should be developed based on sound data and relationships supported by appropriately scaled models, rather than being predicated on the predictions of complicated, overparameterized models.

Finally, there are two very different views of ecosystem management: that of (1) managing people to minimize our collective impact on ecosystem structure and functions; and (2) that of using human actions to manage the ecosystem toward some predefined societal objectives. Managing human actions to reduce impacts on environmental systems at their source (input management) embodies approach (1), whereas managing for particular system outputs, as in approach (2), is risky for complex systems that we cannot completely understand, like ecosystems [*Montgomery*, 1994]. While it is widely recognized that ecosystem management must be done in the larger context of landscape-scale processes, the presumption remains that scientific monitoring will inform us and allow adaptive management. Ecosystem management, although still vaguely defined, should be founded on science-driven performance monitoring and understanding of geomorphic, watershed, and ecological processes to make the consequences of our choices clear. If we are to further degrade our environment, it should be the result of an explicit policy, rather than as a consequence of Xeno's Paradox because implicit choices are either hidden or unacknowledged in the face of economic pressures driving resource exploitation. Achieving this vision will depend on how well we are able to integrate geomorphology and river ecology, not only with each other, but also into our societal and local decision making processes.

CLOSING THOUGHTS

While the fine set of papers in this volume show how we are making strong progress in some areas, we still have much to learn and a long way to go toward realizing the vision discussed above. Although there are many exciting questions on the interface of geomorphology and ecology, there are also institutional and professional impediments to people trying to pursue those questions they find to be of interest or importance in their own region, watershed, or stream reach.

In particular, we need to develop strategies to encourage interdisciplinary work between geomorphologists and ecologists. Although one hears many calls for interdisciplinary research, our institutional systems and journals are respectively not particularly conducive or receptive to such efforts. I recall with some amusement, papers on geomorphology-ecology interactions declined by an ecological journal because the content was too geological and a geological journal because it was not geological enough. Interdisciplinary investigations are often collaborations among many investigators, something devalued by academic promotion and tenure systems that consider the single-author paper the ultimate academic currency. I do not believe that the solution is to simply create new "interdisciplinary" journals and academic departments or programs. Rather, I believe that we need to encourage pushing the boundaries of our own disciplines into the realm of other intellectual traditions and to respect the risks assumed by those who venture into new, and perhaps poorly charted territory.

We also need to encourage closer co-ordination and integration of fundamental and applied aspects of watershed science. Many academic geomorphologists and ecologists consciously shy away from overtly practical projects. Conversely, too many non-federal funding agencies remain hesitant to support fundamental research at universities. In addition to overcoming these problems, we need to develop institutional structures to disseminate research results to professionals working in real-world applications. Interpreting the results of stream channel assessments and developing the understanding of potential and natural conditions needed for designing effective restoration efforts require substantial experience and training. I do not believe that short courses can provide the requisite knowledge and experience. Neither do I believe that our educational system provides enough opportunities for more meaningful experiences to practicing professionals. While the important issue of how to establish standards of professional practice and evaluate those practices and work products is difficult enough for a single discipline, it is far more daunting for work that spans the ecology-geomorphology interface.

In the end, there are no simple cookbooks or manuals that can capture the inherent regional complexity of interactions between geomorphic processes, riverine habitat, and ecological systems. We can translate understanding based on the general physics that underpins fluvial geomorphology to any region; however, it is much more difficult to generalize how regional differences interact with that physics to structure the manner in which river processes influence ecological systems (and vice-versa). Consequently, we need to pursue regional research programs to develop a sound empirical basis for understanding system behavior and for developing models to usefully extrapolate system behavior into the management arena. The papers collected in this volume provide good examples of how geomorphological studies can contribute to different aspects of regional river assessment, management, and recovery programs.

Acknowledgments. Elements of this chapter are based on material recycled from manuscripts started years ago but never completed with Gordon Grant, Bill Dietrich, Mary Power, and Kate Sullivan.

REFERENCES

Beechie, T. J., Collins, B. D., and Pess, G. R., Holocene and recent geomorphic processes, land use, and salmonid habitat in two north Puget Sound river basins, Water Science and Application (this volume), J. B. Dorava, D. R. Montgomery, B. B. Palcsak, and F. A. Fitzpatrick, (eds.). American Geophysical Union, Washington, D. C., 2001.

Brizga, S., and Finlayson, B. (eds.), *River Management: The Australasian Experience*, John Wiley & Sons., Chichester, 301p.

Bull, W. B., *Geomorphic Response to Climate Change*, Oxford University Press, New York, 326p., 1991.

Collins, B. D., and Montgomery, D. R., Importance of archival and process studies to characterizing pre-settlement riverine geomorphic processes and habitat in the Puget Lowland, Water Science and Application (this volume), J. B. Dorava, D. R. Montgomery, B. B. Palcsak, and F. A. Fitzpatrick, (eds.). American Geophysical Union, Washington, D. C., 2001.

Frissell, C. A., Liss, W. J., Warren, C. E., and Hurley, M. D., A hierarchical framework for stream habitat classification: Viewing streams in a watershed context, *Environmental Management,* 10, 199-214, 1986.

Grumbine, E., Protecting biological diversity through the greater ecosystem concept, *Natural Areas Journal,* 10, 114-120, 1990.

Ikeya, H., A method for designation for areas in danger of debris flow, in *Erosion and Sediment Transport in Pacific Rim Steeplands*, edited by T. R. H. Davies and A. J. Pearce (eds.), Int. Assoc. Hydrol. Sci. Pub. 132, 576-588, 1981.

Iverson, R. I., and Major, J. J., Rainfall, ground-water flow, and seasonal movement at Minor Creek landslide, northwestern California: Physical interpretation of empirical relations, *Geological Society of America Bulletin*, 99, 579-594, 1987.

Jacobson, R. B., Laustrtup, M. S., and Chapman, M. D., Fluvial processes and passive rehabilitation of the Lisbon Bottom side-channel chute, Lower Missouri River, Water Science and Application (this volume), J. B. Dorava, D. R. Montgomery, B. B.

Palcsak, and F. A. Fitzpatrick, (eds.). American Geophysical Union, Washington, D. C., 2001.

Johnson, A. C., Swanston, D. N., and McGee, K. E., Landslide initiation, runout, and deposition within clearcuts and old-growth forests of Alaska, *Journal of the American Water Resources Association*, 36, 17-30, 2000.

Keefer, D. K., and Johnson, A. M., *Earthflows: Morphology, Mobilization, and Movement*, U.S. Geological Survey Professional Paper 1264, 56p., 1983.

Kelsey, H. M., Earthflows in Franciscan melange, Van Duzen River basin, California, *Geology*, 6, 361-364, 1978.

Lancaster, S. L., Hayes, S. H., and Grant, G. E., Sediment and wood storage and dynamics in small mountainous watersheds, Water Science and Application (this volume), J. B. Dorava, D. R. Montgomery, B. B. Palcsak, and F. A. Fitzpatrick, (eds.). American Geophysical Union, Washington, D. C., 2001.

Laronne, J., and Reid, I., Very high rates of bedload sediment transport by ephemeral desert rivers, *Nature*, 366, 148-150, 1993.

Larsen, M. C., and Román, A. S., Mass wasting and sediment storage in a small montane watershed: An extreme case of anthropogenic disturbance in the Humid Tropics, Water Science and Application (this volume), J. B. Dorava, D. R. Montgomery, B. B. Palcsak, and F. A. Fitzpatrick, (eds.). American Geophysical Union, Washington, D. C., 2001.

Ludwig, D., Hilborn, R., and Walters, C., Uncertainty, resource exploitation, and conservation: Lessons from history, *Science*, 260, 17&36, 1993.

McDowell, P. F., Spatial variations in channel morphology at segment and reach scales, Middle Fork John Day River, northeastern Oregon, Water Science and Application (this volume), J. B. Dorava, D. R. Montgomery, B. B. Palcsak, and F. A. Fitzpatrick, (eds.). American Geophysical Union, Washington, D. C., 2001.

Montgomery, D. R., Input- and output-oriented approaches to implementing ecosystem management, *Environmental Management*, 19, 183-188, 1994.

Montgomery, D. R., Panfil, M. S., and Hayes, S. K., Channel-bed mobility response to extreme sediment loading at Mount Pinatubo, *Geology*, 27, 271-274, 1999.

Montgomery, D. R., Process domains and the river continuum, *Journal of the American Water Resources Association*, 35, 397-410, 1999.

Naiman, R. J., and Bilby, R. E. (eds.), *River Ecology and Management: Lessons from the Pacific Coastal Region*, Springer-Verlag, New York, 705p., 1998.

Pasternack, G. B., Animal response to river evolution in the tidal freshwater zone, Water Science and Application (this volume), J. B. Dorava, D. R. Montgomery, B. B. Palcsak, and F. A. Fitzpatrick, (eds.). American Geophysical Union, Washington, D. C., 2001.

Reneau, S. L., and Dietrich, W. E., Erosion rates in the southern Oregon Coast Range: Evidence for an equilibrium between hillslope erosion and sediment yield, *Earth Surface Processes and Landforms*, 16, 307-322, 1991.

Schumm, S. A., *The Fluvial System*, Wiley, New York, 338p., 1977.

Selby, M. J., *Hillslope Materials and Processes*, Oxford University Press, 451p., 1993.

Slocombe, D. S., Environmental planning, ecosystem science, and ecosystem approaches for integrating environment and development, *Environmental Management*, 17, 289-303, 1993a.

Slocombe, D. S., Implementing ecosystem-based management, *BioScience*, 43, 612-622, 1993b.

Swanston, D. N., and Swanson, F. J., Timber harvesting, mass erosion, and steepland forest geomorphology in the Pacific Northwest, in *Geomorphology and Engineering*, D. R. Coates (ed.), Dowden, Hutchinson & Ross, Inc., Stroudsburg, Pennsylvania, p. 199-221, 1976.

Swanson, F. J., Kratz, T. K., Caine, N., and Woodmansee, R. G., Landform effects on ecosystem patterns and processes, *BioScience*, 38, 92-98, 1988.

Takahashi, T., Ashida, K., and Sawai, K., Delineation of debris flow hazard areas: in *Erosion and Sediment Transport in Pacific Rim Steeplands*, edited by T. R. H. Davies and A. J. Pearce (eds.), Int. Assoc. Hydrol. Sci. Pub. 132, 589-603, 1981.

Thomas, M. F., and Goodie, A. S., (eds.), *Dambos: Small Channeless Valleys in the Tropics*, Zeitschrift für Geomorphologie, Supplementband 52, 222p., 1985.

Young, A., *Slopes*, Oliver & Boyd, Edinburgh, 288p., 1972.

David R. Montgomery, Department of Earth & Space Sciences, University of Washington, Seattle, WA 98195-1310
dave@geology.washington.edu